Fukushima Daiichi
Nuclear Information Handbook

A Guide to Accident Terminology and Information Sources

H. G. Brack

Davistown Museum

Department of Environmental History

Center for Biological Monitoring Archives

Special Publication 62

ISBN-10: 0-9829951-6-4
ISBN-13: 978-0-9829951-6-7
LCCN: 2011928690
Davistown Museum © 2011

First Edition

Cover design by Sett Balise

Front cover photo: This image made available from Tokyo Electric Power Co. via Kyodo News, shows the damaged No. 4 unit of the Fukushima Dai-ichi nuclear complex in Okumamachi, northeastern Japan, on Tuesday March 15, 2011. White smoke billows from the No. 3 unit. (AP Photo/Tokyo Electric Power Co. via Kyodo News)

Back cover photo: Fukushima Daiichi: Nuclear Facility Damage Overview. Image Collected March 16, 2011. Used with permission from DigitalGlobe Inc.

This publication is sponsored by
Davistown Museum
Department of Environmental History
Special Publication 62
www.davistownmuseum.org

Pennywheel Press
P.O. Box 144
Hulls Cove, ME 04644

Acknowledgements

Many thanks to the following individuals for helping with the compilation of this text:

Sett Balise, Judith Bradshaw Brown, Janet Christrup, Linda Dartt, Laure Day, Beth Sundberg, and the Davistown Museum Department of Environmental History

Dedication

This *Handbook* is dedicated to the thousands of workers at the Fukushima Daiichi complex who have risked their lives improvising the manual cooling efforts that have kept the loss of coolant accidents from evolving into an even larger nuclear disaster.

Preface

This publication is sponsored by the Davistown Museum, a 501 (c) (3) organization. If not otherwise noted and with the exception of the Japan-disaster-related information, most of the contents of this *Handbook* have been extracted from the online archives of RADNET, the Center for Biological Monitoring's (CBM) nuclear information website. The CBM RADNET archives should not be confused with the EPA's online RadNet (http://www.epa.gov/narel/radnet/), which was organized after CBM's RADNET was incorporated within the Environmental History Department of the newly incorporated Davistown Museum (2000). The CBM RADNET archives republished in this *Handbook* include reporting units, definitions, radiological surveillance data, protection action guidelines, and commentary relevant to the ongoing accident at the Fukushima Daiichi nuclear complex. No citations from the RADNET archives date after 2000, though new definitions have been added to this *Handbook*. Comments, corrections, and additional information are solicited, curator@davistownmuseum.org.

Author Biography

A former volunteer fireman (1963-83), Skip Brack holds degrees in English from the University of Massachusetts (1966) and the University of Colorado (1967) and was an English instructor at the University of the Pacific. Skip was also a graduate student at the University of California, Berkeley, where he helped organize the Stop the Draft movement before leaving academia. In 1970, he was an Earth Day organizer, speaker, researcher, and Director of the New England Ecology Center. He organized the die-in at Logan Airport to protest the supersonic transport (SST), which was cancelled by the US Senate in December of 1970. After moving to West Jonesport, Maine, in the summer of 1970, he opened the Jonesport Wood Company, Inc. and has been in the used hand tool business ever since. Brack now operates tool stores in Hulls Cove, Searsport, and Liberty, Maine.

In 1972, Brack organized the Center for Biological Monitoring (1972-2000) and began collating research on chemical fallout and anthropogenic radiation. Skip moved to Hulls Cove in 1983, where he still lives. In 1994, Brack established RADNET: Nuclear Information on the Internet. In 1999, he founded the Davistown Museum, a regional tool, art, and history museum in Liberty, Maine. The Center for Biological Monitoring Archives is now a component of the museum's Department of Environmental History. The museum's extensive website is a major resource for persons, including homeschoolers, interested in New England's Native American, maritime, and industrial history, and the history of hand tools and how they were forged.

Author's publications

<u>**Environmental History**</u>

Radscan: Information Sampler on Long-lived Radionuclides

A Review of Radiological Surveillance Reports of Waste Effluents in Marine Pathways at the Maine Yankee Atomic Power Company at Wiscasset, Maine, 1970-1984: An Annotated Bibliography

Legacy for Our Children: The Unfunded Costs of Decommissioning the Maine Yankee Atomic Power Station

Anthropogenic Radioactivity: Chernobyl Fallout Data: 1986 - 2001

Patterns of Noncompliance: The Nuclear Regulatory Commission and the Maine Yankee Atomic Power Company: Generic and Site-specific Deficiencies in Radiological Surveillance Programs

Phenomenology of Biocatastrophe Publication Series
using the pseudonym Ephraim Tinkham

Volume 1: Essays on Biocatastrophe

Volume 2: Biocatastrophe Lexicon

Volume 3: Biocatastrophe: The Legacy of Human Ecology

<u>**New England Maritime and Industrial History**</u>

Norumbega Reconsidered: Mawooshen and the Wawenoc Diaspora

Davistown Museum *Hand Tools in History* Publication Series

Volume 6: Steel- and Toolmaking Strategies and Techniques before 1870

Volume 7: Art of the Edge Tool: The Ferrous Metallurgy of New England Shipsmiths and Toolmakers from the Construction of Maine's First Ship, the Pinnace *Virginia* (1607), to 1882

Volume 8: The Classic Period of American Toolmaking, 1827-1930

Volume 9: An Archaeology of Tools: A Catalog of the Tool Collection of the Davistown Museum

Volume 10: Registry of Maine Toolmakers, 6th Edition

Volume 11: Handbook for Ironmongers: A Glossary of Ferrous Metallurgy Terms: A Voyage through the Labyrinth of Steel- and Toolmaking Strategies and Techniques 2000 BC to 1950

Phenomenology of Tools: Philosophical Observations on the Nature of Tool Wielding, revised second edition

Table of Contents

Figures

I. Fukushima Daiichi Disaster

Introduction

This *Handbook* results from over four decades of my research and commentary on anthropogenic radiation, the Maine Yankee Atomic Power Company, and the industrial history of a nation that perfected the manufacture of hand tools and atomic weapons but failed to design and build safe nuclear power reactors, including those it exported to other countries such as Japan. This edition of the *Nuclear Information Handbook* has been published as a result of the Fukushima Daiichi nuclear crisis that began on March 11, 2011 after a 9.0 earthquake off the coast of Japan. The tsunami that followed the earthquake destroyed the backup diesel generators at the Fukushima Daiichi complex. The loss of cooling capabilities that followed resulted in seven separate fuel assembly meltdown accidents, three in reactor vessels (loss of reactor coolant accident; LORCA) and four in their adjacent spent fuel pools (loss of coolant accident; LOCA). These seven accidents may be appropriately defined as a multiple interlocking meltdown event (MIME), a new acronym suitable for describing what is a unique occurrence in nuclear industrial history. Excluding all six other Fukushima Daiichi accident sites, the radiation release from the fires and hydrogen explosion in the spent fuel pool in Reactor 4 may be larger than the source term from the Chernobyl accident.

This *Handbook* begins with a synopsis of the Fukushima Daiichi accident cycle and continues with a description of relevant dosage reporting units, definitions, and concepts pertaining to nuclear accidents. It also includes information about baseline data, plume source points, pathways, and radiation protection guidelines. Much of the general public's fear of "radiation" is based on ignorance of its sources, constituents, and potential health physics impact. This *Handbook* hopes to clarify the confusing reporting units of radiation exposure and measurement and provide guidance for non-experts to evaluate the significance of the Fukushima Daiichi or any other nuclear accident.

This *Handbook* provides:

- An overview and time line of the accident at the Fukushima Daiichi reactor facilities, including a description of the fuel assemblies currently in meltdown status.
- A concise introduction to radiation reporting units, exposure and protection action guidelines, pathways, biologically significant radioisotopes, and contamination levels of concern about point sources of anthropogenic radioactivity in the context of background levels of naturally occurring radiation.

- Access to the most important Internet accident information resources, including those of the Japanese government (NISA and MEXT), International Atomic Energy Agency (IAEA), US EPA, US NRC, IRSN (France), and University of California, Berkeley.
- A comprehensive overview of the radiological impact of the Chernobyl accident. Extracted from the online archives of RADNET, the Chernobyl bibliography and the data it contains provide a useful guideline for evaluating the radiological impact of the accident in Japan.
- Historic contamination baseline data pertaining to weapons testing fallout, nuclear power production, and the many nuclear waste plume sources that characterize the atomic age.
- An exploration of the decommissioning of the Maine Yankee Atomic Power Co. at Wiscasset, ME as a paradigm of the problems encountered at many operating nuclear reactors, including the fuel cladding failure accidents at Maine Yankee, which led to its closing.

Caveat

The *Nuclear Information Handbook* and the RADNET archives provide no specific information on the health physics impact of any nuclear accident or source point of anthropogenic radioactivity. In this *Handbook*, and in the RADNET archives, the term "radiological impact" refers to accident deposition levels of long-lived isotopes, especially the indicator isotope Cs-137, and their concentration levels in pathways to human exposure and consumption. No evaluation of the health physics impact of radiation exposure can be made without this data. The FDA-derived intervention levels, the MEXT radiation and daily life graphic, the Wikipedia sievert exposure guidelines, and RADNETs terrestrial contamination levels of concern and the dose assessment criteria in the radiation protection guidelines reprinted in this *Handbook* provide essential information allowing the non-expert to evaluate the significance of data being reported in Japan, the United States, and elsewhere.

Human exposure to ambient radiation is expressed in sieverts and grays, reporting units essential for estimating the severity of an accident and its immediate impact on humans. Accurate accident dose assessment is contingent upon the measurement of the indicator isotopes radiocesium (Cs-137) and radioiodine (I-131), followed by the documentation of their uptake in the food webs that result in exposure to humans, as expressed in the reporting units described below. An evaluation of the health physics impact of any nuclear accident, including the Fukushima Daiichi disaster, cannot be completed without a comprehensive analysis of the source term (radiation release inventory),

2

deposition concentrations, and pathways of the isotopes discharged during this or any other accident. The information contained in this *Handbook* is intended to help the layperson navigate the labyrinths of the world of nuclear information and its many equivocations and rituals of evasion. It is hoped that readers of this *Handbook* will be able to go beyond the confusing and often misleading jargon of ambient exposure as measured in microsieverts and nanograys and find the answers to fundamental questions about contamination deposition and concentration levels expressed in becquerels per square meter, becquerels per kilogram, and becquerels per liter.

Fukushima Daiichi Accident Status

It is now evident that seven nuclear accidents have occurred at the Fukushima Daiichi complex with the potential to surpass the source term (release totals) of the Chernobyl accident. On Tuesday, April 12, 2011 the Japanese government confirmed this possibility by raising the rating of the severity of the accident to a level 7 on the INES nuclear event scale (see below), estimating that radiation releases had already reached 10% of Chernobyl levels and possibly higher. Commenting on the possible future chronic releases of radiation from this accident, TEPCO executive Junichi Mutsmoto notes, "the radiation leak has not stopped completely, and our concern is that it could eventually exceed Chernobyl." (*The New York Times* April 13, 2011, A-5). In fact, the source term of this complex accident-in-progress has already surpassed the Chernobyl releases.

All seven nuclear accidents in progress at the Fukushima Daiichi reactor complex involved fuel assembly melting events followed by hydrogen explosion blowouts, fires associated with burning zirconium fuel rod claddings, and the continuation of nuclear fuel criticality (fission chain reactions). The first indications of major problems at the reactor complex were the rapid drop in water levels in reactor vessels 1 - 3 and in the spent fuel pools of Reactors 1 - 4. This was followed by fires, which both preceded and followed three hydrogen gas explosions that resulted from the continued generation of heat in the reactor vessels of Units 1 - 3. These explosions, graphically depicted by electronic and print media, discharged large quantities of volatile fission products (Cs-137 and I-131) into the atmosphere where they were dispersed in the immediate vicinity of the reactor complex, eastward over the Pacific Ocean, and to many inland locations.

The hydrogen explosion in the spent fuel pool of Reactor 4 on March 15th, 2011 released a substantial percentage of the content of its 1,479 fuel assemblies, each of which contained at least 10,000 curies of Cs-137. Another 2,889 fuel assemblies were involved in meltdown events in Reactors 1 - 3 and their spent fuel pools. Damage to the primary containment structures at Reactors 1 and 2 and melted fuel in their dry wells may prevent construction of a closed loop cooling system, which the Tokyo Electric

Power Company (TEPCO) indicated is the key step in restoring "normal cooling" to the damaged units at Fukushima Daiichi. Only after closed loop cooling systems are constructed at all seven accident locations will the ongoing, though dramatically reduced, fission process in melted fuel assemblies evolve to a "cold shutdown" status. The accident at the complex will be a long-term chronic point source of radiological contamination as long as some degree of fissile activity is still occurring. Radiation releases, including those via secondary and tertiary pathways, may continue for years and will likely far exceed those from the Chernobyl accident.

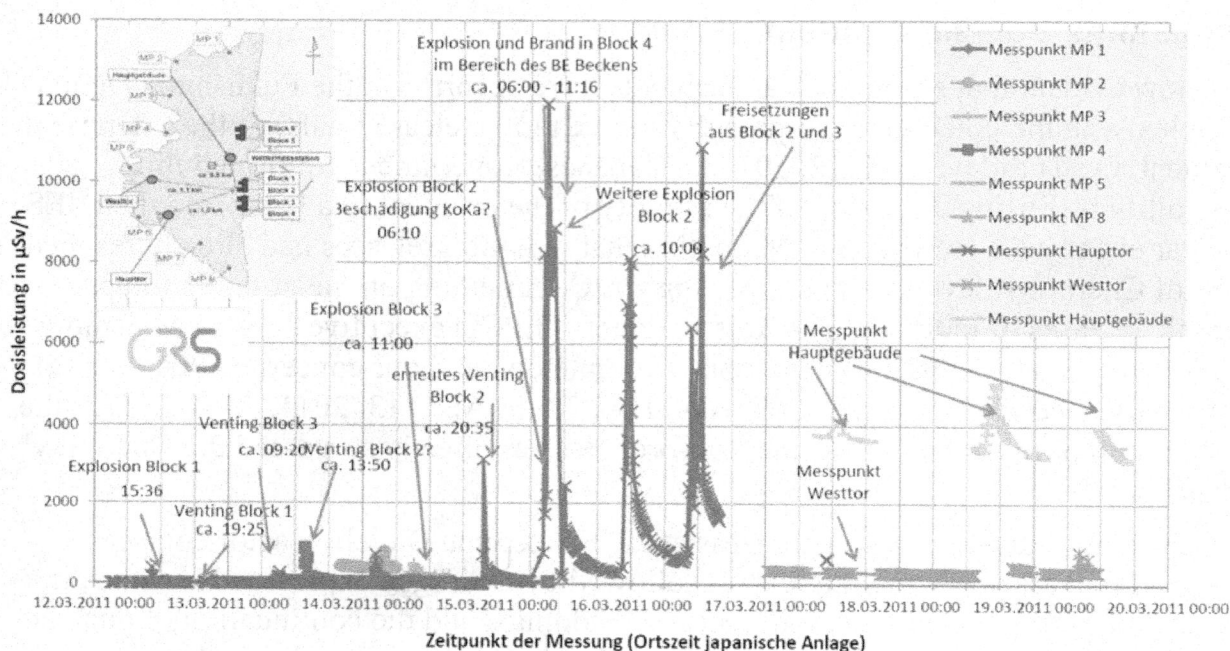

Figure 1. Fukushima Daiichi radiological releases (http://energyfromthorium.com/2011/03/30/areva-fd-presentation/).

At the accident site itself, "peak discharges from the Fukushima Daiichi facility appear to have occurred between March 14th and 16th according to a new assessment… based largely on computer models showing very heavy emissions of radioactive iodine and cesium." (*The New York Times* April 15, 2011). *Figure 1* graphically depicts these emissions though the reporting unit, thousands of microsieverts per hour, tells us nothing about the isotopic concentration levels of these releases. An important component of these peak emission releases occurred when a fire broke out at the spent fuel pool in Reactor 4, which was not operating at the time of the tsunami. A fourth hydrogen explosion was apparently associated with this fire and totally destroyed the outer containment of the spent fuel pool at Reactor 4. The contents of this spent fuel pool had just been removed from this reactor. Peak accident radiation emissions of

4

fission products, including a substantial percentage of the spent fuel pool inventory of Reactor 4, occurred during this fire, which was "successfully contained within a few hours" (TEPCO 2011) only to start up again the next day.

The continuing challenge of cooling seven meltdown events creates the possibility of a future release of large amounts of radioactivity if fuel assembly melting results in the future resumption of high levels of criticality. The evolution of a worst case scenario appears to have been mitigated by ongoing manual cooling efforts. Initially these cooling efforts involved helicopters and fire equipment, an extremely primitive way to control seven ongoing meltdown events. A system of electrically-powered pumps has now been improvised to pump water into the reactor vessels and spent fuel pools, all of which contain damaged fuel assemblies that have suffered loss of cooling. The intense effort by TEPCO to cool the fuel assemblies may be successful in preventing a significant increase in the fission chain reactions essential to the resumption of explosive criticality in the reactor vessels and the fuel assemblies in the spent fuel pools. In the hoped for scenario of continued cooling, there may not be a significant increase in airborne contamination beyond what has already occurred. The continued manual cooling of the reactor vessels and spent fuel pools, whose emissions-containing closed loop cooling systems were destroyed by the hydrogen explosion blowouts, is resulting in large discharges of highly radioactive water into marine pathways. This unexpected washout pathway did not characterize the Chernobyl accident; the last nuclear accident involving discharges to the marine environment occurred at Sellafield in England. The creation, storage, and dispersal/disposal of radioactive water as a result of ongoing cooling efforts may release greater quantities of radiation than the series of hydrogen explosions, which characterized the first stage of the accident. At Fukushima Daiichi, the best case scenario involves the successful construction of seven closed loop cooling systems to capture the emissions from the exposed and melted fuel assemblies, which still exhibit low levels of criticality. The harsh reality of damaged facilities in a highly radioactive environment suggests instead that the struggle to cool the zirconium-clad fuel assemblies in the reactor vessels and spent fuel pools "may continue indefinitely - possibly for three to five years." (*The New York Times* March 30, 2011). The ultimate solution to the problem of disposing of the highly radioactive legacy of this accident will probably be the construction of an entirely new accident waste storage facility.

The accident time line printed in this text provides a snapshot of the sequence of the events during the first five weeks following the accident. Chronic liquid- and steam-associated emissions will continue until the temperature of the melted and/or damaged fuel assemblies are stabilized and are returned to a cold shutdown status. The critical unanswered questions are: how long will accident emissions continue; will their intensity gradually abate; what are their secondary and tertiary pathways and concentration levels; and what is the total accident source term (release inventory)?

Radiological Surveillance Information Availability

The proliferation of information technologies that allow instantaneous posting of data pertaining to nuclear accidents has occurred since the Center for Biological Monitoring (CBM) RADNET website was first established in 1994. The radiological surveillance information links cited in this text provide both bad news and good news about the ongoing accident in Japan. Areas to the west and northwest of the Fukushima Daiichi complex have received fallout comparable to that which resulted from the Chernobyl accident. An extensive network of radiological surveillance information sources has become available on the Internet since the accident began. The online databases of Japan's Nuclear and Industrial Safety Agency (NISA) and Ministry of Education, Culture, Sports, Science, and Technology (MEXT) illustrate the willingness and ability of the Japanese government to provide Internet freedom of information about the environmental impact of the Fukushima Daiichi accident. Accident plumes can be tracked online, for example, at http://www.mext.go.jp/english/radioactivity_level/detail/1303986.htm. France's IRSN website (http://www.irsn.fr/FR/popup/Pages/irsn-meteo-france_19mars.aspx) initially offered an hour by hour video of plume concentrations, which were very helpful in the evaluation of the intensity and pathways of airborne releases (see *Figure 5*). The University of California, Berkeley Nuclear Engineering Dept. has been especially effective at providing timely data about fallout from Japan that is reaching the west coast of the US, as have other US-based online information sources, including the EPA's RadNet monitoring data.

Main stream media also initially reported data previously collected by NISA, MEXT, and US NRC. The *Wall Street Journal* reported that radioiodine has been documented in spinach collected 60 miles southwest of the plant on Friday, March 25th, measuring 54,000 Bq/kg. *The New York Times* reported peak values of Cs-137 had reached 3.7 million Bq/m^2 at a location 25 miles from the crippled reactor complex. The *Times* noted the standard used to remove populations from the Chernobyl site was 1.48 million Bq/m^2 of Cs-137; maximum contamination levels reached in areas as much as a thousand kilometers away from the Chernobyl site were 5.5 million Bq/m^2 (see *Section VIII. Chernobyl Fallout Data* and *Section IV. Baseline Information*)

The most ominous aspect of this ongoing disaster is the danger that it poses to Japanese society and the many communities who harvest the products of terrestrial and marine environments. Also of note is its potential to contaminate food production in locations directly downwind from the accident site, including far distant locations such as the San

6

Joaquin Valley in California, a major source of fresh vegetables for the United States and world grocery stores, and other agricultural production areas in the United States. Isotope-specific data in the reporting units listed later in this text will be essential for concerned citizens to evaluate the significance of contaminated agricultural areas and their food and milk products, as well as the movement of contamination plumes in the marine environment, including tsunami debris and fallout in the great Pacific garbage patch. Rhetorical descriptions of the contamination such as "teeny weenie" (CNN) or "reassuring picture of very low risk" (*Wall Street Journal*) should not serve as a substitute for accurate scientific measurements of contamination reported in easy to understand universal SI reporting units (e.g. becquerels).

Freedom of information about the radiological impact of this accident is accompanied by a lack of information about what exactly has occurred in the reactor vessels in Units 1 - 3 and their spent fuel pools, especially pertaining to their capacity to retain cooling water without leaking. Even more startling is the unwillingness of the media to inquire about the size and movement of the release of radiation that occurred when a hydrogen explosion dispersed a large percentage of the contents of the Reactor 4 spent fuel pool. The US Nuclear Regulatory Commission went on an "extreme alert mode" at its Rockville, MD, headquarters and provided valuable insights into the possible and probable conditions and accident event sequences at the Fukushima Daiichi complex, much of it reported on a daily basis in *The New York Times*. Radiological data surveillance reports compiled in the US between March 11, 2011 and early April indicate that **the radiological impact of fallout reaching the United States has been extremely minimal and poses no current threat to the safety of foods grown anywhere in the United States.** This favorable development should not deter an intense effort to calculate and track an accident source term that has already exceeded that of the Chernobyl disaster.

Accident Site Reactor Design

A brief description of the architectural design of the Fukushima Daiichi complex helps explain the severity and possible complications of accidents that involve fuel assembly melting events at seven locations (3 loss of reactor cooling accidents [LORCA] and 4 loss of spent fuel pool cooling accidents [LOCA]), the first multiple interlocking meltdown event (MIME) to beset a nuclear power complex. As graphically illustrated in aerial photographs, four boiling water nuclear reactors are located in a row (Reactors 1 - 4), with two more reactors separated by several hundred feet of space (Reactors 5 and 6). Each of the four reactors experiencing meltdown events has the same layout. The reactor vessel itself is located within a primary containment structure; this steel containment structure totally envelops the reactor vessel. An outer wall shelters a

7

secondary concrete containment structure enclosing the primary containment structure and its enclosed reactor vessel. Underlying the primary containment is a circular steel "torus" used to collect the steam that powers the electricity-producing turbines. Above and to the right of the primary containment structure, **but within the secondary containment**, lies the vulnerable spent fuel pool of each reactor. Above and outside of the secondary containment structure are highly visible cranes used to move fuel from the reactor vessel to the spent fuel pool. The hydrogen explosions that destroyed the secondary containment structures at all four reactors, including the fuel assembly removal crane at Reactor 3, originated in this underlying torus, which, due to a design flaw, was located underneath instead of within the primary containment structure. The role of improperly located spent fuel pools in these ongoing accidents is discussed in the following essays. Located well away from the four reactors that have suffered extreme damage from hydrogen explosions, the common spent fuel storage facility noted below is currently not at risk of a loss of coolant accident, though it was also damaged by the tsunami, which destroyed the backup diesel generators used to cool Reactors 1 - 4.

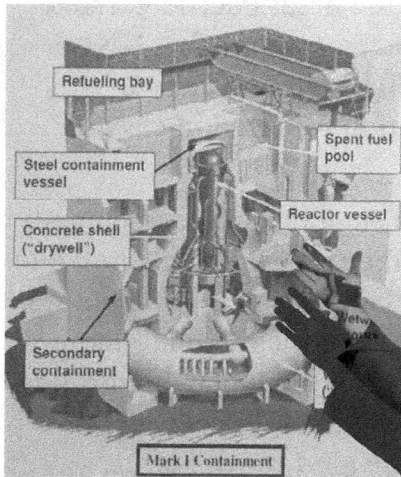

Figure 2. How a nuclear plant works (Reuters 2011).

Fukushima Daiichi: Fuel Assembly Inventory

The ultimate size and impact of the unfolding accident at the Fukushima Daiichi plant in Japan is unknown. Unfortunately, given the complexity of the Daiichi reactor complex and the complicated and interrelated nature of its ongoing sequence of accidents, the worst case scenario could involve radiation releases one order of magnitude greater than those from the Chernobyl accident. Given the number of damaged and/or destroyed fuel assemblies involved in these accidents, the Tokyo Electric Power Company's assertion that the accident source term is 10% of that of Chernobyl is untenable. The radiation released by the explosion in the spent fuel pool of Reactor 4 may approach or equal the Chernobyl source term. The other six ongoing fuel assembly melting events will only add to this record-breaking total. Due to prevailing winds, the marine environment is a recipient of much of these emissions; this meteorological anomaly helps mitigate the impact of these accidents on Japan's crowded urban areas.

On March 18, 2011, *The New York Times* published a summary of the **number of fuel assemblies** in the reactor vessels and spent fuel pools at the Japan facility, the first four of which are the source of accident emissions:

8

- **Reactor 1: 400 in reactor vessel and 292 in spent fuel pool**
- **Reactor 2: 548 in reactor vessel and 587 in spent fuel pool**
- **Reactor 3 with MOX fuel: 548 in reactor vessel and 514 in spent fuel pool**
- **Reactor 4: 0 in reactor vessel and 1,479 in spent fuel pool**
- Reactor 5: 548 in reactor vessel and 826 in spent fuel pool
- Reactor 6: 764 in reactor vessel and 1,136 in spent fuel pool
- A separate common storage spent fuel pool used by all 6 reactors: 6,291

Ongoing fuel assembly melting events are occurring at the reactor vessels of Units 1 - 3 and the spent fuel pools of Units 1 - 4, having a combined inventory of 4,368 assemblies, less than a third of the 13,933 fuel assemblies within the Fukushima Daiichi complex. Each assembly contains at least 10,000 curies of Cs-137, providing an approximate inventory of 43,680,000 curies in the assemblies most at risk. A list of the principal reactor isotopes released at Chernobyl is in *Section III. Additional Definitions and Concepts: Chernobyl Source Terms.* This list highlights the dominance of Cs-137 in the source term of long-lived radioisotopes released during a nuclear reactor accident. The dispersal of biologically significant quantities of Pu-239, Pu-241, and other alpha-radiation emitting isotopes is also an unfortunate component of this accident sequence. Hopefully, a worst case scenario accident involving the intensification of criticality (fissile activity) at one or more of the three reactor vessels and four spent fuel pools still at risk will not occur.

US Nuclear Regulatory Commission Accident Update (*NYT* 4/6/11)

The recent release of the NRC evaluation of the situation at the Fukushima Daiichi complex has brought to light a number of issues pertaining to the future safety of the facility. It is of particular interest that, due to the weakened and compromised structure of the pressure vessels and the stress of the weight of the water needed to control the melting of the fuel assemblies, each reactor is much more vulnerable to both further hydrogen explosions from the cooling of the fuel and further aftershocks. *The New York Times* states, "If the fuel continues to heat and melt because of ineffective cooling, some nuclear experts say that could also leave a radioactive mass that could stay molten for years… The document raises new questions about whether pouring water on nuclear fuel in the absence of functioning electronic cooling systems can be sustained indefinitely… The risks of pumping water on the fuel present a whole new category of challenges that the nuclear industry is only beginning to comprehend." The NRC document "also suggests that fragments or particles of nuclear fuel from spent fuel pools above the reactors were blown 'up to one mile from the units.'" These particles are, in essence, large size hot particles that contain significant quantities of long-lived plutonium and other nonvolatile fission products. Other particulates may consist of CRUD, highly radioactive corrosion and activation products in the reactor core, and thus in spent fuel. The NRC document also makes the anomalous observation that "the worst case solution would be if the water [in the reactor vessel] rose above the fuel level." The *Times* commentary on the NRC report continues "because slumping fuel and salt from seawater that had been used as a coolant is probably blocking circulation pathways, the water flow in number one 'is severely restricted and likely blocked.' Inside the core itself, 'there is likely no water level… It is difficult to determine how much cooling is getting to the fuel.' Similar problems exist in number two and three although the blockage is less severe." The NRC also urged TEPCO to begin filling the reactor vessel with stable nitrogen gas to avoid another hydrogen explosion; this process began on April 7th. The NRC report also indicated that the hydrogen explosion that occurred in fuel pool number 4 resulted in "a major source term release." (*The New York Times* April 6, 2011).

Chernobyl vs. Fukushima Daiichi

The Chernobyl accident involved a single reactor with a radiocesium (Cs-134 and Cs-137) inventory of ±12 million curies. This included as much as 8 million curies of Cs-137, one of the most biologically significant isotopes in an accident plume. Two million seven hundred thousand curies of Cs-137 were discharged into the troposphere as a result of an out-of-control graphite reactor explosion and fire during the accident (Goldman 1987; Aarkrog 1994). Estimates of the Chernobyl source term of I-131 range from 3,180 PBq (85,860,000 Ci) (IAEA) to 5.2 million terabecquerels (140,400,000 Ci) (NISA). In the ongoing accidents at the Fukushima Daiichi facilities, >40 million curies of Cs-137 are at risk of being discharged into the environment in a worst case scenario. The accident is further complicated by the fact that at Reactor 3, MOX fuel, which is partially composed of reprocessed plutonium and uranium oxides, has the potential to release large quantities of plutonium isotopes into the troposphere as the accident unfolds. The fire and explosion at the spent fuel pool at Reactor 4 released a large inventory of fission products and hot particles, also including plutonium, some of which were bulldozed and covered with dirt in the area adjacent to the reactor. The source term sketch below suggests this point source alone may exceed the Chernobyl source term. Ironically, the Fukushima Daiichi emissions are much more invisible than the Chernobyl releases in their pathway movements, as much of the contamination now resides in the marine environment.

An additional complicating factor, not as yet being discussed in the media, is the fact that the Chernobyl accident effectively ended after nine days when the fuel, having melted through the bottom of the reactor structure, which had no containment,

10

solidified. At that time fissile activity ended and the discharge of large quantities of fission products stopped. It is highly unlikely that this solidification process will occur at any of the four reactor facilities now in the process of releasing radioactivity into the environment despite the gradual decline in fissile activity due to cooling efforts. A comprehensive description of the Chernobyl accident did not appear until A.R. Sich completed his MIT PhD thesis (1994) on the accident and began publishing articles in *Nuclear Safety* and *Nuclear Engineering International* (1994-96). A detailed analysis of Cs-137 contamination levels from Chernobyl did not appear until 2006 (Fairlie and Sumner, see *Section IV. Baseline Information*). In the case of the ongoing disaster at the Fukushima Daiichi facility, it may also take a decade to fully document the source term (total release inventory), patterns of dispersion, and accident history. The current assertions that any releases from the accidents will have a minimal environmental and health physics impact have no credibility; the impact cannot be determined until all emissions from the Fukushima Daiichi complex and its spent fuel pools have ended and systematic biomonitoring of all release pathways and contamination levels have been completed. The timeframe for these evaluations will be years, if not decades.

Accident Source Term Sketch

Leak on Reactor No. 2

The Nuclear Regulatory Commission said that some of the core of Fukushima Daiichi's Unit No. 2 had probably leaked from the reactor vessel into the bottom of the containment structure.

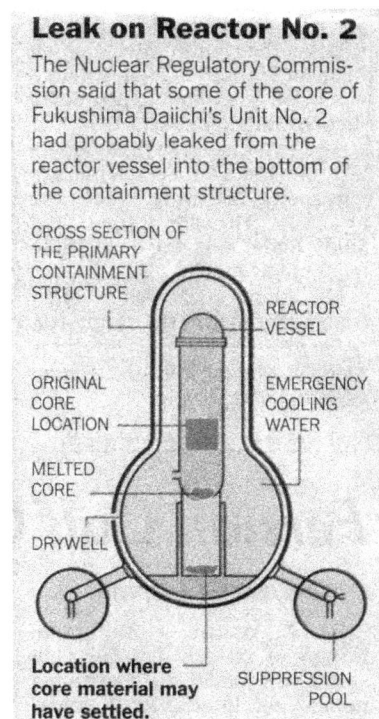

CROSS SECTION OF THE PRIMARY CONTAINMENT STRUCTURE

REACTOR VESSEL

ORIGINAL CORE LOCATION

EMERGENCY COOLING WATER

MELTED CORE

DRYWELL

Location where core material may have settled.

SUPPRESSION POOL

Figure 3. Reactor 2 vessel design. (Glanz April 5, 2011).

The nuclear accidents at the Fukushima Daiichi complex, an ongoing MIME, can be divided into two phases. Phase 1 involved a loss of water coolant in all seven environments. The exposed fuel assemblies began melting; this was followed by the sequence of hydrogen gas explosions, which occurred during the first five days of the accident at Reactors 1 - 3 and the spent fuel pool of Reactor 4. These were followed by five more days of steam and smoke emissions, much of which originated from Reactors 2 and 3. Peak radiation releases occurred during these first ten days. The estimated (modeled) source term of radiocesium during this period are listed below as Phase 1 releases. The large size of these releases is suggested by the terminology "tens of thousands of terabecquerels per hour" (NISA 2011) were released during the early stages of the accident. One terabecquerel = 27 Ci (curies); if 37,000 terabecquerels per hour were released in a series of pulses, which lasted 72 hours, the hourly release rate would be 1 million curies with a total release of 72 million curies. These initial releases contained large quantities of very short-lived radioisotopes,

11

such as I-133 and noble gasses, making it difficult to estimate the release of Cs-137, Cs-134, and other long-lived radioisotopes. While the surveys of ground deposition that followed the Chernobyl accident made it relatively easy to estimate the Cs-137 and Cs-134 source terms, the dispersion of most radiocesium over the Pacific Ocean in the Fukushima Daiichi accident makes calculation of its source term much more difficult. Because the Fukushima Daiichi inventories (7) of Cs-137 are much larger than Chernobyl (± 8 million Ci / 10,000 Ci per fuel assembly = 800 fuel assembly equivalents), a sketch of the Japan accident Cs-137 source term can be made by an evaluation of the condition of the fuel assemblies, all of which have suffered significant damage, melting, and, in the case of the spent fuel pool assemblies, actual dispersion.

Phase 1 Estimated Release

A breakdown of the seven accident source terms can be reasonably estimated utilizing Cs-137 as the indicator isotope. Each fuel assembly contains a minimum of 10,000 Ci of Cs-137. Our modeling of Cs-137 releases at Fukushima Daiichi can then be compared with the Chernobyl release of 2.7 million Ci of Cs-137. In modeling the estimated releases from both the reactor vessels and the spent fuel pool assemblies, it is important to note the severe damage to each group of fuel assemblies once they were uncovered by the loss of cooling water and they began melting. According to the Kyodo News Agency, 70% of the fuel rods in Reactor 1 were damaged. In Reactor 2, the fuel rods were fully exposed for a period of time on Monday, March 14, 2011; 33% of the fuel rods are reported as being damaged. The NRC sketch in *Figure 3* suggests that this might be a conservative estimate as some fuel assemblies have melted into the dry well. The condition of the fuel rods in Reactor 3 is unknown, but they are probably significantly damaged. All the fuel rods in the Reactor 4 spent fuel pool were damaged by the sequence of fires and the hydrogen explosion that dispersed significant portions of the fuel rod fission products onto the plant site as well as out into the ocean on March 15, 2011. Smoke and steam emissions from fuel melting continued for ten days after the peak emissions of March 12 – 15. The fuel assemblies in the spent fuel pools of Reactors 1 - 3 have also experienced significant melting and suffered extensive damage in the hydrogen explosions that destroyed the secondary containment structures in each facility.

Table 1. Fuel assembly synopsis.

Reactor vessel	Reactor assemblies	Spent fuel pool assemblies
1	400	292
2	548	587
3	548	514
4	Empty	(1479)

Reactor vessel	Reactor assemblies	Spent fuel pool assemblies
Total	1496	1393 (+1479)

1. If 5% of the Cs-137 in the reactor fuel assemblies was released out of a total inventory of 14,960,000 Ci of Cs-137, then 748,000 Ci of Cs-137 is the estimated release from these three point sources. This does not include the shorter-lived Cs-134 (1/2 T = 2.1 years).

2. If 10% of the Cs-137 in the spent fuel pools of Units 1 - 3 was released, then 1,393,000 Ci of Cs-137 is the estimated release from these three point sources.

3. If 20% of the Cs-137 in the highly damaged spent fuel pool of Reactor 4 was released, then 2,958,000 Ci of Cs-137 is the estimated release from this source point.

Our modeling of the source term of Cs-137 during the first ten days of these accidents is, therefore, 5,099,000 Ci of Cs-137, compared with the release of 2,700,000 Ci of Cs-137 at the Chernobyl accident. Due to the extreme damage suffered by all seven Fukushima Daiichi Units and the melting and dispersal of reactor and spent fuel, it is highly unlikely that the Fukushima Daiichi source term for Cs-137 is only one tenth that of Chernobyl. Its accurate measurement by ground deposition analyses is problematic as much of the Cs-137 source term was dispersed over the Pacific Ocean.

Phase 2 Estimated Release

The second phase of the accident at the Fukushima Daiichi complex involves the chronic release of low levels of radioactivity. If 37 terabecquerels per hour of radioactivity (not 37,000 terabecquerels) are released for one year, the release rate is 1,000 Ci/hr or 24,000 Ci/day for a total of 8,660,000 Ci of radioactivity. The primary mode of release in phase 2 of this MIME is, and will continue to be, the remobilization and transport of fuel-pellet-derived fission products by ongoing cooling efforts, i.e. the washout pathway. Only if high levels of fissile activity – nuclear chain reactions – reoccur, will airborne emissions exceed those from the washout pathway. If 5% of the total source term of both washout pathway and aerial emissions is Cs-137 (since criticality has subsided but not ended, less short-lived radioisotopes are being released), then chronic releases from all seven fuel assembly melting events would add another 433,000 Ci to the total Cs-137 source term, which would be 5,532,000 Ci. These modeled estimates of the releases from the ongoing accidents in Japan are actually very conservative. There is the possibility that much more than 5% of the reactor vessels now melted fuel assembly fission products have already been released. The same observation could be made about the modeled releases from the spent fuel pools, especially those at

13

the highly damaged Reactor 4 spent fuel pool. The Cs-137 source term at Chernobyl was about 30% of the total of the long-lived isotopic source term. The shorter-lived isotopes, including the radioiodine group, initially dominated the release plumes; after 60 days Cs-137 ($1/2T = 30.174$ years) and Cs-134 emerged as the most important source term isotopes from a health physics perspective.

Accident Event Sequence

After the March 11, 2011 earthquake struck the Fukushima Daiichi complex and triggered the automatic shutdown of Reactors 1 - 3, there was only a 46 minute interregnum before the tsunami destroyed both the backup cooling system and the hydraulic fuel rod insertion equipment in the lower sections of each reactor. The insertion of the control rods slows down and then halts the neutron bombardment of the uranium fuel, which creates the heat that boils the water, as in boiling water reactor. Steam-driven turbines then generate electricity. Due to the severity of the earthquake, the question arises as to whether there was any damage to the fuel assemblies or the control rods that would prevent total control rod insertion to end all fission chain reactions. The same question applies to the valves used to vent accident-related steam pressure (and thus, hydrogen buildup), which failed in the early stages of the accident, a failure that helps explain the hydrogen explosions that occurred in each reactor unit. An extensive account of this debacle is contained in both *The New York Times* and *Wall Street Journal* on May 18, 2011. Was there any lingering fissile activity in the reactor when loss-of-cooling occurred?

Once fuel assemblies begin overheating, the increasing gas pressures within the fuel rods begin to deform the rods. Gas bubbles are formed, the lattice structure of the fuel inside the rods is damaged, and the zirconium fuel rod cladding is embrittled. Mechanical stresses from overheating further damage the fuel rod assemblies. As the fuel rods overheat, the core of the fuel rod expands more than the rim, increasing the surface area of the fuel, and thus, its vulnerability to increasing fissile activity from alpha radiation and neutron bombardment. The increasing heat causes volatile fission products within the fuel rod to burst apart the zirconium fuel cladding, a process that also is enhanced by steam corrosion cracking. The decay heat from the radioactive fuel assemblies or spent fuel, especially that generated by the highly energetic actinides, such as plutonium, causes a drying out of the fuel rod cores as well as fuel rod surfaces, which become too hot for nucleation (water) boiling. A complicated, and not fully understood, sequence of events follows. The intense heat traditionally generated by decaying plutonium and other powerful energy releasing actinides results in radiolysis, which in LOCAs involves the chemical breakdown of water creating highly explosive hydrogen. An intense hot hydrogen reaction flows on and around fuel pellets as well as

14

the now embrittled zirconium cladding. Stress corrosion and cracking of the fuel cladding increases and cladding ductility is lost. As the cladding bursts, the exposed surface area of the fuel is increased allowing accelerated volatile fission product emissions, including radiocesium and radioiodine, which were previously trapped within the fuel rods, 63 of which are in each fuel assembly.

At some point in this process enough heat is generated to breakdown and melt the fuel rod casings (2200 to 3900 °F) and ignite the zirconium fuel cladding, which then burns off, allowing the nucleus of a fuel assembly isotope, such as Pu-239, to more easily absorb a neutron, split into two nuclei, and release its powerful energy (5.1554 MeV) to accelerate the melting process or to begin or continue the fission process. The MIME at Fukushima Daiichi is a previously untried experiment: it is unknown when and where in the fuel melting process melting fuel assemblies will trigger a self-sustaining fission chain reaction and to what extent it can be controlled by manual cooling, especially if melted fuel rods are relocated in hard-to-cool reactor vessel drywells in the form of hot fuel bundles. If the two primary containment vessels in each reactor are damaged and leaking cooling water, how can controlled cooling (a return to consistently lower than boiling water temperatures) be restored in this environment?

Fukushima Daiichi Radiological Surveillance Data Sources

The Nuclear and Industrial Safety Agency (NISA) and Ministry of Education, Culture, Sports, Science, and Technology (MEXT) have provided daily radiological surveillance reports about the extent and impact of atmospheric releases of radioactivity from the meltdown and steam explosions at Reactors 1 - 3; the meltdown, fires, and steam explosion at the spent fuel pool at Reactor 4; and the ongoing steam releases resulting from manual cooling efforts. Unfortunately, information about cumulative deposition levels in the most impacted close-in areas is not available; the Japanese government uses the term "under survey" to avoid reporting the grim news about terrestrial deposition levels that may approach those from the Chernobyl accident. The MEXT website also provides important meteorological information (wind direction, velocity, and precipitation) and "rate of space dose" expressed in nanograys per hour (nGy/h). NISA is reporting offsite background dose rates in un-impacted areas that appear to be close to the IAEA estimate of 0.27 μSv/hr. The extensive discharge of radioactive water to the marine environment and radioactive steam releases from manual cooling will continue indefinitely and will necessitate the compilation of extensive databases pertaining to impacted marine pathways and fisheries, and contaminated terrestrial landscapes and food webs. Currently, there is very little information available about washout pathway-derived contamination in the marine environment. Future monitoring efforts will have to begin with sediment contamination levels, and include phytoplankton, sea vegetables, mollusks, crustaceans, plankton-feeders, and piscivorous (fish-eating) fish. The following links will bring inquisitive visitors to a wide variety of sites providing information on accident contamination levels and dose rates, including an excellent plume cloud video available at the French IRSN website.

Japanese Information Sources

Nuclear and Industrial Safety Agency (NISA); http://www.nisa.meti.go.jp/english/ -- shows countermeasures for the Great East Japan Earthquake

Ministry of Education, Culture, Sports, Science and Technology – Japan (MEXT); http://www.mext.go.jp/english/topics/1303717.htm -- shows radiation data collected in mainland Japan

System for Prediction of Environment Emergency Dose Information (SPEEDI); http://www.bousai.ne.jp/eng/index.html -- shows disaster prevention data

Real time Map of SPEEDI data; http://gebweb.net/japan-radiation-map/jp/ -- shows levels at each monitoring station, color coded

Japan Open Radiation Dashboard; http://www.sendung.de/japan-radiation-open-data/dashboard/ -- shows graphs of radiation data by prefecture

Spreadsheet of Current Reactor Conditions; https://spreadsheets.google.com/ccc?key=0AonYZs4MzlZbdHY4aUJhUlY3Mnd0NVF JRXVidFYtR2c&hl=en#gid=21 -- shows actions being taken on each reactor

U.S. and Other Information Sources

University of California, Berkeley Nuclear Engineering Dept.; http://www.nuc.berkeley.edu/UCBAirSampling -- has data on radiation sampling in milk, rainfall, river water, and air in becquerels and equivalent dose

Institut de Radioprotection Nucleaire (IRSN); www.irsn.fr/EN/Pages/home.aspx -- shows videos of the plume

Where are the Clouds?; http://where-are-the-clouds.blogspot.com/ -- A blog covering the movement and impact of the radioactive plume

EPA's RadNet map; https://cdxnode64.epa.gov/radnet-public/showMap.do

Another RadNet map interface; http://www.epa.gov/japan2011/rert/radnet-data-map.html

Black Cat Systems Online Geiger Counter Nuclear Radiation Detector Map; http://www.blackcatsystems.com/RadMap/map.html -- Amateur network of Geiger counters

RadiationNetwork.com; http://www.radiationnetwork.com/ -- another amateur Geiger counter network

Oregon State Department of Health Monitoring Data; http://public.health.oregon.gov/Preparedness/CurrentHazards/Pages/DailyAirMonitoring.aspx#gamma -- Updated daily

Texas A&M Plume Trajectory Projections; http://csrp.tamu.edu/earthquake/earthquake/Maps.html

MIT Nuclear Information Hub; http://mitnse.com/

AREVA North America: Next Energy Blog; http://us.arevablog.com/

Please email us at: tech@davistownmuseum.org with additional suggestions about other sources of information on Fukushima releases and plume pathways.

Fukushima Daiichi Accident Time Line

Date	Incident
3/11	A 9.0 earthquake is recorded 72 km (45 miles) east of Tōhoku, located northeast of the Fukushima Daiichi nuclear reactor complex with the hypocenter at an underwater depth of approximately 32 km (19.9 miles). A tsunami of historic proportions obliterates the backup generators used to cool the fuel assemblies in the event of the loss of electrical power needed to facilitate the shutdown of an operating reactor.
3/12	A hydrogen gas explosion occurs in Reactor 1, which "tore off the outer wall and roof, but the primary containment vessel surrounding the reactor remained intact."
3/14	A hydrogen gas explosion occurs in Reactor 3, "core melting is presumed. The outer structure of the reactor building was torn away on Monday." TEPCO has also indicated the fuel in the core of Reactor 2 was completely uncovered for a matter of hours.
3/15	A hydrogen explosion occurred in Reactor 2, followed by an explosion and fire in the spent fuel pool of Reactor 4. TEPCO reported that this explosion blew a 26-foot hole in the side of the housing of the spent fuel pool, providing direct venting of any spent-fuel-pool-derived emissions. "Great danger exists if the pool of spent fuel becomes uncovered. The same is true of the other reactors." Perimeter radiation levels rise to 11.9 mSv/hr; onsite levels are reported as 400 mSv/hr. Readings of 0.085 mSv/hr are reported at Toki, 75 miles from the reactor complex. The explosion in Reactor 2 "almost certainly damaged the inner steel containment

Date	Incident
	vessel raising the prospects of a full meltdown of the nuclear fuel inside."
3/16	*The New York Times* reports that yesterday's hydrogen explosion in Reactor 2 originated in "the torus in the cooling area of the number 2 reactor."
3/18	Traces of I-131 and Cs-137 are detected in the air at Sacramento, CA.
3/19	The spent fuel pools at Reactors 5 and 6 are reported to be in "cold shutdown mode."
3/20	Smoke can be seen coming out of Reactor 3 with an unexplained pressure spike.
3/21	Unexplained smoke and steam emissions are coming from Reactors 2 and 3. Fukushima Daiichi reactor-derived radioactive fallout is reported in Charlottesville, VA.
3/22	Pressure in Reactor 1 is recorded as stable. Reactor 4 spent fuel pool is leaking water. "The storage pools… at the number 2 reactor, which holds spent fuel rods, was spewing steam late Tuesday." The IAEA "lacked data about the temperature of the pools holding spent fuel rods at the numbers 1, 3, and 4 reactors." The IAEA reported "radiation levels 1,600 times above normal 12 miles from the plant."
3/25	Four of eight samples of weeds taken at Iitate village between March 18th and 24th had I-131 levels above 1 million Bq/kg with a range of 400,000 to 2,540,000 Bq/kg.
3/26	Eight of nine samples of weeds taken at Iitate village between March 18th and 26th had Cs-137 levels above 1 million Bq/kg; the peak value was 2,870,000 Bq/kg on 3/26.
3/29	TEPCO officials indicate "the isotopic composition of the plutonium found at Fukushima Daiichi suggest the material came from the reactor site." (http://where-are-the-clouds.blogspot.com/).
3/30	Little Rock, Arkansas – 8.9 pCi/l Iodine-131, milk (EPA RadNet lab analysis).
3/31	Energy Secretary Steven Chu reports 70% of the core at Reactor 1 is uncovered and estimates that 33% of the core at Reactor 2 is uncovered. "Large amounts of harmful radioactive elements were found in soil samples of the village Iitate" (*Wall Street Journal* March 31, 2011) located 40 miles northwest of Fukushima Daiichi.
4/1	"Highly radioactive water [is reported] in the turbine buildings attached to Reactors 1, 2, 3, and 4."

Date	Incident
4/4	Water in the pit of Reactor 2 is reported as containing 10^6 (10,000,000) Bq/l of I-131. Dutch Harbor, Alaska – 2.42 pCi/m^3 Iodine-131, air (EPA RadNet).
4/5	Seven tons an hour of highly radioactive water are being discharged into the ocean from the turbine room of Reactor 2. This water "has flooded areas of the plant creating new complications in the effort to stave off full meltdowns of the fuel…" Radiation levels 19 miles off shore are 1/1000th of those 360 yards off shore from the discharge pipe – "the level of radiation at 19 miles offshore was still hundreds to thousands of times as high as levels sampled at the same site in 2005."
4/7	TEPCO workers began injecting nitrogen into the containment vessel of Reactor 1 to help mitigate the possibility of another hydrogen gas explosion.
4/12	"Japan has decided to raise its assessment of the accident at the crippled Fukushima Daiichi nuclear power plant to the worst rating on an international scale, putting the disaster on par with the 1986 Chernobyl explosion… tens of thousands of terra becquerels" of radioactive materials per hour were released from the plant in the aftermath of the tsunami.
4/13	"The radiation leak has not stopped completely, and our concern is that it could eventually exceed Chernobyl," Mr. Matsumoto said. (www.nytimes.com/2011/04/13/world/asia/13japan.html?_r=1)
4/17	TEPCO authorities note the temperature in Reactor 2 "appears to be rising." Professor Tetsuo Matsumoto from Tokyo City University is noted as stating, "How long the decommissioning process would take depended heavily on the state of the nuclear fuel… Will it still be shaped like rods? Or will it have melted and collapsed into a big mass? …It could be ten years, or it could be thirty. You just won't know until you open up the reactor."
4/18	"The presence of highly radioactive water at unit 2 posed a particular challenge… Workers have been cooling nuclear fuel at the reactor's core and in storage pools by pumping in hundreds of tons of water a day, producing dangerous amounts of runoff."
4/19	"Hidehiko Nishiyama, the Deputy Director General of the Nuclear and Industrial Safety Agency, said the authorities were looking for ways to shore up the bottom of the spent uranium fuel-rod storage pool at Reactor 4 to prevent it from collapsing."

Date	Incident
4/20	The *Wall Street Journal* reports "water at the center of the current efforts is in the basement of the number two reactor building and totals some 25,000 tons, according to TEPCO." The *Journal* also reports ocean water outside the plant "has been measured as having 13 million Bq of I-131 per cubic centimeter [13 billion Bq/l], 300 million times the legal limit and 3 million Bq of Cs-137 per cubic centimeter [3 billion Bq/l], which is 30 million times the limit."
	Hidehiko Nishiyama of Japan's Nuclear and Industrial Safety Agency noted that, "heavily contaminated water that has accumulated in basements and trenches at the site is 2 million times as radioactive as the less contaminated water that workers pumped into the ocean from April 4th to April 10th. . . In a further effort to improve cooling . . . a decision had been made to flood the primary containment vessels of the number 1 and number 3 Reactors with enough water to cover up the sides of the reactor pressure vessels up to the level of the uranium fuel rods."
	Matthew L. Wald, writing in *The New York Times*, notes, "Tokyo Electric Power has only a few weeks to patch up the three smashed secondary containments before the coming rainy season, when downpours could wash more contamination into the environment."
4/27	The *Wall Street Journal* reports, "Currently, an estimated 87,500 tons of water has collected in the basements and utility trenches surrounding four of the plant's reactors, according to TEPCO. In addition, about 500 tons of water is being injected into the reactors every day, further adding to the stock of flooding water."
4/29	Due to the Royal Wedding, and the huge outbreak of tornadoes in the southern US, already declining media reports on the status of the Fukushima Daiichi disaster have become nonexistent. No news is good news.
5/2	Osama bin Laden is killed by a Navy Seal strike in Abbottabad. There's not even the slightest amount of room in *The New York Times* and elsewhere for updates about chronic emissions from the MIME in Japan.
5/3	"Japan fades as a cause of concern when the big picture is considered… Drumbeat of nuclear fallout fear doesn't resound with experts… [Exposure to] natural radiation is far bigger than all the manmade emissions, including the current increase from the crippled Japanese reactors… 'It disappears as a contributor to population radiation doses,' said Frank N. von Hippel, a nuclear physicist who advised the Clinton administration and now teaches at Princeton University. But the fear of radiation is different. 'Somehow,' Dr. von Hippel added, 'nuclear things get stigmatized relative to their statistical risks.'" The Fukushima Daiichi disaster

Date	Incident
	is no longer a newsworthy event. Any future reports of exposure to the source term from this MIME will be reported in banana equivalent doses (BEDs).

All quotations are from *The New York Times* unless otherwise noted.

Site and Media Specific Radiological Data

Detailed information about the radiological impact of the Fukushima Daiichi disaster is available online on Japanese government and other websites. Luckily, consistent offshore winds have enabled much of Japan's mainland to escape the most intense terrestrial fallout; in contrast variable winds spread the Chernobyl source term throughout portions of northern and western Europe as far south as Turkey. Not all fallout from the Fukushima Daiichi accident went over the Pacific Ocean. There were sufficient wind shifts from the prevailing easterly transpacific flow to bring significant quantities of radioactive emissions to inland areas. The data listed below under terrestrial fallout illustrate the intensity of the source term releases of biologically significant isotopes in locations in Japan that received a few hours of often-rainfall-associated fallout. The city of Tokyo and its 15 million residents had the luck of the Irish: they were upwind of a nuclear accident that almost certainly released more radioactivity than the Chernobyl accident. In the event of a complete meltdown of the fuel assemblies located in the reactor vessel of unit 2, (the most likely location of any future large accident releases), or with respect to chronic aerial emissions that may persist for years, one wonders if the prevailing winds will continue to transport emissions away from Tokyo and into the Pacific Ocean.

The brief listings below are a small sample of the huge online database of information already compiled about Fukushima Daiichi source term releases.

Japan

Date	Location	Isotope	Media	Quantity	Source
3/15/11	Reactor site	Integrated dose	Air	8,217 μSv/hr	*The New York Times*
3/20/11	60 km NW	I-131	Dust	203 Bq/m^3	MEXT
3/23/11	30 km NW	Integrated dose	Air	8,985 μSv*	MEXT
3/23/11	Tokyo	I-131	Tap water	210 Bq/l	NISA
4/1/11	75 miles NE of Tokyo	I-131 Cs-137	Fish (Kounago or sand lance)	4,080 Bq/kg 526 Bq/kg	Japanese government

*The average dose over 23 hours and 58 minutes was 725 μSv or 30.3 μSv/hr.

United States

Date	Location	Isotope	Media	Quantity	Source
3/18/11	Berkeley, CA	I-131	Rainwater	4.2 Bq/l	www.nuc.berkeley.edu
3/23/11	Berkeley, CA	I-131	Rainwater	20.1 Bq/l	www.nuc.berkeley.edu
3/24/11	Berkeley, CA	Cs-137	Rainwater	0.59 Bq/l	www.nuc.berkeley.edu
3/26/11	Berkeley, CA	Cs-137	Rainwater	0.5 Bq/l	www.nuc.berkeley.edu
3/30/11	Berkeley, CA	I-131	Milk	0.70 Bq/l	www.nuc.berkeley.edu

Other Locations

Date	Location	Isotope	Media	Quantity	Source
3/18/11	Scandinavia	I-131	Air	< 0.30 mBq/m^3	www.epa.gov/japan2011
3/18/11	Netherlands	I-131	Air	0.17 mBq/m^3	www.epa.gov/japan2011

Fukushima Terrestrial Fallout Monitoring Samples

Sampling Location	Sample Type	Sampling Date	Radioactive Concentration in Bq/kg	
			I-131	Cs-137
Itake Village (40 km NW of FD site)	Weed (Leaf Vegetable)	3/18/11 12:20	2,520,000	1,800,000
Itake Village	Weed	3/20/11 12:40	2,540,000	2,650,000
Itake Village	Weed	3/22/11 12:00	1,110,000	1,600,000
Itake Village	Weed	3/24/11 13:05	805,000	1,050,000
Itake Village	Weed	3/26/11 12:00	1,030,000	2,870,000
Itake Village	Weed	3/28/11 11:50	381,000	480,000
Itake Village	Weed	3/30/11 12:25	576,000	1,890,000
Kawamata Town (45 km NW of FD site)	Weed	3/18/11 11:45	173,000	72,800
Kawamata Town	Weed	3/21/11 12:03	315,000	120,000
Kawamata Town	Weed (washed)	3/23/11 11:30	74,400	23,100
Kawamata Town	Soil	3/25/11 15:05	112,000	21,800

Sampling Location	Sample Type	Sampling Date	Radioactive Concentration in Bq/kg	
			I-131	Cs-137
Kawamata Town	Weed	3/26/11 11:20	79,500	54,700
Kawamata Town	Weed	3/29/11 11:00	71,900	67,900
Kawamata Town	Soil	3/31/11 13:40	14,700	949
Tamura City (40 km W of FD site)	Weed	3/18/11 11:35	36,000	40,100
Tamura City	Weed	3/21/11 12:30	30,800	25,000
Tamura City	Weed	3/24/11 11:35	29,400	32,600
Tamura City	Weed	3/27/11 11:45	33,300	19,800
Tamura City	Weed	3/30/11 12:30	18,600	18,800
Minamisouma City (25 km N of FD site)	Weed	3/18/11 13:30	88,600	17,800
Minamisouma City	Weed	3/3/22/11 13:35	140,000	17,200
Minamisouma City	Weed	3/26/11 13:50	83,700	10,500
Minamisouma City	Weed	3/30/11 14:45	113,000	13,100
Ono Town (40 km SW of FD site)	Weed	3/18/11 12:35	181,000	28,300
Ono Town	Weed	3/19/11 12:15	201,000	73,800
Ono Town	Weed	3/30/11 11:08	10,300	6,280
Ono Town	Weed	3/31/11 11:11	9,960	6,00
Iwaki City (45 km S of FD site)	Weed	3/18/11 13:15	690,000	17,400
Iwaki City	Soil	3/19/11 13:15	12,600	288
Iwaki City	Weed	3/24/11 15:00	154,000	6,210
Iwaki City	Soil	3/25/11 13:45	23,900	519
Iwaki City	Weed	3/27/11 12:30	126,000	7,470

Sampling Location	Sample Type	Sampling Date	Radioactive Concentration in Bq/kg	
			I-131	Cs-137
Iwaki City	Soil	3/31/11 12:51	8,370	150
Kawamata Town (35 km NW of FD site)	Weed	3/25/11 15:07	663,000	497,000
Kawamata Town	Weed	3/31/11 13:40	227,000	465,000
Date City (50 km NW of FD site)	Soil	3/24/11 12:10	41,200	6,850
Date City	Weed	3/25/11 16:18	77,100	40,700
Date City	Weed	3/31/11 14:25	22,500	24,500
Date City	Soil	3/31/11 14:25	27,200	6,740
Nihonmatsu City (45 km NW of FD site)	Weed	3/25/11 11:40	73,400	235,000
Nihonmatsu City	Soil	3/25/11 11:35	32,900	9,330
Nihonmatsu City	Weed	3/31/11 10:50	17,700	131,000
Nihonmatsu City	Soil	3/31/11 10:50	24,400	14,200
Ookuma Town (5 km SW of FD site)	Soil	3/31/11 13:00	423,000	98,100

Isometric surveys of the ground deposition of accident-derived Cs-137 or Cs-134/137 composites expressed in Bq/m^2 are not yet available. No accident impact evaluations can be made without this data, which will hopefully become available to the Japanese public at some time in the future. Information about source term ground deposition will be an essential component of future decisions pertaining to resettlement or further evacuation of impacted areas in Japan.

Fukushima Daiichi Subprime Boiling Water Reactor Design

Four important design characteristics differentiate the boiling water reactors at the Fukushima Daiichi complex from pressurized water reactors, such as those used at Three Mile Island and the now-decommissioned Maine Yankee plant at Wiscasset, and make them much more vulnerable to accident scenarios. These characteristics differentiate the "subprime" design of boiling water reactors from the more durable pressurized water reactors.

24

- The spent fuel pool is located adjacent to and above the reactor vessel primary containment structures but within the secondary containment structure. In pressurized water reactors, the spent fuel pool is located well away from the reactor vessel and outside of the secondary containment structure.

- A large donut-shaped tank for containing excess steam, known as a "torus," is located at the base of the reactor vessel but is not enclosed within the primary containment structure.

- The efficiency of the torus in collecting and transferring heat generated in the reactor vessel inadvertently resulted in the design flaw of plant vulnerability to hydrogen explosions. The torus was the repository of hydrogen generated by the breakdown of water molecules by neutron bombardment from radiolysis, an integral component of the renewed fissile activity. This vulnerability is due to the fact that the force of any hydrogen explosion in the torus would be directed in part to the exterior of the primary containment vessel, impacting the integrity of the secondary containment structure and its vulnerable tenant, the spent fuel pool. In contrast, the "very large very strong but very costly containment" allowed the Three Mile Island facility to survive a hydrogen blast (*The New York Times* March 20, 2011).

- The control rod insertion equipment was located underneath the reactor vessel, in contrast to pressurized water reactors, which have control rod insertion equipment on top of the reactor vessel. At Fukushima Daiichi, this control rod equipment was destroyed by the tsunami.

After the cooling of three reactor vessels and four spent fuel pools was halted by the tsunami, resulting fuel assembly overheating and melting caused the accumulation of hydrogen in the underlying torus wet well. In a pressurized water reactor with a gravity fed control rod mechanism at the top of the reactor vessel, this outcome could have been avoided. Instead, a sequence of hydrogen explosions occurred in each of the torus wet wells that underlie Reactors 1 - 3. These explosions destroyed the two outer layers of the secondary containment structure that enclosed the primary containment structure that held the reactor vessel. They also resulted in severe damage to the spent fuel pools that are located within the secondary containment structure of each reactor. The spent fuel pool at Reactor 4 was also severely damaged by a hydrogen explosion that also probably originated in the underlying torus, possibly due to hydrogen that leaked from Reactor 3. At Fukushima Daiichi, these secondary containment structures were turned into rubble. Damage to the steel primary containment structure surrounding the reactor vessel is a possibility, but information about the extent of such damage is not yet available. The severity of damage to the reactor buildings is graphically illustrated in media coverage of this accident. This damage includes the destruction of the fuel assembly crane at Reactor 3, which may have damaged its reactor vessel (Spotts 2011). During the loss of reactor coolant accident (LORCA) at Three Mile Island, the robust nature of the steel containment vessel contained the hydrogen explosion, which would otherwise have led to a full meltdown event. *The New York Times* article (March 20, 2011) on the design flaws of boiling water reactors also notes the disadvantage of locating the spent fuel pool in an elevated position so that it can easily receive spent fuel; it is very vulnerable to shaking during earthquake events and is much more easily damaged than a conventional underground spent fuel pool. David Lockbaum of the Union of Concerned Scientists, a critic of boiling water reactor designs, is quoted in this article as noting the interconnected vulnerability of both reactor pressure vessel and spent fuel pool to accidents that automatically involve both in unfolding accident scenarios: "If you have a problem with the reactor or the spent fuel pool, you will have a problem with both." In this context, the Fukushima Daiichi accident provides the paradigm for a novel form of nuclear accident: a **multiple interlocking meltdown event (MIME),** in which each loss of coolant accident accelerates the deterioration of cooling capabilities of closely adjoining spent fuel pools and neighboring reactors. Loss of coolant in the spent fuel pool of any boiling water reactor can quickly translate into problems in cooling fuel assemblies in the reactor vessel. The chances for a total meltdown event at any of the seven accident locations and its evolution into an even larger MIME are diminishing as melt event temperatures and, thus, emissions are gradually falling.

General Electric, current owner of NBC News, provided the design for construction of the reactor group installed at the Fukushima Daiichi site. Thirty-five of these boiling water reactors (BWR) are located at nuclear power plants in the United States. Essentially a low budget "subprime reactor design," the construction of the Fukushima Daiichi complex, as well as BWRs throughout the world, included the negligent installation of spent fuel pools adjacent to and slightly above the reactor vessel, a poorly designed heat recovery system (the torus) that resulted in the hydrogen explosions in Japan, and a vulnerable hydraulic control rod insertion mechanism under the reactor vessel. The design of the Fukushima Daiichi reactor complex, the construction of which provided extensive profits to Wall Street investors, gives an ominous new meaning to the term "subprime real estate."

Conundrum: Reactor Vessel Integrity

A question arises with respect to the manual cooling of the three reactor vessels at Fukushima Daiichi that are experiencing meltdown events. Reactor pressure vessel integrity is critical in preventing release of fission products to the environment and containing the gases emitted as a result of the zirconium cladding failures and fires that characterize the ongoing accidents at this location. Hydrogen explosions have blown the roofing off all four spent fuel pools. Manual cooling, which has prevented an increase in melted fuel temperatures, has and will continue to result in the washout of spent fuel pellets and other fissile material from the four spent fuel pools at risk. Manual cooling via a network of intake valves that are also probably damaged is being used to cool the three reactor vessels. This raises the following questions: to what extent has the damage to the primary containment structure and its reactor vessel hindered ongoing cooling efforts? What will be the long-term chronic reactor vessel source term releases in the form of both radioactive water and radioactive steam? If the steel encased reactor vessels can be easily cooled by streams of water, how effective could they be in retaining the fission products that are a natural product of the ongoing meltdown events? If emergency workers are injecting water into the pressure vessels through existing valves and intake piping, aren't these same damaged reactor water cooling system components a pathway for the rapid accumulation of radioactive water in the basements and tunnels of the reactor complex and ultimately in the marine environment?

Cooling Event Releases

Significant source term releases resulting from Fukushima Daiichi meltdown events are associated with manual cooling efforts, which initially used Tokyo fire equipment. These cool-down releases are associated with plumes of evaporated water (steam), which transport fission products into the atmosphere. The term "feed and bleed" refers to the steam pressure releases that result from these cooling attempts. Cooling event releases also take the form of washout events, whereby fission products being released from damaged and/or melted fuel rods in the spent fuel pool are flushed into reactor cellars, tunnels, and other structural locations before being released to the marine environment or recovered in temporary radioactive water storage units such as barges and tanks. In the ongoing battle to prevent the resumption of criticality, or the acceleration of pockets of ongoing criticality in the masses of "hot" fuel, feed and bleed steam releases will characterize the Fukushima Daiichi accident scenario, until all seven fuel assembly groupings are relocated in a secure enclosure that will capture all steam, water vapor, and volatile gas emissions, as is the case in a normally-operating nuclear reactor. It may be appropriate to describe the accident at the Fukushima Daiichi complex as a **continuum** of radiation releases where low-levels of criticality-derived emissions will join steam, water vapor, and radioactive water releases in a series of chronic source term discharges to the environment of historic duration. One of the major challenges of the Fukushima Daiichi recovery process will be how to store and dispose of radioactive water captured from ongoing fuel assembly cooling efforts. Buildup of sea salt residues and the continuing deformation and possible combustion of the zirconium-clad fuel assemblies within the reactor vessels and spent fuel pools will further complicate accident remediation activities. It is already obvious that the marine environment will become the ultimate repository of much of this radioactive water. Once in the marine environment, tracking all source releases is much more difficult than when contamination is deposited in terrestrial environments and thus in more easily measured food webs. A key question about the accident source term will be: how much radiation was released to the environment by the atmospheric and washout pathway releases and where did it go?

Fukushima Daiichi: A Black Swan Event

When Nassim Nicholas Taleb wrote *The Black Swan: The Impact of the Highly Improbable* in 2007, he was anticipating unexpected events such as the subprime real estate crisis, which lead to the financial meltdown of 2008. The earthquake and tsunami of March 11, 2011 resulted in the mother of all black swan events, the loss of cooling at three nuclear reactors and their associated spent fuel pools. The meltdown events that followed already constitute the most complicated nuclear accident that has ever occurred. This particular black swan event is characterized by an unanticipated and ongoing series of steam vapor releases to the atmosphere and radioactive water discharges to the marine environment. The next worst case scenario involves the evolution of one of the current meltdown events into a full meltdown at either a reactor vessel or a spent fuel pool, in which case the Fukushima Daiichi reactor complex will become too radioactive to allow further effective accident mitigation efforts. The ultimate worst case scenario would be the gradual evolution of full meltdown events at all three at-risk reactor vessels and at all four at-risk spent fuel pools. David Lochbaum's comments about the vulnerability of co-dependent reactor vessels and spent fuel pools are particularly relevant to this scenario (*The New York Times* March 20, 2011). The world now prays that this will not occur. A more remote possibility is the spread of loss of coolant capabilities to the spent fuel pools at Reactors 5 and 6, as well as to the large adjacent common spent fuel pool storage area containing 6,291 fuel assemblies, raising the specter of a world catastrophe, including contamination of the world's food supply. This penultimate nuclear disaster is very unlikely, but the vulnerability of boiling water reactors to fuel assembly meltdown events should be an ongoing concern wherever they may operate.

Disaster Timeframe

The duration of the Fukushima Daiichi disaster is a key unknown that will greatly influence the amount of radioactivity released. After nine days, the "corium" of the

Chernobyl graphite reactor melted and flowed into the lower regions of the reactor building and then solidified. "The decay heat dropped due to the uptake of surrounding materials (the stainless steel and serpentine corners of the lower biological shield), combined with rapid spreading of the melted fuel up to 40 meters from the epicenter of the melted corium." (Sich 1996). The unfolding nuclear disaster at the Fukushima Daiichi reactor site in Japan is unlikely to be characterized by the accidental and unexpected sudden decline of the decay heat, which is the key element in halting any loss of reactor coolant accident (LORCA) or any loss of coolant accident (LOCA). Rather, the Japan MIME is a two-phase accident with peak emissions reaching "tens of thousands of terabecquerels per hour" (NISA April 12, 2011) during the first ten days of the accident. **Chronic relatively low levels of radioactive steam and radioactive washout water emissions, possibly measured in terabecquerels per hour, will continue for months, if not years.**

The best case scenario for the ongoing accidents at Fukushima Daiichi is a gradual diminishing of the temperatures in the seven reactor assembly groups now experiencing melting events. If the ideal of the construction of a closed loop cooling system can be accomplished in the future, this will be a major step in preventing future emissions via both the atmospheric transport and washout pathways. Unfortunately, huge releases of radioactivity have already been discharged during this accident. Much of this source term release has been localized in and around the plant, including the discharge of as much as 25% of the inventory of the spent fuel pool at Reactor 4. The remobilization and secondary transport of hydrogen explosion, steam, smoke and washout pathway fission product emissions will continue for years and possibly for decades. Rainfall events, which can mobilize and transport terrestrial Fukushima Daiichi-derived contamination, will be an ongoing cause of contamination movement. The huge number of workers at the accident site and their transportation equipment will be an important secondary pathway for contamination movement. Another secondary pathway, which may become a very important future source of public concern, involves a transoceanic movement of the long-lived contamination emitted from all seven point sources at the Fukushima Daiichi complex. At the time of the explosion of the spent fuel at Reactor 4, luckily, the prevailing winds were blowing into the Pacific, where a significant portion of the contents of the fuel assemblies of Reactor 4 now reside. An unknown component of the disaster timeframe involves the degree to which emissions from Fukushima Daiichi are deposited and transported from the waters near the accident site to other locations. Much of the contamination, which prevailing winds bring over the Pacific Ocean, will reside in sediments at the ocean bottom. Some, however, will be transported on tsunami debris or by tidal currents to the North Pacific garbage patch and food webs of the Pacific Ocean. The actual percentage of the source term that travels via ocean currents and eventually arrives on the shorelines and beaches of North America, including Hawaii, Alaska, Washington, Oregon, and California, may never be known, but, nonetheless, may provide significant exposure for fishermen, recreational boaters, seaweed harvesters, and beachgoers for decades into the future. The same concern exists for Korean and some Chinese shorelines and fishermen.

Fukushima Daiichi vs. Three Mile Island Cleanup

The April 20, 2011 edition of *The New York Times* included an article by Matthew Wald with commentary about the ordeal of cleaning up the Three Mile Island facility.

> Lake Barrett, the senior Nuclear Regulatory Commission engineer during the early phases of the cleanup, said, by comparison, "It was a walk in the park compared to what they've got... In Japan, four separate reactors are damaged and fixing each one is complicated by the presence of its leaking neighbors. . . It will also require a major infusion of equipment to replace parts far from the reactor's core, like pumps and switchgear that were destroyed by the tsunami... At Three Mile Island, the water in the reactor building and the primary auxiliary building gave radiation doses as high as 1,000 REM per hour," said Ronald L. Freemerman... the Project Manager of the cleanup... "That meant a worker would hit the NRC's annual limit in about a minute." Engineers from Three Mile Island laid out the next three steps... first, decontaminate the floors and walls to hold down the potential radiation dose, second, rebuild the secondary containments of Units 1, 2, and 4 and fix or replace the heavy cranes just beneath their ceilings. That would allow workers to defuel the reactor. That step alone took five years at Three Mile Island, where no building had to be rebuilt... third, peek inside the reactor vessel to figure out what tools would be needed to remove the wrecked fuel in the core. "Three Mile Island was a surprise," Mr. Freemerman said, "because so much of the core had melted and flowed beneath a support made of five plates of thick steel."... Only then did they realize they would need new remote-controlled tools to cut through the metal, to get to the material below... They also used long handled picks and scoops to break apart the fused mass of ceramic fuel pellets and metal... Eventually, about 150 tons of radioactive rubble was shipped to an Energy Department laboratory in Idaho Springs, ID, where it still sits, waiting as all used American fuel does, for a final resting place.

Cleanup after the Three Mile Island meltdown took 14 years and cost 1 billion dollars. The big difference between the Three Mile Island accident and the seven ongoing accidents (MIME) at Fukushima Daiichi, other than the much larger inventories of radioactive waste available for release in the Japan complex, is that the more robust containment building at Three Mile Island remained intact after the hydrogen explosion, which is an integral part of most meltdown accident scenarios. The result of the explosion at Fukushima Daiichi was and continues to be the widespread dispersal of radioactive contamination in and around the damaged buildings at the reactor sites, which can no longer safely contain their now semi-molten contents. The future risk of worker exposure will be a major impediment to repair efforts and the construction of closed loop cooling systems, including the reconstruction of the Reactor 4 secondary containment and spent fuel pool structures.

The Legacy of Plutonium

The nuclear age was engendered when the first transuranium element, neptunium-239 (1/2 T = 2.355 days), was synthesized in 1940. 2.355 days later, its daughter product, plutonium-239 was created. The age of nuclear fission chain reactions was underway. From the proliferation of nuclear weapons to nuclear power plants, including those at Fukushima Daiichi, Pu-239 is the enabler of the nuclear age. Plutonium-239 is a handy piece of technological innovation: upon its decay it produces intense energy, radiation, and fast neutrons, which once slowed are the key element in self-sustaining fissile activity. The energy that heats the water that creates the steam to drive the turbine originates in this fissile activity. Fuel assemblies contain copious amounts of Pu-239, a product derived-from the fission of uranium. The high energy alpha particle radiation of Pu-239 is the driving force of self-sustaining nuclear fission chain reactions. For 40 years, Pu-239 (1/2 T = 24,000 years) produced the energy at the Fukushima Daiichi reactor complex. The legacy of plutonium is that loss-of-cooling in reactor fuel assemblies, all of which contain large amounts of Pu-239, equals loss of control of the nuclear fission chain reaction process. As the progeny of captured neutrons, Pu-239s intense alpha radiation will, itself, induce neutron release, and thus, near fissile or fissile activity in the multiple globs of Fukushima Daiichi melted fuel assemblies until the melted fuel is relocated in a closed loop high-level waste storage facility at cold shutdown temperatures outside of the damaged reactor environs.

Any onsite closed loop cooling system improvised at the damaged Fukushima Daiichi reactors is a temporary solution to a complex decommissioning conundrum. All melted fuel assemblies will have to be moved to a specially designed and built high-level waste storage facility most likely located within the Fukushima Daiichi complex. Unfortunately, Pu-239 and other long-lived actinides are integral components of the washout pathway and soil deposition inventories. Storing several hundred thousand tons of radioactive water laced with Pu-239 has never been previously attempted. The removal of square miles of topsoil and thousands of tons of debris contaminated with plutonium as well as radiocesium and other isotopes into an improvised onsite radioactive waste storage site is one of the many challenges of the safe remediation of the Fukushima Daiichi disaster zone.

A Closed Loop Cooling System

The only possible future solution to ending the chronic releases of radiation now occurring as a result of the seven fuel assembly meltdown events at Fukushima Daiichi involves the construction of a closed loop cooling system, probably in a specially designed accident waste storage facility. As Matthew Wald (*The New York Times* March 20, 2011) notes "cooling with the recirculated water could end releases of radioactive materials, but will require new pumps and possibly new piping." Once a closed loop cooling system is built, radioactive water "can be pumped through filters that will strain out the radioactive elements." Japanese authorities, including the Prime Minister Naoto Kan, announced plans to construct closed loop cooling systems at all four reactors on Friday, April 15, 2011. Their announcements included a hoped-for timetable that would result in viable functioning cooling systems within nine months. The problem with constructing such a closed loop cooling system at any of the four reactor buildings in Japan is how to repair the extensive damage to the piping and reactor water systems in buildings that were severely damaged by the hydrogen explosions that occurred within the first week of the accident. The high levels of radioactivity in the plant environment, especially within the damaged reactor buildings, make this an extremely complicated and challenging task. This task will be further complicated by the extreme damage suffered by the spent fuel pool at Reactor 4, which is nearing collapse. Some of the contents of this spent fuel pool had already been ejected by the hydrogen explosion of March 15, 2011. Any work in the vicinity of this debris will result in high radiation exposure to the personnel attempting to build a new closed loop cooling system. It is also questionable whether any such closed loop system could be constructed in this environment. While it is possible that closed loop systems could be constructed at Reactors 1 and 3 because the primary containment vessels have not been significantly damaged (a possibility, not a certainty), "the primary containment vessel at the number 2 Reactor is damaged and leaking gases and the leaks need to be plugged before it can be flooded." (*The New York Times* April 20, 2011). Chronic releases of radiation in the form of both gas and water will continue indefinitely at the Fukushima Daiichi complex until all seven of the ongoing fuel melting events are completely isolated, cooled, and no further radiation can be emitted into the environment. Public relations efforts to the contrary, there is no possible way to estimate how long this will take until the actual condition of all the reactor vessels, spent fuel pools, and the fuel assemblies they enclose is documented. The traditional procedure of achieving cold shutdown status by inserting the control rods into the fuel assemblies is not an option given the fact that all the fuel assemblies have been deformed and partially melted during these accidents and all control rod insertion equipment has been destroyed.

Prevailing Wind Directions: Luck of the Irish

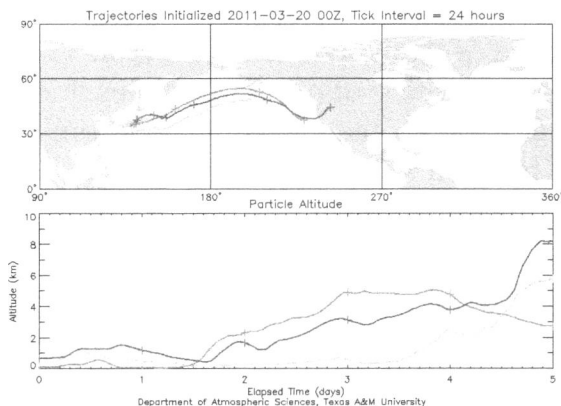

Figure 4. Transpacific plume dispersal (http://csrp.tamu.edu/earthquake/earthquake/20110320T00Z.html)

Several unique circumstances make the Fukushima Daiichi disaster different from the Chernobyl disaster. During the early stages of the accident (10 days) and during the beginning of its second phase of chronic releases, prevailing winds have brought most, but not all, contamination out over the Pacific Ocean as airborne emissions or into shoreline waters as radioactive water washout. This fallout pattern contrasts with the Chernobyl accident, where most source term fallout was over terrestrial environments. The relatively heavy nonvolatile hot particle emissions from the hydrogen explosions at all four spent fuel pools suggests that a significant percentage of accident source term releases now reside in the nearby marine environment, helping to explain the very high radiation levels reported in sea water. That well-publicized water leak from the basement of Reactor 2 is only a small part of the often wind-driven source term releases to the Pacific Ocean. The graphic in *Figure 4* clearly illustrates the transpacific nature of prevailing winds along with the tendency of airborne emissions to reach greater altitudes as they dissipate over the Pacific Ocean.

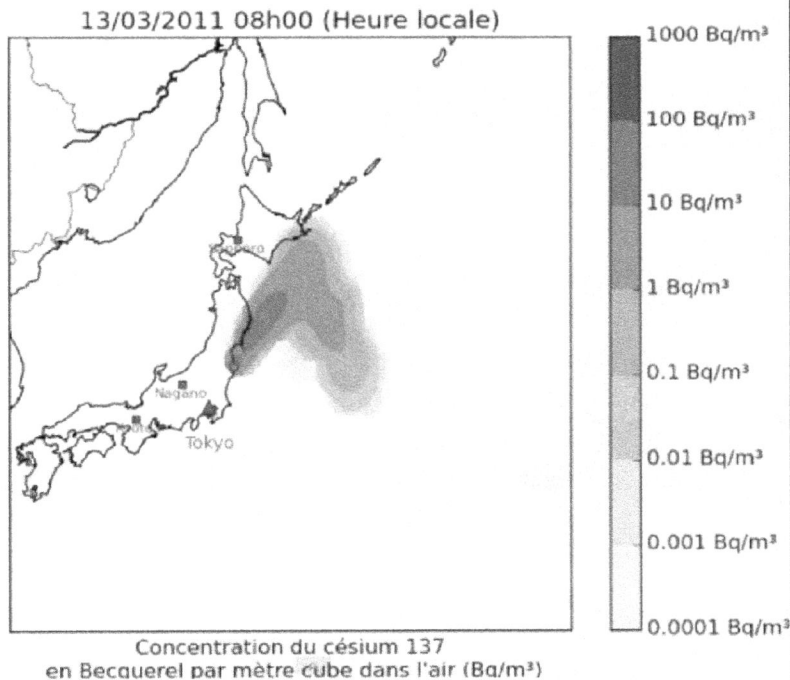

Figure 5. Cesium concentration plume (http://www.irsn.fr/FR/popup/Pages/animation_dispersion_rejets_19mars.aspx)

Graphical depictions of washout pathway emissions in the western Pacific are not yet available.

The weather map in *Figure 4* is from the Department of Atmospheric Sciences at Texas A&M University. It shows the movement (top) and altitude (bottom) of particles during the five day period starting March 20, 2011. The altitude for the lines on the bottom map started at 70 and 660 meters and ends at just below 3 km and just above 8 km. Most of the airborne source term releases from the Japan accident were deposited in the ocean or otherwise dissipated. Deposition levels in North America have been minimal.

Figure 5 and *Figure 6,* available via an online plume video courtesy of France's Institut de Radioprotection Nucleair (IRSN), illustrate the luck of the Irish: prevailing winds brought the majority of accident emissions to the northeast and then to the east of the Fukushima Daiichi complex. The reporting units in *Figure 5* showing air concentrations up to 1,000 Bq/m^3 provide important information about the intensity of the radiation releases from Fukushima Daiichi. The IRSN is one of several sources that provide information about the air concentration of fallout from Chernobyl. While some close-in locations had brief air concentrations at or above 100,000 Bq/m^3, in general, the most severely impacted areas in Ukraine and Belarus had air concentrations of 100 to 1,000 Bq/m^3. Other severely impacted areas in

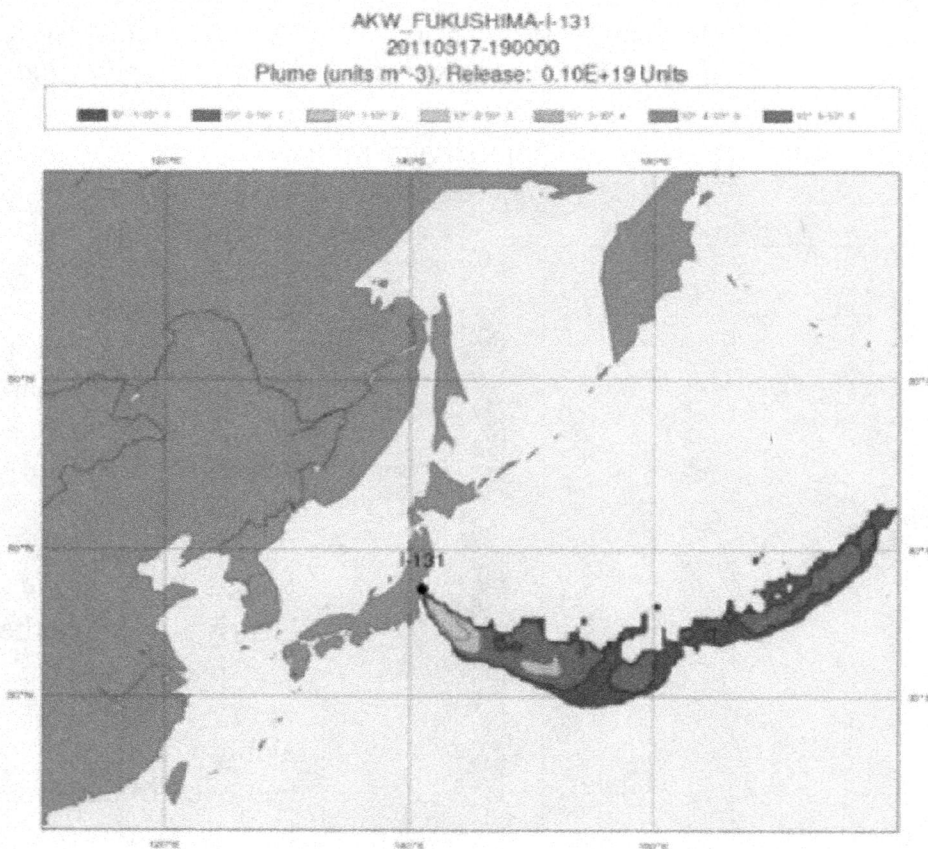

Figure 6. March 19th plume dispersal (http://www.irsn.fr/FR/popup/Pages/irsn-meteo-france_19mars.aspx)

Finland, Sweden, England, Germany, and Turkey had air concentrations in the 10 to 100 Bq/m^3. France and other less impacted areas in Europe had maximum air concentrations in the 1 to 10 Bq/m^3. In France, the IRSN maps, which are easily viewed on their online video, note current background gross air concentration levels at 10 $\mu Bq/m^3$ (0.00001 Bq/m^3). The IRSN map provides clear evidence that the Fukushima Daiichi releases, which will continue for a much longer duration than the Chernobyl releases, were at least briefly in the same order of magnitude in their initial stages as the Chernobyl releases. An area near the Fukushima Daiichi reactor complex and to the east over the Pacific Ocean had air concentrations in the 100 to 1,000 Bq/m^3 range with an even larger area having concentrations of 10 to 100 Bq/m^3. The reactor site itself and the one mile radius noted by the US NRC may have had air concentrations that also approached 100,000 Bq/m^3.

Figure 6 indicates a significant change in the fallout pattern four days after that illustrated in *Figure 5*. In the interim between these two dates, prevailing winds brought some fallout over the mainland of Japan. An important project for the accident analysts of the future will be more detailed compilation of fallout maps and their comparison with the now well-documented fallout patterns that resulted from the Chernobyl accident. Unfortunately, the Fukushima Daiichi multiple interlocking meltdown event (MIME) is much more complex than the relatively straightforward graphite fuel explosion that occurred at Chernobyl. The seven point sources involved in the Japan disaster have much greater inventories of long-lived isotopes such as radiocesium available for dispersal. Chernobyl had approximately 8 million curies of Cs-137 in the reactor that exploded; after over a decade of analysis, its release has been pinpointed at 2.7 million curies. The seven locations at Fukushima Daiichi that have experienced blowout and meltdown events have approximately 43 million curies of Cs-137 (1/2 T = 30.1 yr) available for release, along with tens of millions of curies of Cs-134 (1/2 T = 2.1 yr). Because much of the Japan accident release ended up in the Pacific Ocean, aerial surveillance of ground deposition won't be available to reconstruct air concentration levels or the accident's source term. Rather, an accurate accounting of the source term will have to await a complete analysis of the amounts of fission products remaining in each of the seven damaged fuel assembly groupings. This task is greatly complicated by the fact that the accident duration in Japan is already much longer than the duration of the Chernobyl accident. While radiation releases are clearly declining since the first two weeks of the accident, it appears releases will continue for years simply because it may take at least this long to cool, isolate, and remove the tens of millions of curies of fission products in the 4,368 fuel assemblies involved in this mother of all loss of coolant accidents. Accident mitigation efforts will also have to deal with the challenge of disposing of tens of thousands of tons of radioactive water, a problem not encountered at Chernobyl. With the recent estimate that at least 3,000 workers will soon be employed trying to improvise a closed loop cooling system and otherwise stabilize the four reactor sites that have been severely damaged by the hydrogen explosion blowouts, continuous remobilization of the accident source term will be an ongoing problem.

Nuclear Mis-information?

Two key issues bedevil the accurate reporting of the extent of contamination from any nuclear accident and the attempt to evaluate the health physics significance of accident emissions. The first issue involves the confusing terminology of the many reporting units used by various governmental, corporate (e.g. TEPCO - the Japanese electric power company), and environmental organizations. Easy access to modern information technology media now facilitates the ability of the lay person to make an informed judgment about the risks of exposure to accident-derived plumes of radioactive emissions. The second issue is much more difficult to alleviate, namely the lack of knowledge of the reporters and commentators who provide news about the Fukushima Daiichi disaster and other accidents. A March 23, 2011 article in the *Wall Street Journal* graphically illustrates this conundrum. The *Journal* reported Tokyo Electric Power Company's main gate gamma radiation level reading at 240 microsieverts per hour (μSv/hr), a fairly low reading considering that seven nuclear accidents were occurring all at once. It then reports spinach collected "60 miles southwest of the plant last Friday contained 54,000 becquerels of the radioactive element iodine-131… All the numbers add up to a reassuring picture of the very low risks from the radiation emitted from Fukushima so far, which is less than the amount people typically get from common sources such as the sun, medical tests, and air travel." (Bailik 2011). Given the FDA derived intervention level of 167 Bq/kg for radioiodine-131 for a one year old child and the annual, daily, and hourly background exposure rates noted in the following section, the *Wall Street Journal* report becomes a paradigm for the misinformation and generalizations that characterize most mass media reporting on the ongoing Fukushima Daiichi accidents. In the weeks after the spectacular first phase of the accident, the hydrogen blowouts and fires became less newsworthy chronic emissions, and coverage of the Japanese disaster gradually waned. Accident plume intensity and movement have received very little media coverage despite the fact that radiation releases may be larger than those of the Chernobyl accident.

Public and media unfamiliarity with the reporting units and protection action guidelines is being compounded by a growing lack of information about the status of the ongoing fuel assembly melting events at the Fukushima Daiichi complex. *The New York Times* and the British Broadcasting Corporation (BBC; www.bbc.com) have provided consistently accurate reporting about the initial stages of the accident that culminated in the hydrogen explosion at the spent fuel pool at Reactor 4 on March 15, 2011. The Public Broadcasting Service (PBS; www.pbs.org) and Cable News Network (CNN; www.cnn.com) have also provided some credible coverage of the ongoing crisis in Japan. The *Wall Street Journal* is the source of the most frequent daily updates about the accident. Accurate media coverage of ongoing releases and their concentration levels as well as the contamination levels in terrestrial and marine environments is nonexistent.

Information Availability

The government of Japan and the Tokyo Electric Power Co. (TEPCO) can be complimented on their willingness to provide accurate, up-to-date, radiological surveillance data related to some, but not all, components of the impact of the ongoing nuclear accidents at the Fukushima Daiichi nuclear power plants. Unfortunately, accurate information about the condition of the primary containment structures at Reactors 1 - 3 is not available, nor is it possible to evaluate the amount of melted fuel now lying at the bottom of the reactor vessel dry wells. It was also impossible to evaluate the degree to which fissile activity (criticality) had subsided in Reactors 1 - 3 during the brief interregnum of cooling between the time of the earthquake and the tsunami-derived loss of cooling capacity when the backup generators were flooded and destroyed. It was also not possible to determine the extent to which fissile activity resumed in any of these environments after the loss of reactor coolant accidents (LORCAs) and spent fuel pool loss of coolant accidents (LOCAs) began. Accurate information about ongoing efforts to control and reduce fuel assembly heating and melting is also not available. After emissions peaked during the first ten days of the accident, onsite radiation levels abated enough to allow hundreds of emergency workers to return to the accident site. Information about what they were doing was also not available. A key unknown is how normal cooling activities will be restored in reactor buildings that are literally piles of rubble. A disturbing question is arising as the third month of the accidents at Fukushima Daiichi begins: why is there little or no visual documentation of ongoing accident mitigation activities given the initial graphic images of fire equipment being used to cool the seven accident sites?

Bulldozing

Shortly after the hydrogen explosion at the recently filled-with-hot-fuel pool at Reactor 4 tore off its roof and released "a large source term" (US NRC), possibly more than 25% of its inventory of fission products, TEPCO authorities began an intensive effort to mitigate the resulting fallout. Some of the contents of this spent fuel pool, if not blown into the ocean, fell "within a mile" of the fuel pool and were bulldozed by TEPCO to mitigate worker exposure to the high levels of radioactivity that resulted from the ground deposition of hot particles, CRUD, and other spent fuel fragments. Presumably, TEPCO workers have been bulldozing surface contamination in the form of radioactive waste-laden soil into pits and then covering them with less radioactive materials. This would explain why the reactor site, which was so radioactive that even the presence of 50 workers involved extremely hazardous exposure to ionizing radiation, could then be the location where almost a thousand workers labored to pipe water into the three reactor vessels and spray the now unroofed fuel pools with water from fire trucks, which now have been replaced with a network of hoses and electrical pumps. TEPCO was not shy about showing Tokyo fire equipment cooling the reactor meltdown events, but has been less forthcoming about the frenetic bulldozing that accompanied these efforts. Coincidentally, the United States National Imagery Management Agency (NIMA) has ±2,500 employees devoted to tracking the radioactive emissions from every North Korean, Iranian, etc. point source. NIMA "big government" radiological surveillance specialists are having a field day tabulating the emissions from each fuel assembly melting event at the Japan accident site.

After 40 years of Cold War and post-Cold War spying efforts, the advanced technology developed for aerial radiological surveillance (ARS) has reached a level of perfection such that heat emitted by campfires can be easily recorded and quantified as to intensity. The rapid analysis of ARS fallout data prompted the US Defense Department to initiate a 50-mile safety perimeter for US Defense Department employees in the days after the accident. It was also ARS technology that enabled the IRSN (France) to create the graphic video images of the Fukushima Daiichi plume movement. As soon as the accident occurred, ARS technology could immediately document both the spread of radioactivity and the sudden increases in fuel temperature. Reactor 2, which seemed to have the largest percentage of its fuel sitting in the reactor dry well, will be the most important at-risk environment and of obvious interest to ARS specialists. If there are no further sudden increases in the temperature of now molten fuel assembly conglomerations, NIMA aerial radiological survey technicians will still be able to track and closely tabulate ongoing chronic accident emissions even as they fall (hopefully) to a terabecquerel or two per hour (1 terabecquerel = 27 curies). In the age of information technology, the results of googling bulldozer activity and radioactive emissions are hard to keep secret despite governmental, TEPCO, and media information blackouts. Bulldozing is just not as "cooling" (cool) as heroic firefighters manning their equipment. The online plume videos formulated by France's IRSN are a testament to the efficacy of the marriage of information technology with aerial radiological surveillance technology, bulldozing and hoses notwithstanding.

Water In/Water Out: Leak Rates

The key to deciphering the environmental impact of the seven interrelated fuel assembly melting events at Fukushima Daiichi involves the destinations of the water used to cool down the reactor fuel. Reactor vessels and spent fuel pools do not usually leak water during normal operations nor emit radioactive steam into the atmosphere. The current effort to avoid a full scale meltdown involves the constant pumping of water into three reactors and four spent fuel pools. *The New York Times* noted, "as the water boils in the reactors, pressure rises too high to pump in more water, so workers have to vent to the atmosphere" (April 18, 2011) to relieve the pressure before feeding in more water. "At some reactors the arms of the trucks that deliver the water have been placed over the damaged walls of the buildings, enabling water to be shot more directly **at** the reactors and pools and reducing runoff." TEPCO announced a plan to build an entirely new closed loop cooling system, a myopic goal that would take months to halt the release of radioactive steam and water (Belson 2011, A1). Radioactive emissions at the accident site have two pathways associated with cooling efforts: radioactive water washout, now the intense object of recovery, and radioactive steam, which is more difficult to recover. More data leaks about water leaks is needed – where is WikiLeaks when we need it?

Evaluation of the total Fukushima Daiichi source term (release inventory) at all seven facilities now experiencing fuel assembly melting events is contingent upon documenting the rate at which cooling water "leaks into the environment" either as radioactive water or radioactive steam. Since all four reactor buildings have been severely damaged by hydrogen explosions (blowouts), restoration of "normal" cooling rates is highly unlikely. Accurate measurement of the variations in radioactive water or steam releases will be essential to determine the ongoing source term releases of this accident. Ultimately, the question will arise as to what percentage of the radioisotope inventory was released from each unit and what percentage of the original core and fuel pool inventories remained onsite? Obviously, the resumption of criticality at one or more units will create a plume of highly radioactive gaseous emissions, of which radioactive steam will only be one component. If such emissions occurred early in the accident due to ongoing criticality, a possible cause of the hydrogen explosions that were graphically illustrated on TV, neither the Japanese government, TEPCO, the IAEA, nor the US NRC have mentioned criticality-derived emissions as a component of these ongoing accidents.

Fukushima Daiichi Lingering Questions

There are a number of unanswered questions that need to be addressed as the accident at Fukushima Daiichi evolves from a potentially world-threatening catastrophic disaster to the nuclear power industry's largest quagmire of multiple point sources of chronic radiation emissions.

1. Did the reactor vessel and spent fuel pool cooling failures that occurred at the beginning of the accident result in the resumption of criticality at any of the units or was there lingering low-level criticality in the reactor vessels at the time the tsunami struck and cooling capabilities were lost?

2. Since radiation emissions have diminished since the four hydrogen gas explosions occurred, are there any remaining fuel assemblies that are still critical, i.e. that are undergoing fission chain reactions such as in the damaged reactor vessel of Units 1 and 2, which have melted fuel in the lower sections of their reactor vessel and in their underlying dry wells? (See *Figure 3*.)

3. Will TEPCO be able to prevent any future rise in temperature within the melted fuel assemblies as appeared to be happening in Reactor 2 as recently as April 17, 2011?

4. Will continued manual cooling efforts succeed in the further reduction of heat generation in the fuel assemblies that melted as a result of loss of reactor vessel and spent fuel coolant?

5. Will the ultimate goal of a "cold shutdown" where no more boiling occurs be successfully implemented at all seven accident locations and when will this occur?

6. To what extent will damage to the secondary reactor containment buildings and the reactor and water cooling systems prevent construction of improved closed loop cooling systems?

7. Will a return to cold shutdown status have to await construction of a specially designed onsite melted fuel storage facility and how long will this construction project take?

8. How long will manual cooling efforts generate radioactive steam emissions, given the fact that the roofing over all four spent fuel pools has been destroyed?

9. How long will manual cooling efforts result in the washout of fuel assembly-derived fission products to the terrestrial and especially the marine environments adjacent to the reactor complex?

10. To what extent will TEPCO be successful in the recovery of radioactive water and where will it be stored?

11. Will the radioactive water component of the accident source term exceed that of atmospheric gaseous and particulate emissions?

12. Will the Japanese government, TEPCO, and the International Atomic Energy Agency be able to reconstruct the source term from **each** of the environments experiencing loss of coolant accidents, and for each source term pathway, especially including the washout pathway. How many years will this analysis take?

The answers to these questions will help all concerned spectators of the unfolding tragedy in Japan, including the citizens of Japan who are so severely impacted by the radiological contamination of their homeland, better understand what has taken place. The accidents at Fukushima Daiichi involve fuel cladding failures on a grand scale. The source term from the hydrogen explosion in the spent fuel pool of Reactor 4, which may have released ±25% of its inventory, may itself exceed the total source term of the Chernobyl accident. Any third grade honor student can do the math and make a rough estimate, for example, of the source term of cesium-137. It is public knowledge, courtesy of *The New York Times*, that the spent fuel pool at Reactor 4 contained 1,479 fuel assemblies. It is also well-known that a fuel assembly in every nuclear reactor has accumulated at least 10,000 curies of Cs-137. Ten thousand multiplied by 1,479 equals 14,790,000 curies as the minimum content of Cs-137 in this recently-loaded spent fuel pool. The Chernobyl source term for Cs-137 was 2.7 million curies. If 25% of the content of the spent fuel pool of Reactor 4 was ejected from the confines of the destroyed building by a hydrogen explosion and associated fires, deposited locally as fuel assembly fragments, windblown steam and smoke emissions, or washed out fission products by manual cooling efforts, the source term release from **only** the Reactor 4 spent fuel pool was _____ curies of Cs-137. Okay kids, do the math and fill in the answer, and then mail the answer to your local newspaper or electronic media outlet. Who will be the first to mention that the source term release from just one of the seven accident locations in Japan is probably greater than that from Chernobyl, even though it was not as widely deposited in terrestrial environments as the tropospheric dispersion of emissions from the Chernobyl explosion? Fox News, CNN, PBS, NYT, WSJ, Charlie Rose, Amy Goodman, etc.: Can you rise to the occasion, or do you think this information should be kept secret? (Answer: ±3,697,500 Ci of Cs-137 may have been released just from the explosion that destroyed spent fuel pool 4.) This accident will continue indefinitely because there are six other radioactive contamination point sources in this multiple interlocking meltdown event (MIME). It is highly unlikely that the current estimate by TEPCO and the Japanese government that the accident release is only 10% of that at Chernobyl is accurate.

As this book was going to press, all media seem obsessed with the Royal Wedding, tornados in the American midlands, the death of Osama bin Laden, and a Mississippi River flood disaster that will transport more ecotoxins into the Gulf of Mexico than those discharged by the Gulf oil spill. As a result of the combination of too much news to report and ongoing psychic numbness, will the spectacle of the ongoing agony of the citizens of Japan no longer be deemed newsworthy?

II. Basic Definitions and Concepts

The following guidelines and reporting units have been extracted from the Center for Biological Monitoring RADNET archives and contemporary online information sources such as the United States Food and Drug Administration (US FDA), United States Department of Energy (US DOE), United States Environmental Protection Agency (US EPA), Centers for Disease Control (CDC), the International Atomic Energy Agency (IAEA), Institute for Energy and Environmental Research (IEER), Greenpeace, and Wikipedia.org. Additional definitions and explanations continue in *Section III. Additional Definitions and Concepts.* The following delineation of reporting units is prefaced by comments on the problem of evaluating dose assessments from exposure to ambient radiation in the context of multiple pathways of radiation exposure, including ingestion, inhalation, and secondary/tertiary pathway exposures.

The Problem of Exposure to Ambient Radiation

The fly in the ointment pertaining to any nuclear accident or nuclear discharge exposure scenario is that the radiation-detecting equipment that measures ambient air contamination or the surface contamination of radioactivity on skin and clothing (in reporting units of absorbed dose, e.g. microsieverts/hour, or in rate of space dose, e.g. nanograys per hour) provides only a fragment of the information needed to evaluate actual exposure. Accidents such as the ongoing disaster in Japan create plumes of radiation, which are then deposited on terrestrial landscapes and in marine environments by rainfall events as well as by dry deposition. The accidents at the Fukushima Daiichi facility are also characterized by the discharge of large volumes of contaminated seawater as a result of the improvised attempts to cool the fuel assemblies.

Contamination derived from these and other accidental releases is then taken up by the biogeochemical cycles of the earth's biosphere, exposing all living creatures, including humans, to anthropogenic (manmade) radiation. The accidents at the Fukushima Daiichi facility may well turn out to be historically difficult-to-remediate long-term chronic contamination point sources, with multiple exposure pathways ranging from ground shine and re-suspended and remobilized deposition to contamination of terrestrial and marine food webs. Anomalous pathways exposure may include wash-up of contaminated tsunami debris on the shores of North America.

Once an accident plume has passed, measurements of ambient radioactivity exposure levels provide little or no information about total accident exposure for communities living within principle fallout zones, or for population groups whose principle exposure is through contaminated food products. The tragedy of the ongoing Japan disaster is that radioactivity will be emitted continuously for a long period of time; its terrestrial deposition will depend on accident duration and intensity, wind direction, ocean current dispersal of liquid contaminants, and the re-concentration of radioisotopes after their deposition in marine sediments. Rhetorical commentary on the accident plume, such as "Radioactivity is low," or "Such and such location will have a minimum impact from this accident" should have zero credibility. The actual evaluation of human exposure to accident-derived radioactivity from Fukushima Daiichi must await detailed isotope and media-specific analysis of the contaminant load in all impacted abiotic and biotic environments. In places such as California, Oregon, and Washington, which began receiving tiny amounts of radioactivity on Saturday, March 19, 2011, an evaluation of the significance of the radiological impact of the Fukushima Daiichi accidents can only be measured by analysis of the soil deposition of indicator isotopes, such as I-131 (1/2 T = 8 days) and Cs-137 (1/2 T = 30 years), **as measured in becquerels per square meter or per liter or per kilogram of abiotic and biotic media**. Low or non-existent concentrations of radioactivity in air, as measured in microbecquerels per cubic meter ($\mu Bq/m^3$) or microsieverts per hour ($\mu Sv/hr$) provide no information about plume passage and the deposition that has already occurred. If new cesium deposits can be documented (on top of old weapons testing and Chernobyl radiocesium baseline deposits), new deposits of other biologically significant isotopes, including MOX fuel-derived plutonium-239 (1/2 T = ±24,000 years), will also be present. Due to its short radiological half-life, I-131 is always an indicator of the recent discharge of newly created fission products. No I-131 contamination was detectable a few months after the Chernobyl accident ended, due to its short half-life. Nonetheless, hundreds of millions of citizens in Russia and Europe were exposed to biologically significant concentrations of thyroid-seeking I-131 before it decayed. The analytic techniques that measure gamma radiation, and beta-emitting radiocesium and radioiodine isotopes, cannot be used to measure contamination of biotic and abiotic media by long-lived alpha-emitting plutonium, curium, and americium, which will require time-consuming laboratory analysis to evaluate.

Exposure to accident plumes cannot be evaluated until the following information is available:

- How many becquerels per square meter of the accident indicator isotope cesium-137 have been deposited in my community?
- How many becquerels per kilogram of cesium-137 are in the food that my family and community are ingesting?
- How many becquerels per liter of cesium-137 are in the milk, including breast milk, that my children are drinking?

- How many becquerels of iodine-131 are in the public and private drinking water supplies that my family is drinking?
- How many becquerels of iodine-131 are in the milk, including breast milk, that my children are drinking?
- How many becquerels of iodine-131 are in the leafy vegetables, mushrooms, and other foods that my family is eating?

The presence of these indicator isotopes in amounts approaching protection action guidelines or levels of concern also mean that individuals, families, and communities are being exposed to the wide variety of other isotopes released by nuclear accidents. The radiological fallout data compiled after the Chernobyl accident, much of it summarized in *Section VIII. Chernobyl Fallout Data*, provide a glimpse into the many isotopes contained in any accident source term release. The Chernobyl ground deposition monitored in Skustar, Sweden (see Sweden: Hardy 1986) included important nuclides of concern associated with all accidents: Zr-95, Nb-95, Ru-103, Te-132, Cs-134, Ba-140, La-140, and Ce-141. At this location, background radiation rose to 900 μR/hr (9 μSv/hr) in comparison to Stockholm where it remained near background levels at 30 μR/hr (0.3 μSv/hr). Another Swedish researcher notes background radiation as 10 to 15 μR/hr (0.1 to 0.15 μSv/hr) (Sweden: Reizenstein 1987). Given the high levels of ground deposition at Skustar, the elevated ambient or background radiation level graphically illustrates how little information about the seriousness of a nuclear accident can be derived from background readings that often vary widely under normal conditions and may not appear excessively high even after a major fallout event, the duration of which may only last a few hours.

The following definitions are intended as a guide to the basic terminology essential to understanding the environmental impact and health physics implications of the unfolding disaster at the Fukushima Daiichi facility or any other nuclear accident. The fundamental question that all concerned world citizens may have, or may soon have, is: how much radioactivity is my community and my family being exposed to?

Reporting Units for Radiation Exposure and Environmental Contamination

The principal reporting units now used throughout the world for the measurement of radioactivity in biotic and abiotic environments are "sievert" (dose), "becquerel" (activity), and "electron volt" (energy). **One becquerel equals one disintegration per second of a radioactive substance.** Each isotope has its own unique electron voltage (eV) or energy level. The activity and energy levels of each radioisotope and their decay modes (gamma, beta, or alpha) are the basis for any health physics impact dose assessment. The dosage of human exposure to external radiation is measured in micro- or millisieverts for ground shine, cloud shine, and deposits on shoes and clothes. Volumetric and surface contamination is measured in becquerels per kilogram, liter, or square meter. The power (energy) of each disintegration is reported in electron volts. Commonly encountered reporting units for human exposure to ionizing radiation and nuclear accident radionuclide concentration, deposition, and pathway movements are listed below. Relevant definitions follow this list.

Radiation Reporting Units

Ambient Radiation Exposure: Dose Reporting Units

- microsieverts per hour (μSv/hr)
  - The International Atomic Energy Agency defines background radiation from natural sources as averaging 0.27 μSv/hr; this does not include hourly average exposure to manmade sources (see below), which raises the average exposure from all sources to 0.7 μSv/hr (US NRC)
- millisieverts per hour (mSv/hr)
- millisieverts per year (mSv/yr)
- nanograys per hour (nGy/hr)
- micrograys per hour (μGy/hr)
- microrems per hour (μR/hr)
- aerial monitoring of ambient radiation: millirems per hour (mR/hr)
- 1 rem (R) = 0.01 sieverts (Sv); 0.27 μSv/hr = 27 μR/hr
- 1,000 millionths of a sievert (1,000 μSv) = 1 thousandth of a sievert (1 mSv)

Activity Reporting Units Expressed in Becquerels

- 1 curie = 37 billion becquerels = 37 GBq; to convert curies to becquerels, multiply by 37 billion
- air: microbecquerels per cubic meter (μBq/m^3)
  - airborne contamination may be in the form of dust, other particulates, or water vapor, including evaporated steam
  - after the Fukushima Daiichi disaster, the Japanese government reported ambient air concentrations in "tens of thousands of terabecquerels per hour"; one terabecquerel equals one trillion becquerels.
  - to convert becquerels per cubic meter to microbecquerels per cubic meter, multiply the Bq/m^3 times 1 x 10^{-6}

- aerial surveillance of gross beta/gamma ground contamination: becquerels per square meter (Bq/m^2)
- surface contamination: becquerels per square meter (Bq/m^2)
- surface contamination as measured in bequerels are also reported using prefixes:
  - 1 MBq (M = mega = million) = 1,000,000 bequerels = 27 microcuries = 27 µCi
  - 1 GBq (G = giga = billion) = 1,000,000,000 Bq = 27 millicuries = 27 mCi
  - 37 GBq = 37,000,000,000 Bq = 1 Curie = 1 Ci
  - 1 TBq (T = tera = trillion) = 1,000,000,000,000 becquerels = 27 Curies = 27 Ci
  - 1 PBq (P = peta = quadrillion) = 1, 000,000,000,000,000 becquerels = 27,000 Curies = 27,000 Ci = 27 kCi
- milk and water: becquerels per liter (Bq/l)
  - high contamination levels in water samples are sometimes reported in becquerels per cubic centimeter (Bq/cm^3). A liter contains 1,000 cubic centimeters
- food: becquerels per kilogram (Bq/kg)

Other Activity Reporting Units

- the US has a tradition of reporting radioactivity in disintegrations per minute (d.p.m.) especially for low levels of radioactivity associated with weapons testing fallout
- to convert d.p.m. (disintegrations per minute) to becquerels, divide by 60 (e.g. 180 d.p.m. = 3 Bq)
- to convert becquerels to d.p.m., multiply by 60
- older protection action guidelines use the reporting unit microcurie (µCi) to delineate intervention levels (1 µCi = 1,000,000 picocuries = 37,037 Bq) (United States Department of Health and Human Services 1982)
- up until recently, the United States used the now antiquated reporting system of picocuries; 1 Bq = 27 picocuries
  - to convert picocuries to becquerels, divide by 27 (27.027)
  - to convert becquerels to picocuries, multiply the number of becquerels by 27 (27.027)

o to convert picocuries per cubic meter to becquerels per cubic meter, multiply pCi/m^3 times 0.037

Energy Reporting Units

- 1,000,000 electron volts (eV) = 1 MeV (1 mega-electron volt)

- 1,000 electron volts (eV) = 1 meV (1 milli-electron volt)

Commonly Used Prefixes

Also see *Table of Prefixes*.

f = quadrillionths = femto = 10^{-15}	P = quadrillion = peta = 10^{15}
p = trillionths = pico = 10-12	T = trillion = tera = 10^{12}
n = billionths = nano = 10-9	G = billion = giga = 10^9
μ = millionths = micro = 10-6	M = million = mega = 10^6
m = thousandths = milli = 10-3	k = thousands = kilo = 10^3

Sievert (Sv)

- An international unit of radiation dosage, which measures the amount of radiation that is absorbed by a person, usually expressed in microsieverts or millisieverts. One sievert is equal to 100 REMs, a dosage unit of X-ray and gamma ray exposure. A millisievert equals 0.1 REM.

Becquerel (Bq)

- One disintegration per second of a radioactive material, also defined as "the activity of a quantity of a radioactive material in which one nucleus decays per second." (wikipedia.org/wiki/becquerel).

Radiation Equivalent Man (REM)

- The conventional unit of **dose equivalent**. The corresponding International System (SI) unit is the **sievert (Sv)**; 1 Sv = 100 REM.

- "A measure of the biological damage of a given absorbed dose of radiation. The rem takes into account the varying ways in which ionizing radiations transfer their energy to human tissue. Rems are derived from rads by multiplying rads by the 'quality factor' of the type of radiation in question. **For gamma and most beta radiation, the quality factor is one; in other words, rems equal rads. For alpha radiation, the quality factor is 20, that is, rems equal 20 times rads.**

Neutron radiation quality factors vary according to neutron energy."
(http://www.ieer.org/fctsheet/radiationhealthfactsheet_2011.pdf).

- Additional information pertaining to dose assessments is available in *Sections VII. Radiation Protection Guidelines* and *III. Additional Definitions and Concepts* (see Derived Air Concentration through Effective Action Level).

Gray (Gy)

Equal to 100 rad, Gy is the SI unit of absorbed radiation dose of ionizing radiation (for example, X-rays) and is defined as the absorption of one joule of ionizing radiation by one kilogram of matter (usually human tissue). Radiation exposure is sometimes expressed in the reporting units nGy (nanograys, billionths of a gray) or micrograys (μGy), millionths of a gray.

- 1,000 nanograys (1,000 nGy) = 1 microgray (1 μGy)

- 1 rad = 0.01 Gy

- 1 μGy = 1 μSv

Ionizing Radiation

Biologically significant radiation with an energy level above 155 eV (0.000155 MeV), which has the capacity to cause carcinogenic, mutagenic, and teratogenic effects by knocking the electrons out of the orbits of the molecules in irradiated cellular tissues.

Electron Voltage

The measure of the energy of electromagnetic radiation is reported in units of electron volts (eV). Dose reporting units express human exposure to biologically significant radioactivity. Ionizing radiation is universally measured in millions of electron volts (MeV). The power of the ionizing radiation emitted by each becquerel (disintegration per second) of an isotope is dependent on the electronic voltage of that particular isotope. The energy released by the disintegration of individual radioactive isotopes from naturally-occurring electromagnetic radiation sources such as the sun, and from manmade non-ionizing electromagnetic radiation sources, is reported in electron volts (eV).

Selected Electron Voltages:

Naturally Occurring Radiation

    Sunlight = 3 to 5 eV

    *Potassium-40 = 10 to 40 eV

    *Radon-222 = 5.48966 MeV (5,489,660 eV)

Uranium-235 = 200MeV (200,000,000 eV)

Anthropogenic (Manmade) Radiation

Radio waves = 1.2398×10^{-10} to 1.2398×10^{-5} electron volts

Iodine-131 = 0.6065 MeV (606,500 eV)

Cesium-137 = 0.51163 MeV (511,630 eV)

Plutonium-239 = 5.1554 MeV (5,155,400 eV)

* The radical difference in the energy of these two naturally occurring isotopes explains why tens of thousands of people die annually from exposure to radon, whereas the death rate from eating bananas is extremely low.

Electromagnetic Radiation (EMR)

Energy radiated in the form of a wave which can accelerate charged particles. Electromagnetic radiation can travel through a vacuum. Its energy varies greatly; radio waves have the longest wavelengths and the lowest frequency and energy (1.2398×10^{-10} to 1.2398×10^{-5} electron volts). X-rays and gamma rays have the shortest wavelengths and highest frequencies and energies (up to and above 6×10^6 electron volts, or six million electron volts). For a comprehensive explanation of the public health consequences of ionizing radiation, i.e. electromagnetic radiation above 155 eV see Section 10 in Gofman (1981) *Radiation and human health; a comprehensive investigation of the evidence relating low-level radiation to cancer and other diseases.*

Background Radiation

Total background radiation exposure averaged over large population groups (i.e. the entire population of the United States) is reported as 6.2 mSv/yr. This hypothetical exposure level assumes no person is being exposed to the plumes of any ongoing nuclear accident. "Background radiation" can be divided into two components.

- Exposure to naturally occurring radiation, such as radon gas, cosmic radiation, and the ubiquitous naturally-occurring radioactive potassium (K-40) (see the notorious banana equivalent dose noted below). Approximately half or more of background radiation is from naturally occurring sources.

- Exposure to routine anthropogenic sources such as radiological imaging (mammograms, CT scans, dental x-rays, radiation therapy, and nuclear medicine facilities). Background radiation also includes exposure to the cumulative fallout from past nuclear weapons tests, residual accumulations of Chernobyl contamination, other nuclear power and fuel reprocessing accidents, the ongoing operation of nuclear power plants, and fossil fuel burning, especially coal.

The estimated exposures from background radiation noted below represent the average exposure of hundreds of millions of people. In reality, such "background exposure" varies widely with geographical location, exposure to sunlight via sunbathing, use of medical technologies such as CAT scans (see below), and proximity to other sources of anthropogenic radiation.

Average annual exposure from all sources of background radiation, including anthropogenic sources, is frequently reported at different orders of magnitude (milli vs. micro) and for different lengths of time (year, day, or hour):

- 0.0062 sieverts per year (Sv/yr)
- 6.2 millisieverts per year (mSv/yr)
- 620 millirems per year (mR/yr)
- 6200 microsieverts per year (μSv/yr)

This level of exposure translates to

- 1.7 millirems per day (mR/day)
- 17 microsieverts per day (μSv/day)
- 0.017 millisieverts per day (mSv/day)
- 0.7 microsieverts per hour (μSv/hr)
- 0.07 millirems per hour (mR/hr)

Slightly more than half of these **averaged** annual, daily, and hourly exposure rates are from anthropogenic sources. Variations in exposure to the many naturally-occurring and anthropogenic radiation sources can vary by up to two orders of magnitude, depending on proximity to highly radioactive geological formations, altitude, and exposure to anthropogenic sources such as CT scans and nuclear accidents.

Exposure Guidelines

Estimates of radiation exposure on an hourly basis vary widely, depending on the source of the information. As noted above, the US Nuclear Regulatory Commission (NRC) estimate of 6.2 millisieverts (mSv) annual exposure from all sources of radiation averages to 0.7 microsieverts (μSv) per hour. The International Atomic Energy Agency (IAEA) estimates hourly exposure to background radiation at 0.27 μSv per hour. The Massachusetts Institute of Technology (MIT) Department of Nuclear Engineering website gives a range of 0.05 to 0.1 μSv per hour. The Japanese government reports background radiation in both micrograys (μGy) per hour and nanograys (nGy) per hour.

On Japanese government accident site print outs, near normal readings appear to be 0.024 to 0.032 µGy per hour (24 to 32 nGy/hr). They also report a rate of space dose with a range of readings in 29 locations in Japan from 117 to 463 nGy per hour (0.117 to 0.463 µGy per hour) reflecting the probable impact of the Fukushima Daiichi releases. **Variations in rate of space dose up to one order of magnitude therefore provide almost no information about the presence of airborne emissions, which can quickly subside, nor much information about the ground deposition of accident-derived radioactivity**. More pertinent are environmental monitoring samples that clearly differentiate highly contaminated areas from those having less accident-derived fallout (see below).

- Naturally occurring background radiation exposure: 3 millisieverts/year (varies widely with geographical location and may be as much as 20 times higher in sections of India, Iran, and Europe) (wikipedia.org/wiki/Sievert)

- Total average radiation dose received by a person living in the US: 6.2 millisieverts/year (http://www.nrc.gov/reading-rm/doc-collections/fact-sheets/bio-effects-radiation.html)

- Mammogram: 3 millisieverts

- Chest CT scan (computed tomography or CAT scan): 6 to 18 millisieverts (Van Unnik 1997)

- Limit of exposure for nuclear workers for one year: 50 millisieverts (mSv)

- Exposure levels of concern for Japanese schoolchildren: 20 millisieverts/year (54.8 microsieverts/day or 2.28 microsieverts/hr); this recently raised protection action guideline is currently generating a lot of controversy in Japan

- Peak radiation dose at the boundary of the Fukushima Daiichi nuclear power station on March 16, 2011: **400 millisieverts per hour** (mSv/hr) (400,000 microsieverts per hour); this extraordinarily high reading reflects the fleeting presence of a super hot plume zone (±2 hours) where radiation levels quickly fall after a quick release accident (e.g. hydrogen explosion) due to the decay of short-lived isotopes (1/2 T = < 1 hr)

- Note the graph in Figure 1 reports peak ambient radiation exposure levels at 12,000 microsieverts per hour (**12 millisieverts per hour**), an estimate probably derived from aerial radiological surveillance at some location other than the main gate

Classification of Radioactive Waste

The classification of radioactive waste is defined in the NRC regulation 10 CRF § 61.55 Waste Classification (http://www.nrc.gov/waste.html).

- Low-level radioactive waste (LLRW) includes radioactively contaminated protective clothing, tools, filters, rags, medical tubes, and many other items
- Waste incidental to reprocessing (WIR) refers to certain waste byproducts that result from reprocessing spent nuclear fuel, which the US Department of Energy (DOE) has distinguished from high-level waste (described below)
- High-level waste (HLW) is "irradiated" or used nuclear reactor fuel
- Uranium mill tailings are the residues remaining after the processing of natural ore to extract uranium and thorium

Wikipedia has provided the service of simplifying the NRC's waste classifications (Class A, Class B, and Class C) in the following table. The high levels of biologically significant fission products such as Cs-137 permitted in a cubic meter of Class C low-level wastes are particularly notable. 4,600 curies times 37 billion becquerels is a very large concentration of radioactivity; that it might be disposed of in a shallow landfill, such as that at Barnwell, SC, highlights the controversial nature of radioactive waste disposal.

Radionuclide	Class A (Ci/m^3)	Class B (Ci/m^3)	Class C (Ci/m^3)
Total of all nuclides with less than 5 year half-life	700	No limit	No limit
H-3 (Tritium)	40	No limit	No limit
Co-60	700	No limit	No limit
Ni-63	3.5	70	700
Ni-63 in activated metal	35	700	7000
Sr-90	0.04	150	7000
Cs-137	1	44	4600
C-14	0.8		8

Radionuclide	Class A (Ci/m^3)	Class B (Ci/m^3)	Class C (Ci/m^3)
C-14 in activated metal	8		80
Ni-59 in activated metal	22		220
Nb-94 in activated metal	0.02		0.2
Tc-99	0.3		3
I-129	0.008		0.08
Alpha emitting transuranic nuclides with half-life greater than 5 years	10 nCi/g		100 nCi/g
Pu-241	350 nCi/g		3500 nCi/g
Cm-242	2000 nCi/g		20000 nCi/g

LLW should not be confused with Greater-Than-Class C wastes (GTCC), high-level waste (HLW), spent nuclear fuel (SNF) or transuranic waste (TRU) (http://en.wikipedia.org/wiki/Low_level_waste).

Transuranic Waste: Waste containing more than 100 nanocuries of alpha-emitting transuranic isotopes per gram of waste with half-lives greater than 20 years, except for high-level waste (http://en.wikipedia.org/wiki/Transuranic_waste).

Greater-Than-Class C Wastes (GTCC): "Most hazardous of LLRW - dangerous to inadvertent intruder beyond 500 years. Must be disposed in geologic repository unless alternate method proposed by DOE and approved by NRC." (http://www.gtcceis.anl.gov/guide/gtccllw/index.cfm). The United States Department of Energy (DOE), Office of Environmental Management (EM) also notes "A majority of the GTCC-like waste in this waste type consists of transuranic (TRU) waste that may have originated from non-defense activities… these wastes may contain concentrations of alpha-emitting TRU radionuclides with half-lives greater than 20 years in concentrations exceeding 100 nanocuries per gram (nCi/g). These TRU wastes may not have a path to disposal if they are determined to have been generated from non-defense activities. The long-term hazard associated with these wastes is similar to that posed by other GTCC LLRW."

Checklist of Biologically Significant Radionuclides

Cesium-137 is the nuclide of choice used in this *Handbook* to characterize changing patterns of the dietary intake of artificial radionuclides.*

*Prior to the first nuclear explosion at Alamogordo, New Mexico on July 16, 1945, the dietary intake of radiocesium was zero.

Other biologically significant radionuclides and their sources of production include the following:

Isotope Name	Half-life	Principal Decay Mode	Maximum Energy	Product of
Naturally Occurring Radionuclides:				
Radium-226	1599.0 y	alpha	4.78450 MeV	Natural Source ^{238}U decay scheme
Radon-222	3.82351 d	alpha	5.48966 MeV	Same
Polonium-210	138.3763 d	alpha	5.30451 MeV	daughter ^{210}Bi in radium decay scheme
Artificially Produced Radionuclides which also Exist Naturally:				
Tritium	12.346 y	beta	0.018610 MeV	^{6}Li
Carbon-14	5730 y	beta	0.155 MeV	^{14}N
Krypton-85	10.701 y	beta	0.672 MeV	^{84}Kr
Artificial Radionuclides Produced by the Fission Process:				
Strontium-89	50.55 d	beta	1.488 MeV	^{88}Sr
Strontium-90	28.82 y	beta	0.546 MeV	fission
Iodine-129	157 x 10^{7} y	beta	0.150 MeV	fission
Iodine-131	8.040 d	beta	0.6065 MeV	fission
Cesium-134	2.062 y	beta	1.454 MeV	^{133}Cs
Cesium-137	30.174 y	beta	0.51163 MeV	fission

Isotope Name	Half-life	Principal Decay Mode	Maximum Energy	Product of
Transuranic Nuclides Produced by the Fission Process:				
Neptunium-237	2.14×10^6 y	alpha	4.2 MeV	^{241}Am
Plutonium-238	87.71 y	alpha	5.49921 MeV	^{238}Np
Plutonium-239	2.4131×10^4 y	alpha	5.1554 MeV	^{235m}U
Plutonium-241	14.355 y	alpha	5.17 MeV	Multiple n-capture from ^{238}U, ^{239}Pu
Americium-241	432.0 y	alpha	5.48574 MeV	^{241}Pu
Curium-242	162.76 d	alpha	6.1129 MeV	same as ^{241}Pu
Curium-244	18.099 y	alpha	5.80496 MeV	same as ^{241}Pu
Other Important Fission Products Include:				
Molybdemum-99	67 hr	beta	1.23 MeV	
Technetium-99	2.12×10^5 y	beta	0.292 MeV	
Ruthenium-103	39.8 d	beta	0.70 MeV	
Ruthenium-106	1 y	beta	0.039 MeV	
Silver-110m	252 d	gamma	0.74 MeV	
Tellurium-132	78 hr	beta	0.22 MeV	
Barium-140	12.8 d	beta	1.02 MeV	
Cerium-144	290 d	beta	0.31 MeV	
Europium-154	8.2 y	beta	0.70 MeV	
Important Activation Products Include:				
Nickel-63	100 y	beta	0.067 MeV	
Nickel-59	76,000 y	electron capture	1.06 MeV	

Isotope Name	Half-life	Principal Decay Mode	Maximum Energy	Product of
Cobalt-58	0.194 y	electron capture and gamma	0.474 MeV	
Cobalt-60	5.2719 y	beta	0.31788 MeV	
Iron-55	2.68 y	electron capture		
Manganese-54	0.855 y	gamma	0.835 MeV	
Niobium-95	35 d	beta	0.160 MeV	
Zirconium-95	0.175 y	gamma	0.396 MeV	

Note: Most beta emitters have gamma emissions as a secondary mode of decay and *vice versa* (exceptions: tritium, strontium-90, and ruthinium-106).

Basic Radiotoxicity for Ingestion of some Radionuclides

The following guide to the radiotoxicity of commonly encountered radionuclides was compiled by the Organisation for Economic Co-operation and Development (OECD) and European Nuclear Energy Agency (ENEA). Biologically significant radionuclides include not only anthropogenic (manmade) isotopes such as Pu-239 and Am-241 but also a number of naturally occurring radioisotopes such as Ra-226, which is the source of radon, the most carcinogenic of all naturally occurring radioisotopes.

High Toxicity*

Ac-227	Am-241	Am-243	Cf-251	Cm-243	Cm-244
Cm-245	Cm-246	Cm-247	Cm-248	I-129	Np-237
Pa-231	Pb-210	Po-210	Pu-238	Pu-239	Pu-240
Pu-242	Ra-223	Ra-224	Ra-226	Ra-228	Th-230
Th-232					

Medium Toxicity**

Ag-110m	Am-242	Ba-140	Bi-210	C-14	Cd-109
Ce-141	Ce-143	Ce-144	Cl-36	Cm-242	Co-56
Co-57	Co-58	Co-60	Cs-134	Cs-135	Cs-136
Cs-137	Eu-152 (13 years)		Eu-152 (9 hours)		Eu-154
Eu-155	Fe-52	Fe-55	Fe-59	Hg-197	I-130
I-131	I-132	I-133	I-135	K-43	La-140
Mn-54	Mo-99	Na-22	Na-24	Nb-93m	Nb-95
Nd-147	Np-239	P-32	Pa-233	Pm-147	Pm-149
Pr-143	Pu-241	Rb-86	Rb-87	Rh-105	Ru-103
Ru-105	S-35	Sb-124	Sb-125	Se-75	Sm-147
Sr-85	Sr-89	Sr-90	Sr-91	Sr-92	Tc-99
Te-125m	Te-127m	Te-127	Te-129m	Te-131m	Te-132
Th-227	Th-231	TI-204	U-232	U-233	U-234
U-235	U-236	U-238	Y-90	Y-91	Y-93
Zn-65	Zr-93	Zr-95	Zr-97		

Low Toxicity***

Ca-45	Cr-51	Cs-131	Cu-64	H-3	I-134
In-113m	Ni-59	Ni-63	Rh-103m	Sm-151	Tc-99m
Te-129					

---

*   ALI < 1 μCi

**   1 μCi < ALI < 1 mCi

*** ALI > 1 mCi

1. Transcribed from "Review of the Continued Suitability of the Dumping Site for Radioactive Waste in the North-East Atlantic" N.E.A., OECD, Paris, April 1980. p.42 (B-214)

Figure 7. Radionuclide isotope toxicity (Appendix V. from Brack 1984 pg. 64).

Selected Radiation Level Guidelines

Government radiation safety limits.

United States

Note: Also see FDA Derived Intervention Levels below as well as the many other protection action guidelines listed later in this *Handbook*. To convert pCi to Bq divide by 27.

Safety Limit	Isotope	Media	Source
300 pCi/l	I-131	Drinking water	1961 FRC Protection Action Guidelines
100 pCi/m^3	I-131	Air	1961 FRC Protection Action Guidelines
8,000 pCi/kg	I-131	General use foods	Federal Register 1986
1,500 pCi/kg	I-131	Infant foods	Federal Register 1986

Japan

Safety Limit	Isotope	Media	Source
30 Bq/l	I-131	Drinking water	Japan Health Ministry
300 Bq/l or kg	I-131	Milk, dairy products	Japan Health Ministry
2,000 Bq/kg	I-131	Vegetables (except root vegetables)	Japan Health Ministry
2,000 Bq/kg	I-131	Fish	Japan Health Ministry
200 Bq/l	Cs-137	Drinking water	Japan Health Ministry
200 Bq/l or kg	Cs-137	Milk, dairy products	Japan Health Ministry
10 Bq/kg	Cs-137	Grains	Japan Health Ministry
10 Bq/kg	Cs-137	Meat, eggs, fish, etc.	Japan Health Ministry
5000 Bq/kg	Cs-137	Soil for growing rice	Japan Health Ministry
500 Bq/kg	Cs-137	Rice	Japan Health Ministry
20 mSv/year	All accident sources	Annual exposure threshold action level for evacuation	Japan Health Ministry

Banana Equivalent Dose (BED)

The famous banana equivalent dose (BED) is derived from the fact that bananas contain the ubiquitous naturally occurring radioactive isotope potassium (K-40): (1/2 T = 1 billion years; energy level 10 to 40 eV). Of all the potassium in the world, 0.0117% consists of this nearly harmless naturally occurring radioisotope. One gram of potassium has an activity level of 30 Bq; an average banana contains a half gram of potassium; it will have an activity level of 15 Bq. The relative insignificance of radioactive potassium in bananas should not be used to divert attention from the presence of biologically significant quantities of isotopes such as radioiodine or plutonium: 15 Bq of iodine-131 is a biologically significant activity level; 2.5 Bq/kg of plutonium-239 in any food mandates extreme precautionary measures.

As we enter the third phase of nuclear-industrial society under the pressure of declining fossil fuel reserves (the first was the era of nuclear weapons and the test explosions; the second was the first generation of now aging and often subprime nuclear power generating reactors), perhaps we will begin expressing risk of exposure to radiation in banana equivalent dose values even though K-40 is not powerful enough to qualify as ionizing radiation. Nuclear technology enthusiasts love these banana equivalent doses. How many BEDs are in a 6 oz. cup of tea containing 200 Bq/l of I-131?

Exposure Pathways

There are four basic pathways of human exposure to radiation resulting from nuclear accidents of any kind:

- External: accident location radiation shine, plume cloud shine, ground shine, shine from contaminated clothing and shoes
- Absorption by dermal deposition
- Inhalation: plume inhalation and inhalation of re-suspended ground deposition
- Ingestion:
  o primary (from foliar and surface contamination)
  o secondary (via direct pathways to human consumption such as the forage-cow-milk pathway or the sediment-phytoplankton-mollusk-crustacean-fish pathway)
  o tertiary (via indirect pathways to human consumption, e.g. the incorporation of contaminated whole foods such as milk, whey, wheat, corn, or soy into processed foods and their redistribution to markets in areas unaffected by ground deposition)

For additional information and pathway citations see *Section VI. Pathways.*

The Risks of Exposure to a Passing Plume

- Inhalation of biologically significant radioisotopes, including radioiodine, plutonium, cesium, and strontium particles
- External exposure to cloud and ground shine
- Exposure to contaminated surface water and contaminated surfaces, which will continue long after plume passage
- Immediate exposure to contaminated air or the contamination that might be on clothing or exposed skin, as measured in microsieverts, does not provide information about total exposure via absorption, inhalation, and ingestion
- After airborne plume particles have been deposited or have dissipated, ambient air and skin surface measurements provide almost no information about exposure
- The key question is, what is the "long-term exposure" of an individual to contaminated food products via ingestion? Close-in and/or downwind accident exposure may include inhalation, absorption, and ingestion, but the ingestion pathway provides most of the exposure for individuals not living near the accident site. Transfer of radioactive contamination into the food chain, especially in vegetables, wheat, soy, corn, milk, meat, and high fructose corn syrup, as well as in seaweed, mollusks, crustaceans, and fish, can provide exposure to individuals and communities that have experienced no radioactive fallout whatsoever from the Fukushima Daiichi or any other nuclear accident.

MOX Fuel Pathway Alert

MOX fuel-derived plutonium, curium, and americium isotopes will be a component of Fukushima Daiichi Reactor 3 emissions, including those originating in its spent fuel pool. Pathways of importance are inhalation of plume cloud particles and inhalation of re-suspended particulates after deposition by rainfall events. If criticality in the Fukushima Daiichi Reactor 3 reaches full meltdown status, MOX fuel-derived contamination patterns could be hemispheric in distribution.

Fukushima Daiichi Washout Pathway

The high volume manual water cooling efforts now underway at the Fukushima Daiichi complex may control or prevent resumption of fissile activity and criticality at the seven sites now experiencing meltdown events, but these cooling efforts also constitute a never-before-experienced pathway for nuclear effluents to enter the environment. Given the damage to Reactors 1 – 3, their spent fuel pools, and the spent fuel pool at Reactor 4 that resulted from the hydrogen explosions and fires, manual cooling efforts mobilized large quantities of fuel pellet-derived and fission products, which, if not retained in makeshift radioactive water storage facilities, will eventually be discharged into the marine environment. The long-term impact of ongoing cooling efforts will result in massive source term releases from the washout pathway that may equal or even surpass those from the Chernobyl accident.

Sediment as a Repository of Discharges to the Marine Environment

Some of the radioactive contamination released to the marine environment at Fukushima Daiichi may be immediately transferred from seawater to fish via phytoplankton, as has already been noted with respect to contamination discovered in the fish species kounago (sand lance): 4,080 Bq/kg I-131 and 526 Bq/kg Cs-137. The majority of source term releases that occur either as washout or atmospheric deposition quickly accumulate in the sediment. Marine concentration ratios pertaining to anthropogenic radioactivity deposited in sediments were collated by the United Nations Scientific Committee on the Effects of Atomic Radiation (UNSCEAR) in their 1982 report (see *Section III. Additional Definitions and Concepts*). The remobilization of Fukushima Daiichi-derived fallout in marine sediments will provide exposure pathways of concern for generations of future fishermen, recreational boaters, beach goers, and seafood consumers.

Deposition Mechanisms

In any nuclear accident, the radioactive plume results in two types of deposition: **dry deposition** and **wet deposition**. Atmospheric nuclear weapons testing resulted in the radioactive contamination of the stratosphere that involved relatively uniform hemispheric fallout from both wet and dry deposition. Chernobyl and the ongoing accident at the Fukushima Daiichi facility involved tropospheric contamination; the **highest levels of ground deposition are associated with rain and snowfall events (wet deposition)** and are erratically distributed by wind direction and rain/snowfall intensity. In the case of Fukushima Daiichi events, the village of Iitate, located 25 miles northwest of the accident site, has received very high levels of rainfall-associated fallout, possibly due to its location, which is surrounded by hills and mountains. An accurate isometric map of terrestrial contamination in Japan utilizing aerial radiological surveillance of the ground deposition of Cs-137 is not yet available even after twelve weeks of gradually declining emissions. This raises a question: is this restriction of easily quantified data not available to the general public in Japan because biologically significant concentrations of Fukushima Daiichi-derived contamination occur in areas outside the evacuation zone?

MEXT Sketch of Radiation Exposure

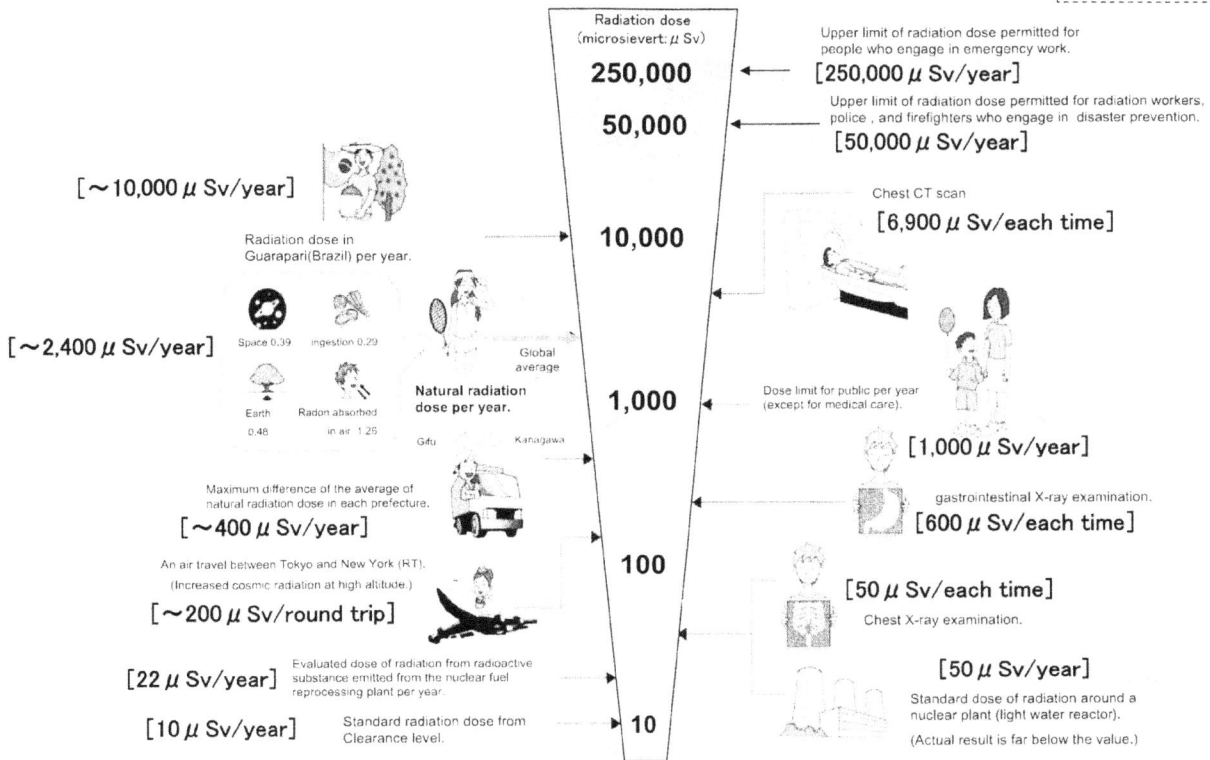

Radiation in Daily-life

※Unit : μ Sv

Radiation dose (microsievert: μ Sv)

250,000 — Upper limit of radiation dose permitted for people who engage in emergency work. [250,000 μ Sv/year]

50,000 — Upper limit of radiation dose permitted for radiation workers, police, and firefighters who engage in disaster prevention. [50,000 μ Sv/year]

[~10,000 μ Sv/year] — Radiation dose in Guarapari(Brazil) per year.

Chest CT scan [6,900 μ Sv/each time]

10,000

[~2,400 μ Sv/year] — Natural radiation dose per year. Space 0.39, ingestion 0.29, Earth 0.48, Radon absorbed in air 1.26. Global average

1,000 — Dose limit for public per year (except for medical care). [1,000 μ Sv/year]

Maximum difference of the average of natural radiation dose in each prefecture. Gifu — Kanagawa [~400 μ Sv/year]

gastrointestinal X-ray examination. [600 μ Sv/each time]

An air travel between Tokyo and New York (RT). (Increased cosmic radiation at high altitude.) [~200 μ Sv/round trip]

100

[50 μ Sv/each time] Chest X-ray examination.

[22 μ Sv/year] — Evaluated dose of radiation from radioactive substance emitted from the nuclear fuel reprocessing plant per year.

[50 μ Sv/year] Standard dose of radiation around a nuclear plant (light water reactor). (Actual result is far below the value.)

[10 μ Sv/year] — Standard radiation dose from Clearance level. **10**

Figure 8. Radiation in daily life (Ministry of Education, Culture, Sports, Science and Technology of Japan (MEXT)
http://www.mext.go.jp/component/english/__icsFiles/afieldfile/2011/04/07/1304690_040619.pdf**).**

EPA Radiation Exposure Action Level

The level of exposure that the EPA considers actionable is 1000 mR, or 1 R, received over four days. Measurements of radiation at these levels would trigger actions contained in the *Protective Action Guidelines for Nuclear Incidents* (United States Environmental Protection Agency 1992).

One thousand mR over four days can be broken down into the following daily and hourly exposure levels that should be a useful guide for concerned citizens and can easily be compared to radiation levels being reported pertaining to the Fukushima Daiichi fuel melting events.

- 250 millirems per day (mR/day)
- 0.0025 sieverts per day (Sv/day)
- 2.5 millisieverts per day (mSv/day)
- 2,500 microsieverts per day (μSv/day)
- 10.415 millirems per hour (mR/hr)
- 0.10415 millisieverts per hour (mSv/hr)
- **104.15 microsieverts per hour (μSv/hr)**

Individuals experiencing radiation exposure near or above these EPA action levels should immediately take precautionary measures such as sheltering, avoiding rainfall exposure, and avoiding intake of contaminated foods, water, and milk. Evacuating to a safer location would be a practical response to this level of radiation exposure.

Terrestrial Contamination Levels of Concern for the Indicator Isotope Cs-137

Sudden increases in the environmental presence of the indicator isotope Cs-137 are indicative of new or reactivated point sources of anthropogenic radioactivity, e.g. nuclear weapons test explosions or nuclear accidents. The fissile activity that results in the production of Cs-137 also produces the short-lived isotope Iodine-131 ($\frac{1}{2}T = 8$ days), among the most biologically significant of all anthropogenic radioactive isotopes. Increasing deposition of Cs-137 automatically mandates intensive monitoring for the presence of I-131 in pathways to human consumption.

- **3.7 Bq/m^2**: Begin monitoring of ambient exposure radiation levels (microsieverts per hour), terrestrial contamination levels above background (Cs-137, Bq/m^2), and contamination in rainwater (I-131, Bq/l) via the Weather Channel and the EPA, CDC, DOE, FDA, IAEA, state, and local websites, as well as other online information sources.
- **37 Bq/m^2**: Expand monitoring to include measurements of local contamination of food, water, and milk by indicator nuclides I-131 and Cs-137 as measured in Bq/kg or Bq/l. Begin protective actions: avoid exposure to rainfall events, remove shoes and clothing before entering domestic environments, and shower after exposure to rain.
- **370 Bq/m^2**: Expand protective actions: stay indoors whenever possible, close windows, seal openings, cover gardens with tarps, shelter livestock, avoid ingestion of leafy vegetables and fruits harvested in fallout areas. Expand monitoring of food, water, and milk; discard or avoid ingesting foods and water contaminated by the indicator nuclides I-131 and Cs-137 above 370 Bq/kg or

Bq/l (10,000 picocuries, the protection action level used by the US FDA to dispose of imported foods contaminated by Chernobyl-derived fallout). Continue monitoring of ambient radiation levels.

- **3,700 Bq/m^2**: Expand protective actions to restrict movement of children, outside workers, nonessential travel, and recreational activities such as hiking and sunbathing. Closely monitor food, milk, and water intake, sources, and radiation levels. Begin using respirators in soil cultivation work. Begin maximum use of information technology to keep informed of ongoing emissions.
- **37,000 Bq/m^2**: Prepare to evacuate to a safer zone, e.g. the southern hemisphere or another planet.

Without the systematic compilation of the distribution patterns of the indicator radioisotopes Iodine-131 (1/2T = 8 days) and Cs-137 (1/2T = 30 years), no reasonable evaluation of the health physics impact of the Fukushima Daiichi disaster or any other nuclear accident or point source can be tabulated.

Plume Hot Zones

Numerous short-lived nuclides characterize any quick release accident. Plume hot zones can be divided into the following three major categories:

- Super hot zone (±2 hours): initial ambient radiation levels following any accident will be highly elevated due to the presence of short-lived isotopes such as Xe-137 (1/2 T = 3.81 m; daughter product is Cs-137), I-132 (1/2 T = 2.3 h), I-133 (1/2 T = 2.8 h), Te-129 (1/2 T = 69.6 m), Te-131 (1/2 T = 25 m), and Ce-143 (1/2 T = 33 h). These isotopes are often associated with hot particles. The health physics impact of these short-lived isotopes is entirely contingent upon weather conditions. A worst case scenario involves calm weather conditions accompanied by rainfall. In the event of a hydrogen gas explosion, as occurred in Japan, high levels of radiation would remain at ground level; survival would be dependent upon the availability of self-contained breathing apparatus in combination with some sort of fallout shelter such as an interior bedroom or cellar. Close-in super hot zone fallout would also be associated with drizzle and fog. The US NRC guideline of a ten mile evacuation zone would be relevant in the context of calm weather where the plume super hot zone was distributed more or less evenly around the accident site. Both the Chernobyl accident and the Fukushima Daiichi hydrogen explosions involved wind speeds of 10 to 15 miles per hour, which helpfully dispersed the plume hot zone above ground level. In the case of the accident in Japan, much of the plume super hot zone was carried over the Pacific Ocean by prevailing winds. In an accident scenario at a US boiling water reactor, such as the Pilgrim Nuclear Generating Station in Plymouth, MA, the super hot plume zone could conceivably extend downwind from the reactor site 10 to 25 miles, extending to communities such as Kingston, Marshfield, Scituate, Cohasset, Hingham, Rockland, and Pembroke when prevailing winds are from the south.

- Hot zone (2 days): An accident-derived plume hot zone is associated with the dispersal and deposition of Xe-135 (1/2 T = 9.14 h), Te-132 (1/2 T = 78 h), Mo-99 (1/2 T = 67 h), Sb-127 (1/2 T = 3.85 d), Te-132 (1/2 T = 3.2 d), and Np-239 (1/2 T = 2.35 d). The indicator isotope I-131 and its uptake in pathways to human consumption is also characteristic of plume hot zones. Most hot zone deposition would occur downwind from the accident site and would be highly associated with rainfall events. While the source term of a quick release accident may only last a few hours or less, hot zone deposition will be a function of wind direction. As illustrated by the Chernobyl plume pathway, impacted communities receiving high levels of I-131, Cs-137, and the many other isotopes associated with a nuclear accident can be several hundred kilometers or more from the accident site. In the case of the Pilgrim Nuclear Generating Station in Plymouth, MA, an accident occurring when prevailing winds are blowing from the southeast, as is typical in the spring and summer, plume hot zone fallout levels above 37,000 Bq/m^2 (1 curie/km^2) of Cs-137 or I-131 or both could extend throughout eastern New England as far as northern Vermont, New Hampshire, and central Maine. With respect to the accident in Japan, prevailing winds brought most of the tropospheric fallout over the Pacific, but brief shifts in wind direction also contaminated areas up to 75 km in all directions from the facility with contamination in excess of the 37,000 Bq/m^2 level of concern.

- Extended fallout zones: The extent of the impact of the Chernobyl accident was a surprise to many nuclear accident experts who had predicted that significant fallout would be limited to within ten miles of the accident site. Revised US NRC guidelines now postulate a 50 mile radius of biologically significant amounts of radiation dispersal, which might be the subject of exposure via the ingestion pathway. In the case of the Chernobyl accident, biologically significant quantities of accident fallout were documented in excess of 1,000 miles from the accident site. Global transport of accident-derived nuclides was documented in the 9 to 14 day range. See *Section VIII. Chernobyl Fallout Data* for a detailed listing of accident surveillance reports from many nations.

66

I-131 as an Indicator of Fissile Activity

In a normal nuclear reaction involving the fission process, I-131 is produced as a volatile component of fissile activity (the fission process) at a rate of 35 to 50 times that of Cs-137. In the case of a loss of coolant accident in a spent fuel pool that had no fissile activity for sixty days, there would be no detectable I-131 activity in comparison to Cs-137 emissions, which would be the prevailing volatile indicator isotope in any source term release. The presence of I-131 in any accident plume is indicative of either ongoing fissile activity (continued criticality or resumed fissile activity) or of recent fissile activity. The Chernobyl Cs-137 source term was 2.7 million Ci or 100,000 terabecquerels; according to the IAEA, the Chernobyl I-131 source term was 85,860, 000 Ci or 3.12% that of the I-131 source term (Fairlie 2006). Of interest is the recent NISA assertion that the Chernobyl I-131 source term was 5.2 million terabecquerels or 140,400,000 Ci. While the source of this information is unknown, it was cited to illustrate the fact that Fukushima Daiichi releases of I-131, which Japanese authorities estimated was between 370,000 and 630,000 Ci, were much less than that at Chernobyl (*Wall Street Journal* April 12, 2011). Of particular note with respect to the Fukushima Daiichi MIME is that a significant portion of the accident source term involved releases of fission products from spent fuel assemblies that had no fissile activity for 60 days or longer. I-131 releases, therefore, were proportionately less in comparison to Cs-137 releases due to the radioactive decay of I-131 in the fuel assemblies in the spent fuel pools that suffered loss of coolant accidents. Resumption of fresh activity within the melted spent fuel pools would be accompanied by renewed I-131 emissions.

Health Physics Levels of Impact

A thumbnail sketch of the health physics impact of exposure to ionizing radiation:

- 0 - 0.25 Sv (0 - 250 mSv): None
- 0.25 - 1 Sv (250 - 1000 mSv): Some people feel nausea and loss of appetite; bone marrow, lymph nodes, and spleen damaged
- 1 - 3 Sv (1000 - 3000 mSv): Mild to severe nausea, loss of appetite, infection; more severe bone marrow, lymph node, and spleen damage; recovery probable, not assured
- 3 - 6 Sv (3000 - 6000 mSv): Severe nausea, loss of appetite; hemorrhaging, infection, diarrhea, peeling of skin, and sterility; death if untreated
- 6 - 10 Sv (6000 - 10,000 mSv): Above symptoms plus central nervous system impairment; death expected
- Above 10 Sv (10,000 mSv): Incapacitation and death

  (wikipedia.org/wiki/Sievert).

1997 Revised FDA Radioactive Contamination Guideline: Part 1

See *Section VII. Radiation Protection Guidelines* for additional selections from this publication.

United States Food and Drug Administration. (March 5, 1997). Draft: *Accidental radioactive contamination of human food and animal feeds: Recommendations for state and local agencies*. Center for Devices and Radiological Health, US FDA, Washington, D.C.

- This draft was issued on 3/5/97 and represents a radical revision of the 1982 FDA recommendations, which are rescinded by these proposed standards.
  - Derived intervention levels (DILs) are far stricter (more conservative) than the 1982 regulations. Derived intervention levels for the radiocesium group (1,160 Bq/kg for 15 year old = 31,320 picocuries/kg) are far closer to the "levels of concern," which resulted in seizure of food containing 10,000 picocuries/kg of radiocesium following the Chernobyl accident (see Chernobyl-derived radiocesium in US imported foods).
  - The most radical change in these guidelines is the inclusion of numerous additional radionuclides for consideration following a nuclear reactor or other type of nuclear accident (United States Food and Drug Administration 1997, Appendix E). The derived intervention level for transuranic nuclides such as ^{238}Pu, ^{239}Pu, and ^{241}Am range from 2.0 to 2.5 Bq/kg for a 3 month old infant. These more inclusive guidelines are an acknowledgment of the lessons of the Chernobyl accident, i.e. a major nuclear accident includes many different radionuclides whose health physics impact cannot be delineated by a single protection action guideline standard such as 10,000 picocuries (370 Bq).
  - The one significant unfortunate lapse in this draft is the use of the "number of samples contaminated above regulatory limits" to summarize contamination levels derived from the Chernobyl accident without reference to the specific levels of contamination in the samples analyzed (see tables C-1, C-2, and C-3). This continues the FDA inclination to withhold nuclide-specific data after incidents of widespread contamination of foodstuffs. The substitution of an arbitrary action limit to replace nuclide-specific data illustrates that the FDA is still inclined to withhold information about rising levels of radioactive contamination in the food chain. In the event of another accident, the use of this arbitrary limit raises the question: will the FDA withhold data if contamination trends up towards the derived intervention level? All levels of contamination below the DIL are, after all, "below regulatory limits."

- "Recommendations on accidental radioactive contamination of human food and animal feeds were issued in 1982 by the Food and Drug Administration (FDA) (Shleien 1982). Since then, there have been enough significant advancements related to emergency planning to warrant updating the recommendations." (United States Food and Drug Administration 1997, 1).
- "DILs [Derived Intervention Levels] are limits on the concentrations permitted in human food distributed in commerce. ... Comparable limits were not provided in the 1982 FDA recommendations. DILs apply during the first year after an accident." (United States Food and Drug Administration 1997, 3).

Table D-5 (pg. D-13)						
DERIVED INTERVENTION LEVELS (Bq/kg)						
(individual radionuclides, by age group, most limiting of either PAG)						
Radionuclide	**3 months**	**1 year**	**5 years**	**10 years**	**15 years**	**Adult**
Sr-90	308	362	616	389	160	465
I-131	196	167	722	1200	1690	2420
Cs-134	1600	2190	1940	1530	958	930
Cs-137	2000	2990	2810	2180	1370	1360
Cs group[a]	1800	2590	2380	1880	1160	1150
Ru-103	6770	8410	12200	16400	25000	28400
Ru-106	449	621	935	1340	2080	2360
Pu-238	2.5	21	17	14	12	10
Pu-239	2.2	18	14	13	10	9.8
Am-241	2.0	17	13	11	9.1	8.8
Pu+Am group[b]	2.2	19	15	13	9.6	9.3
[a] Computed as: (DIL for Cs-134 + DIL for Cs-137) /2						
[b] Computed as: (DIL for Pu-238 + DIL for Pu-239 + DIL for Am-241) /3						

- "The 1982 FDA recommendations were developed from the prevailing scientific understanding of the relative risks associated with radiation as described in the 1960 and 1961 reports of the Federal Radiation Council (Federal Radiation Council 1960-61). Since 1982, FDA and the other federal agencies in the United States have adopted the methodology and terminology for expressing radiation doses provided by the International Commission on Radiological Protection (ICRP) in 1977 (International Commission on Radiological Protection 1977; ICRP 1984a; United States Environmental Protection Agency 1987)" (United States Food and Drug Administration 1997, 5).
- "The equation given below is the basic formula for computing DILs.

$$\text{DIL (Bq/kg)} = \frac{\text{intervention level of dose (Sv)}}{f \text{ x Food Intake (kg) x DC (Sv/Bq)}}$$

  Where:
  DC = Dose coefficient; the radiation dose received per unit of activity ingested (Sv/Bq).
  f = Fraction of the food intake assumed to be contaminated.
  Food Intake = Quantity of food consumed in an appropriate period of time (kg)." (United States Food and Drug Administration 1997, 8).

- "The food monitoring results from FDA and others following the Chernobyl accident support the conclusion that I-131, Cs-134 and Cs-137 are the principal radionuclides that contribute to radiation dose by ingestion following a nuclear reactor accident, but that Ru-103 and Ru-106 also should be included (see Appendix C)." (United States Food and Drug Administration 1997, 10). "DIL is equivalent to, and replaces the previous FDA term Level of Concern (LOC)." (United States Food and Drug Administration 1997, 12).
- "The types of accidents and the principal radionuclides for which the DILs were developed are:
  - nuclear reactors (I-131; Cs-134 + Cs-137; Ru-103 + Ru-106),
  - nuclear fuel reprocessing plants (Sr-90; Cs-137; Pu-239 + Am-241),
  - nuclear waste storage facilities (Sr-90; Cs-137; Pu-239 + Am-241),
  - nuclear weapons (i.e., dispersal of nuclear material without nuclear detonation) (Pu-239), and

- o radioisotope thermoelectric generators (RTGs) and radioisotope heater units (RHUs) used in space vehicles (Pu-238)." (United States Food and Drug Administration 1997, 13).
- "For each radionuclide, DILs were calculated for six age groups using Protective Action Guides, dose coefficients, and dietary intakes relevant to each radionuclide and age group. The age groups included 3 months, 1 year, 5 years, 10 years, 15 years and adult (>17 years). The dose coefficients used were from ICRP Publication 56 (International Commission on Radiological Protection 1989)." (United States Food and Drug Administration 1997, 14).

For more information on these guidelines, see *Section III. Additional Definitions and Concepts*

Ionizing Radiation: Sources and Biological Effects

In 1982, the United Nations Scientific Committee on the Effects of Atomic Radiation (UNSCEAR) issued a 773-page report to the General Assembly, which remains the definitive reference on the sources and health physics impact of radiation. A particularly important subject, and one not covered in this *Handbook*, is the detailed description of dose assessment models in Appendix A of the report, which includes environmental, atmospheric transport, terrestrial, and aquatic dosimetric models. The following summary of the table of contents suggests why the UNSCEAR text remains a basic reference for anyone wishing to learn about ionizing radiation despite the rapid expansion of medical technologies using radiation (e.g. CAT Scans), the proliferation of nuclear power reactors and the Chernobyl accident.

A. Dose assessment models

B. Exposures to natural radiation sources

C. Technologically modified exposures to natural radiation

D. Exposures to radon and thoron and their decay products

E. Exposures resulting from nuclear explosions

F. Exposures resulting from nuclear power production

G. Medical exposures

H. Genetic effects of radiation

I. Non-stochastic effects of irradiation

J. Radiation-induced life shortening

K. Biological effects of radiation in combination with other physical, chemical, or biological agents

International Nuclear Event Scale (INES) For Nuclear Installations

The IAEA has published the following scale for rating accidents-in-progress. This is reprinted from the Swiss Nuclear Safety Inspectorate (HSK) website (http://www.ensi.ch/index.php?id=31&L=2) and has been utilized by the government of Japan to evaluate the accident at the Fukushima Daiichi complex, which has been elevated to Status Level 7.

This international scale for the expression of incident severity in nuclear installations has been in use since early 1990. Based on their relevance to plant safety, it distinguishes the following seven levels of incidents:

Level	Descriptor	Criteria	Examples
7	Major accident	External release of a large fraction of the reactor core inventory typically involving a mixture of short- and long-lived fission products (in quantities radiologically equivalent to more than tens of thousands of terabecquerels of iodine-131). > 2,700,000 Ci I-131. Possibility of acute health effects. Delayed health effects over a wide area, possibly involving more than one country. Long-term environmental consequences.	Chernobyl, USSR (1986) Fukushima Daiichi, Japan (2011)
6	Serious accident	External release of fission products (in quantities radiologically equivalent to the order of thousands to tens of thousands of terabecquerels of iodine-131) Full implementation of local emergency plans probably needed to limit serious health effects.	

Level	Descriptor	Criteria	Examples
5	Accident with off-site risks	External release of fission products (in quantities radiologically equivalent to the order of hundreds to thousands of terabecquerels of iodine-131). Partial implementation of emergency plans (e.g. local sheltering and/or evacuation) required in some cases to lessen the likelihood of health effects. Severe damage to a large fraction of the core and major plant contamination.	Windscale, UK (1957) Three Mile Island, US (1979)
4	Accident without significant off-site risks	External release of radioactivity resulting in a dose to the most exposed individual off-site of the order of a few millisieverts. Need for off-site protective actions generally unlikely except possibly for local food control. Some damage to reactor core as a result of mechanical effects and/or melting. Worker doses likely to have acute fatal consequences.	Saint Laurent, France (1980)
3	Serious incident	External release of radioactivity above authorized limits, resulting in a dose to the most exposed individual off-site of the order of tenths of a millisievert. High radiation levels and/or contamination on-site as a result of workers likely to lead to acute health effects. Incidents in which a further failure of safety systems could lead to accident conditions, or a situation in which safety systems would be unable to prevent an accident if certain initiators were to occur.	Vandellos, Spain (1989)

Level	Descriptor	Criteria	Examples
2	Incident	Incidents with major failure of safety provisions, but still leaving sufficient safety margins to cope with additional faults. Radiological incident with members of the personnel receiving doses in excess of the annual limit Significant contamination of the installation which was not to be expected on the design basis.	Sosnowy Bor, Russia (1992)
1	Anomaly	Functional or operational anomalies which do not pose a risk but which indicate a lack of safety provisions. This may be due to equipment failure, human error or procedural inadequacies.	
0	No safety significance	Situations where operational limits and conditions are not exceeded and are properly managed in accordance with adequate procedures belong here. Examples: Individual failure in a redundant system. Single operational mistake without consequences. Faults (no multiple simultaneous failure) detected in periodic inspections or tests. Automatic reactor scram with normal plant behavior. Reaching of limiting operation conditions, while adhering to the proper regulations.	

III. Additional Definitions and Concepts

The following definitions are extracted from RADNET: Information about Source Points of Anthropogenic Radioactivity, the online archives of the Center for Biological Monitoring, now part of the Department of Environmental History of the Davistown Museum, as well as from other print and online sources as previously noted.

The Fukushima Daiichi nuclear disaster is an ongoing accident whose significance and extent cannot yet be evaluated. RADNET's annotated bibliography of Chernobyl fallout patterns and intensity and the RISO National Laboratories weapons testing fallout database both provide information essential to evaluating the impact of the Japan disaster. The Oak Ridge National Laboratory database for cumulative fuel waste inventories provides the basic information necessary to estimate Cs-137 inventories in the fuel assemblies of aging nuclear reactors (±10,000 Ci of Cs-137 per fuel assembly), including those at the Fukushima Daiichi complex.

Radioactivity

Radioactivity is the spontaneous decay of the nucleus of an atom by the emission of particles, usually accompanied by electromagnetic radiation. It is also defined as the mean number of nuclear transformations occurring in a given quantity of radioactive material per unit time, expressed in sieverts (Sv – human exposure), becquerels (Bq – quantitative measurement), or curies (Ci – activity levels). Most radionuclides (radioactive nuclides in contrast to stable nuclides) have multiple forms of radioactive emissions, and are classified according to their **principal decay modes**. The most common **types of radiation** and examples of their sources are:

Alpha Radiation: Emitted by plutonium-239: a nucleus of a helium atom; large in mass, unable to penetrate more than a few microns of biological tissue (e.g. cannot penetrate a piece of paper).

Beta Radiation: Emitted by tritium: a high speed electron, small in mass, moderate penetrating abilities, e.g. unable to penetrate more than a few millimeters of biological tissue. It is also emitted by cesium and many other radioisotopes.

Gamma Radiation: Emitted by zirconium-95: highly penetrating, very energetic x-rays emitted by an excited nucleus. Gamma radiation will often but not always exit living tissues without depositing its biologically significant electron voltage (eV).

Curie

The fundamental measurement of radioactivity in the environment: the amount of radioactive material giving off 3.7×10^{10} d.p.s., or 37 billion disintegrations per second. **One disintegration per second is called a becquerel, now the universal reporting unit for radioactive contamination.** In the United States, the picocurie (1 pCi = 0.037 d.p.s. or 1×10^{-12} of a curie) was once the unit used for many measurements of radioactive contamination, and may again appear as a reporting unit with respect to the unfolding catastrophe in Japan.

Radioactive Half-Life (1/2 T)

The time required for one half the atoms in a radioactive substance to decay. For example, the radioactive half-life of cesium is 30.174 years, 1/2T = 30.174 y. Radionuclides with short half-lives are hot, emitting large amounts of radiation but decaying quickly and contrast with radionuclides with longer half-lives whose energy is emitted over a longer period of time. The **biological half-life** is the time required for the body to eliminate half of a radioactive substance by regular physiological processes of elimination. This definition differs slightly from **effective half-life** which is the time required for half of the radioactive contamination to be diminished by both radioactive decay and biological elimination.

Daughter Products

A synonym for decay products, daughter products result from the radioactive disintegration of a radionuclide. Daughter products can either be stable or radioactive. Many important radionuclides are components of other nuclides' decay series: e.g. niobium-95 is a decay product of zirconium-95; neptunium-237 is a decay product of americium-241; americium-241 is a decay product of plutonium 241. Plutonium-238, the third most common constituent in spent fuel, is a decay product of neptunium 238. All curium nuclides decay to plutonium isotopes. Also called "growing in." An important daughter product of ubiquitous gaseous stack releases of nuclear reactors is ^{134}Cs, a daughter product of ^{133}Xe.

Fission

Fission is "the splitting of a heavy nucleus into two roughly equal parts (which are nuclei of lower-mass elements), accompanied by the release of a relatively large amount of energy in the form of kinetic energy of the two parts and in the form of emission of neutrons and gamma rays." (http://www.lbl.gov/abc/wallchart/glossary/glossary.html). A fissionable nucleus in an atom such as plutonium absorbs a neutron, becomes unstable, and splits into two nuclei, releasing a large amount of energy. Nuclear power is a controlled, self-sustaining fission process; nuclear explosions are an uncontained chain reaction version of the fission process. In the detonation of thermonuclear (fusion or hydrogen) bombs, the fission process is the trigger for the more powerful fusion event. **Fission products are the artificial radioactive offspring** of nuclear industries and accidents; their inventories and pathways in the environment are the subject of this *Handbook*.

Ionizing Radiation

Radiation with energy above 155 eV which has the ability to knock other electrons out of the orbits of atoms and molecules, often creating more ionizing radiation and adversely affecting living tissues. Biologically significant radiation is an ionizing dose of radiation above 155 eV which may have carcinogenic, mutagenic, or teratogenic health effects in humans.

Effective Action Level (FDA)

Following the Chernobyl accident, the Food and Drug Administration implemented an unofficial protection action guideline when it observed high levels of Chernobyl-derived radiocesium contaminating imported foods approximately one year after the accident. The FDA seized and destroyed foods contaminated in excess of 10,000 pCi/kg (370 Bq/kg) thereby setting an **effective action level** which was significantly more conservative (lower) than the protection action guidelines promulgated by various US government agencies before and after the Chernobyl accident. See *Section VII. Radiation Protection Guidelines* for a more complete description of the wide variety of protection action guidelines.

Exposure Pathway

The route that links radioactive contamination from a specific source point to a receptor population in a specific ecosystem.

Hot Particles

Airborne particles of partly volatilized fuel from nuclear accidents or from defective fuel cladding, which can also be carried by liquid effluents. Hot particles from leaking reactor fuel are also known as **"fuel fleas"** because they become electrically charged as a result of radioactive decay and "hop" from one surface to another. Typical hot particles are ~10 μm in size and can contain nuclides ranging from activation products to reactor-derived fission products (e.g. ^{95}Nb, ^{95}Zr, ^{103,106}Ru, ^{141,144}Ce, etc.) which were widely dispersed after the Chernobyl accident. CRUD is another type of hot particle. A significant percentage of the radioactive emissions from the ongoing accident at the Fukushima Daiichi complex are hot particles ejected by the hydrogen explosion from the four spent fuel pools experiencing loss of coolant accidents (LOCAs).

Indicator Nuclides

They are the principal radioactive products of nuclear industries or accidents. In the first few days of a nuclear accident, radioiodine-131 dominates the **activity release profile**. Other longer-lived radionuclides such as ^{106}Ru, ^{137}Cs, and ^{239}Pu dominate the later time compartments of the release pulse, producing exposure long after media coverage of a nuclear accident has faded and radioiodine-131 levels have subsided (see *Section VI. Pathways*). Cesium-137 is the most significant of the many nuclides which remain after the short-lived radionuclides have decayed. With respect to the Fukushima Daiichi meltdown events, the Department of Nuclear Engineering at the University of California is testing air samples in reporting units of becquerels per liter of rainwater for the indicator isotopes I-131, I-132, Cs-134, Cs-137, Te-132, and Be-7, the last of which is naturally-occurring cosmogenic radiation.

Peak Concentration / Mean Concentration

The peak concentration is the highest reading in a series of samples; the mean concentration is the average of readings in a series of samples.

Cold Shutdown

Cold shutdown is the maintenance of reactor and spent fuel pool water temperatures below the boiling point. If declining water levels expose the fuel rods, as happened at Three Mile Island and the Fukushima Daiichi nuclear power complex, the consequential heating creates radioactive steam. If heating is not controlled, the zirconium cladding on the fuel rods within the fuel assembly can ignite and burn, resulting in further radioactive emissions. A full scale meltdown, just barely avoided at the Fukushima Daiichi nuclear reactors, results when the reactor vessels or spent fuel assemblies become too hot to be cooled by water and fissile activity resumes.

Nuclear Accident Scenarios

Nuclear accidents and mishaps are commonplace events at nuclear reactors, weapons production facilities, and other nuclear installations - the Three Mile Island accident was not the "last" nuclear accident that occurred in the United States. The following is a quick synopsis of accident scenarios.

- **Reactor water system leaks**: A very common form of mishap at many nuclear reactor facilities. The release of tritium to groundwater is an indicator of ongoing reactor water system leaks, and has been documented at dozens of US reactors. More serious water system leaks will also release fission products.
- **Fuel cladding failure accidents**: A fuel cladding failure accident occurred at the Maine Yankee Atomic Power Company in the late 1990s and resulted in the

78

closure of the plant. According to licensee records, 66 failed fuel assemblies released fuel pellets into the reactor water system. Some of these were removed by vacuums and placed in the spent fuel pool, but the majority of fission products remained in the reactor vessel as "low-level radioactive waste" and were buried with the reactor vessel in South Carolina when the facility was decommissioned in 1999 (see *Section IX. Nuclear Dada: The Maine Yankee Atomic Power Plant*). The most notorious fuel cladding failure accident occurred at the Haddam Neck, CT, reactor site and released large quantities of fission products to the atmosphere in the early 1990s.

- **Spent fuel pool loss of coolant accident (LOCA)**: Four LOCAs are now underway at the Fukushima Daiichi complex. Loss of spent fuel pool cooling results in fuel assembly overheating, followed by fuel pellet expansion, fuel assembly swelling, cracking, and deformation. Fuel assemblies may then burst open, releasing fuel pellets as well as fission products into the remaining coolant. A total loss of coolant can result in the gradual melting of the entire fuel pool fuel assembly matrix. The large quantity of fission products being washed out of the Fukushima Daiichi spent fuel pools and into basements, drains, and the marine environment are an indication of an ongoing loss of coolant accident. A LOCA in a spent fuel pool, if cooling is lost or fails, can result in the resumption of partial or full criticality and the release of large quantities of fission products. The timeframe for the duration of a partially uncontrolled LOCA, such as is now occurring at the Fukushima Daiichi complex, is unknown but could be measured in months and possibly even years.

- **Loss of reactor coolant accident (LORCA)**: The accident at Three Mile Island was a LORCA that was eventually resolved by the resumption of the cooling of the reactor vessel spent fuel assemblies. LORCAs appear to be underway at three of the six Fukushima Daiichi reactor units. The other three units were not critical, that is undergoing fission chain reactions, at the time the tsunami destroyed the backup cooling systems. A LORCA occurs when a reactor vessel is breached or otherwise damaged and fuel assembly coolant (usually water) is no longer available. The continuing fission chain reaction can intensify resulting in the melting of fuel followed by the melting of the reactor vessel internals, bursting of pipes, and damage to pressure-operated relief valves. This is called a "serious core event" and can intensify as a function of time, coolant dispersal, and damage to the pressure vessel as the primary containment boundary. Once the pressure vessel is breached, large releases of fission products will occur. Continued heating of the fuel assemblies can result in a total meltdown; the extreme heat generated by an ongoing meltdown will eventually result in the melting of the pressure vessel and the dispersal of the molten fuel into the ground or groundwater underneath the reactor vessel. This is the worst case scenario accident for an operating nuclear reactor other than vaporization.

- **Vaporization**: Vaporization of a nuclear reactor complex, including both the spent fuel pool and the reactor vessel and its fuel assemblies, can occur during a nuclear attack. During the Cold War, the United States and Russia both targeted each other's nuclear facilities. Vaporization of a nuclear reactor site or weapons production facility as a result of nuclear war or terrorist attack would result in the dispersal of the entire cumulative inventory of the facility or facilities. The world catastrophe that would follow would result from the efficient vaporization of all onsite fission products, with a release of 25 to 500 times the contamination discharged at the Chernobyl accident, depending on how many reactors were at the vaporization accident site.

Radiometric Survey

A radiological survey of a contaminated site, especially sediments, soil, or other media containing sufficient data points to characterize the spread of contamination from a particular source point isometrically, i.e. via contour maps using isopleths which express the values of the data points. Aerial radiometric surveys have been utilized since the 1950s to characterize oil-bearing geological formations, by the defense department for analyses of Russian and other weapons production facilities, and after the Chernobyl accident to characterize fallout in Russia, Sweden, and England. Radiometric surveys of the ground deposition resulting from any nuclear accident are stated in becquerels per square meter. A radiometric survey of Fukushima Daiichi-derived C s-137 ground contamination is not yet available.

Biologically Significant Radionuclides

Radioactive substances such as plutonium, cesium, strontium, radioiodine, and tritium provide the most significant health hazards to humans among all nuclides released from anthropogenic sources. Biological significance is a result of a combination of high decay energy, biogeochemical availability, efficient energy transfer to biological systems, and ubiquitous production during nuclear accidents and from nuclear industries. Biologically significant radionuclides, such as I-131 and Cs-137, are called **indicator nuclides** and are used to characterize inventories and pathways of nuclear effluents in the biosphere.

The biological significance of radiation results from the enormous amount of energy contained in each emission. Visible light has an energy range of 1.77 to 4.13 **electron volts (eV)**. Most chemical changes occur within a range of 5 to 7 electron volts (eV).

Biologically significant radiation levels range from **18,610 eV (0.01861 MeV)** for the weak beta emitting tritium (1/2T = 12.346 yr) to **511,630 eV (0.51163 MeV)** for the ubiquitous cesium-137 (1/2T = 30.174 yr) to **5,155,400 eV (5.1554 MeV)** for the highly radiotoxic plutonium-239 (1/2T = 24,131 yr). These highly energetic emissions carry enough energy to tear electrons from neutral atoms and molecules. In delicate biological tissues, the impact of introducing radiation containing hundreds of thousands to millions of electron volts "can only be described as chemical and biological mayhem." (Gofman 1981, 22). For example, the alpha radiation resulting from the decay of plutonium-239 has little penetrating power due to its large mass, but, if inhaled and deposited in the lung, is among the most radiotoxic of nuclides since its 5,155,000 eV (5.155 MeV) will be distributed within the area of only a few cells.

The weaker beta radiation of **tritium** (H^3) is slightly more penetrating than alpha radiation; its biological significance comes from its ubiquitous production during the fission process, its tendency to follow the water cycle in nature, and its ability to become **tissue bound** in humans and the biotic environment. **Cesium-137**, a beta emitter with a gamma component, is biologically significant due to its energy level, its long half-life, its ubiquitous production during the fission process, and its tendency to follow the potassium cycle in nature, giving a whole body dose to those who ingest it.

Relative Biological Effectiveness (RBE)

A key component of the biological significance of radiation, RBE expresses the phenomenon that one kind of radiation is more effective (damaging) than another. Gofman (1981, 47) notes, "the RBE for alpha particles may be 10 for one biological effect, whereas it may be 1 or 2 for some other biological effect... the RBE for one radiation compared to another is not a fixed quantity."

Source Term

The quantity, chemical and physical form, and time history (release duration) of contaminants released to the environment from a facility (Centers for Disease Control 1999).

Source Term Release

Radioactive waste inventories discharged from a particular nuclear accident or source point, e.g. Chernobyl, Sellafield, weapons tests, etc. Each plume is characterized by a unique fingerprint of radioactive emissions which can be identified by a particular series of isotopic ratios. Weapons testing fallout was high in radiostrontium, low in cesium-134, and, thus, differed from the Chernobyl source term release which had much less radiostrontium and a higher ratio of cesium-134 to cesium-137 than weapons test fallout. Eisenbud (1987) and most early reports on the Chernobyl accident, in a classic example of misinformation, based the source term for Chernobyl upon Russian data which only included inventories of radionuclides deposited on Russian soil. Further research indicated that the source term release for Chernobyl included larger quantities of radioactive emissions than initially estimated and much higher levels of contamination than expected in locations which were a great distance from Chernobyl. An important study of the pre-Chernobyl sources of radioactivity, including naturally occurring, industrial, atomic power, weapons testing, and fuel reprocessing sources is the UNSCEAR Text (1982).

Source Term Release Duration

The source term release duration can vary from a few seconds for a weapons test explosion to hundreds of years or more for chronic discharges from source points such as military weapons production facilities. In the case of the Japan disaster, the source term release duration, and thus ongoing contamination of the biosphere, has the potential to be measured in years. The January, 1968, crash of a United States bomber carrying nuclear weapons, into the ocean near Thule, Greenland, released an estimated inventory of 1 TBq ^{239,240}Pu as well as smaller quantities of ^{238}Pu and ^{241}Am (Aarkrog 1994). The duration of this source term release was a matter of a few minutes; the duration of the plume movement is a function of the long radioactive half-lives of the isotopes in the source term release. The geographic magnitude of the plume pulse is a function of the chemical forms of the released isotopes and the biogeochemical cycles which may aid their spread in the biosphere: in the case of the Thule accident, the plutonium will tend to remain localized on the ocean sea bed unless it undergoes a change in chemical form from plutonium oxide to a form of plutonium more susceptible to bioaccumulation and transport by natural processes. The source term release duration from Chernobyl was measured in weeks; the biogeochemical cycling of the longest lived radionuclides within the source term pulse will be measured in millennia. The source term release duration will continue for as long as anthropogenic radioactivity is released from that particular location. This raises a question for accidents such as Chernobyl or Fukushima Daiichi: while the principal source term release occurred over a period of a few weeks, what is the duration of the secondary, chronic leakage of radioactivity from these unsecured source points?

Critical Mass

The minimum mass of fissionable material which can achieve a nuclear chain reaction with a specified geometrical arrangement and material composition (Centers for Disease Control (CDC), *Savannah River Site (SRS) dose reconstruction*, 1999).

Bioindicators

Biological media which are the most susceptible to the accumulation of **biologically significant radionuclides**. Many bioindicators are in **pathways** to human consumption, allowing rapid transfer of radioactivity from the **abiotic** environment (air, precipitation, freshwater, sea water, soil and marine sediments) to **sentinel organisms** as well as crops and crop products such as milk, cheese and meat. Most pathway analyses for the ecological cycling of radionuclides begin with soil or sediment as the repository of radioactive contamination. The process by which living organisms absorb radioactive contamination is called **bioaccumulation**; bioaccumulation may also be defined as the assimilation of contamination prior to its movement up the food chain. Among the most significant bioindicators are:

- **sea vegetables: Fucus vesiculosus,** brown algae and other benthic algae are among the most sensitive bioindicators and are often used to gauge weapons fallout contamination and nuclear reactor pollution from many radionuclides which these organisms will readily absorb. The terrestrial counterparts to sea vegetables as sentinel organisms are **lichens, moss, mushrooms,** and **grass. Leafy vegetables** such as spinach are examples of bioindicators which humans consume directly and which quickly absorb foliar deposition of radiocesium as well as the short-lived radioiodine-131 ($1/2T = 8.04$ d). Milk and milk products, as food crop products of the forage-livestock pathway, are bioindicators which concentrate the rapid transfer of radioactive contamination following nuclear accidents and releases. The presence of iodine-131 in milk is a **key indicator** of the magnitude of a nuclear accident.
- **benthic invertebrates: Mussels** (*Mytilus edulis*), oysters (*Crassostrea virginica*), etc. are another group of sensitive bioindicators and are also used to evaluate the impact of other types of chemical fallout.
- **fish:** Less sensitive than benthic algae (sea vegetables) as bioindicators, fish are an important indicator of the level of human consumption of radioactive contamination. Freshwater fish often show much higher levels of the bioaccumulation of radionuclides and other forms of chemical fallout than marine specimens.
- **grazers: reindeer, sheep, goats, and livestock**: Products from these participants in the forage-livestock pathway, meat, cheese, and milk, often exhibit rapid bioaccumulation of radioactive contamination.

Concentration Factors

Radionuclide	Marine concentration factor (m^3 t^{-1})					Freshwater concentration factor (m^3 t^{-1})	
	Fish	Crustacea	Molluscs	Sediments	Seaweed	Sediments	Fish
^{54}Mn	500	10000	10000	10000	10000	10000	300
^{58}Co	100	1000	1000	10000	1000	30000	300
^{60}Co	100	1000	1000	10000	1000	30000	300
^{65}Zn	2000	5000	100000	10000	1000	1000	1000
^{89}Sr	1	10	10	500	10	2000	30
^{90}Sr	1	10	10	500	10	2000	30
^{106}Ru	1	500	2000	10000	2000	40000	10
^{110}mAg	1000	5000	50000	10000	1000	200	3
^{125}Sb	500	300	100	10000	100	300	1000
^{131}I	10	100	100	100	1000	200	30
^{134}Cs	50	30	30	500	30	30000	1000
^{137}Cs	50	30	30	500	30	30000	1000
^{144}Ce	10	1000	1000	10000	1000	30000	1000

(UNSCEAR 1982; http://www.unscear.org/docs/reports/1982/1982-F2_unscear.pdf).

Marine Concentration Ratios for the Transuranics		
Material	**Plutonium**	**Americium, curium, and neptunium**
Sediment	100,000	100,000
Plankton	5,000	50,000
Benthic algae and macrophytes	5,000	50,000
Benthic invertebrates	1,000	10,000
Fish		
Bottom feeders	250	2,500
Plankton feeders	25	250
Piscivorous (fish eaters)	5	50

(Hanson 1980, 620).

Additional Alphabetized Definitions

Absorbed Dose: The energy imparted by ionizing radiation per unit mass of irradiated material. The units of absorbed dose are called the gray (Gy) (United States Department of Health & Human Services 1997, 305).

Actinide: The fifteen manmade chemical elements with atomic numbers 89 (actinium) to 103 (lawrencium). Plutonium is the most important actinide element and is the principal power source in nuclear reactors and nuclear weapons. Only the actinides uranium and thorium occur naturally in small quantities; all are both highly radioactive and radiotoxic. Other biologically significant actinides include americium-241, the curium isotopes, and the neptunium isotopes. The daughter product of neptunium-239 is plutonium-239.

Action Level: A derived media-specific radionuclide-specific concentration or activity level of radioactivity that triggers a response such as seizure of contaminated foodstuffs following a nuclear accident (see FDA DIL). In the MARSSIM protection action guideline, the action level is called the **investigation level** and would trigger the response of further investigation or site cleanup if the **release criterion** is exceeded.

Activation Products: Nuclides formed through transformation of stable reactor components into radioactive isotopes after intense bombardment with fission products. Radioactivity is thus induced through neutron bombardment or other types of radiation in reactor vessel components and corrosion products (and also in weapons casings) which were stable before the reactor vessel went on-line. The transuranic nuclides plutonium, americium, curium, etc., are also neutron activation products, originating from neutron capture in uranium nuclides rather than from the fission of these nuclides. Other important activation products include carbon-14 and tritium as well as activation products derived from activated stainless steel and carbon steel, activated sludge, corrosion deposits and concrete, and contaminated building products, e.g. ^{55}Fe, ^{54}Mn, ^{65}Zn, ^{58}Co and ^{60}Co. An additional listing of activation products and corrosion products can be found in the *Checklist of Biologically Significant Radionuclides*.

Air and Water Dose Conversion Factors: The University of California, Berkeley, Department of Nuclear Engineering is an important source for reports on Fukushima Daiichi accident-derived radioactive fallout in the United States. The Nuclear Engineering Department utilizes the following conversion factors to relate "the activity of a radionuclide (in Becquerels (Bq) or microcuries (μCi)) to the equivalent dose received by the person (in millirem or microSieverts (μSv))." (http://www.nuc.berkeley.edu/node/1897). In their following conversion table, Be-7 is a naturally occurring cosmogenic isotope that routinely appears in rainwater.

Dose Conversion Factors for water

Units	I-131	I-132	Cs-134	Cs-137	Te-132	Be-7
millirem/μCi	6.849E+01	6.849E-01	7.610E+01	6.849E+01	7.610E+00	1.142E-01
uSv/μCi	6.849E+02	6.849E+00	7.610E+02	6.849E+02	7.610E+01	1.142E+00
millirem/Bq	1.851E-03	1.851E-05	2.057E-03	1.851E-03	2.057E-04	3.085E-06
μSv/Bq	1.851E-02	1.851E-04	2.057E-02	1.851E-02	2.057E-03	3.085E-05

Dose Conversion Factors for air

Units	I-131	I-132	Cs-134	Cs-137	Te-132	Be-7
millirem/μCi	1.042E+02	1.042E+00	1.042E+02	1.042E+02	2.315E+01	6.944E-01
μSv/μCi	1.042E+03	1.042E+01	1.042E+03	1.042E+03	2.315E+02	6.944E+00
millirem/Bq	2.815E-03	2.815E-05	2.815E-03	2.815E-03	6.256E-04	1.877E-05
μSv/Bq	2.815E-02	2.815E-04	2.815E-02	2.815E-02	6.256E-03	1.877E-04

Airborne Multi-Sensor Pod System (AMPS): The AMPS is a recent technological innovation for the collection of multi-sensor data for a variety of national security purposes. One component of remote sensing efforts, the AMPS is of particular interest because one of the pods to be used in aircraft utilizing this system will have high resolution spectral analysis capabilities pertaining to ground deposition of radioactivity deriving from nuclear accidents and nuclear waste plume source points. The following acronyms are of interest in defining and understanding remote sensing technologies which relate to radiological surveillance programs.

- AMS: Airborne Multispectral Scanner
- CASI: Compact Airborne Spectrographic Imager
- ESI: Effluent Species Identification (pod)
- GRIS: Gamma Ray Imaging System
- MSI: Multi-Sensor Imaging
- RSL: Remote Sensing Laboratory
- R-TARAC: Real-Time Airborne Radionuclide Analyzer and Collector

Annual Limit on Intake (ALI): The derived limit for the amount of radioactive material taken into the body of an adult worker by inhalation or ingestion in a year. For a given radionuclide, ALI is defined as the smaller of the intakes that would result in a committed effective dose equivalent of 5 rems and a committed dose equivalent of 50 rems to any individual organ or tissue (United States Department of Health & Human Services 1997, 306).

AP 1000 Nuclear Power Plants: The proposed modern design for the next generation of nuclear power plants to be built in the US and elsewhere, the AP 1000 will be a mass-produced facility using Lego construction for the secondary containment structure. The AP 1000 will be built with a massive "passively safe" gravity-operated cooling system with three days capacity located on the top of each reactor vessel.

As Low as Reasonably Achievable (ALARA): The reduction of exposure to ionizing radiation so as to reduce collective doses as far below regulatory limits as is reasonably possible.

Baseline Environmental Management Report (BEMR): A congressionally mandated report prepared by the Secretary of Energy in June of 1996 to estimate the cost and schedule of cleaning up the nation's nuclear weapons complex. It is also referred to as the "Baseline Report."

Blowout: The hydrogen explosions that result from fuel assembly loss of coolant, overheating, fuel melting, and the consequential reduction of water molecules to hydrogen and oxygen. At Fukushima Daiichi, the excess hydrogen that accumulated in the reactor vessel torus due to fuel assembly melting was the cause of the explosions that destroyed the secondary containment structures of Reactors 1 - 4.

Characterization Survey: A type of survey that includes facility or site sampling, monitoring, and analysis activities to determine the extent and nature of contamination (MARSSIM 1996, GL-3).

Chelation: The process by which both naturally occurring and artificial agents can be used as sequestering agents, thereby making radionuclides and other chemicals in a particular media available for transfer to another environment. Artificial chelating agents such as EDTA have a wide variety of industrial uses and are often used to remove radioactive contamination. Unfortunately, natural chelating agents and chelating processes make plutonium oxide from stratospheric fallout and nuclear accidents, which is usually in a biologically unreactive state in soils and sediment, much more biologically available for uptake in pathways to human consumption. Almost no information is available about the long-term mobilization of plutonium isotopes by naturally occurring chelating agents (Hanson 1980, Section 10).

Cold Shutdown: The return of nuclear reactor water temperatures to consistently lower than boiling water temperature levels.

Comprehensive Environmental Response, Compensation, and Liability Act (CERCLA): A federal law enacted in 1980 that governs the cleanup of hazardous, toxic, and radioactive substances. The Act and its amendments created a trust fund, commonly known as **Superfund**, to finance the investigation and cleanup of abandoned and uncontrolled hazardous waste sites (UNITED STATES DEPARTMENT OF ENERGY 1996, GL-2).

Concentration Ratios: The tendency for many radionuclides to uniformly migrate in one proportion or another in various media in the biotic and abiotic environments. Sediment is the repository of radioactive fallout in abiotic media and thus the point of origination for many pathway analyses. Biological media either concentrate radionuclides as they pass through water to sediment or concentrate radionuclides after they have been remobilized from sediment by various biogeochemical processes (water = 1).

Contamination: The presence of residual radioactivity in excess of levels which are acceptable for release of a site or facility for unrestricted use (MARSSIM 1996, GL-4). This is a particularly controversial definition of contamination in that it is predicated upon arbitrary release criteria that, in effect, allow significant levels of contamination to remain in a remediated or decommissioned site.

88

CRUD: "An acronym for 'Chalk River Unidentified Deposits.' ...black, highly radioactive substances found on the inside of piping and components at the Chalk River nuclear reactor... CRUD has now become a standard industry term referring to minute, solid, corrosion products that travel into the reactor core, become highly radioactive, and then flow out of the reactor into other systems in the plant... CRUD can settle out in crevices or plate-out on the inside of piping in considerable quantities... The major components of CRUD are iron, cobalt, chrome, and manganese... CRUD is a concentrated source of radiation and represents a significant radiological risk because of its insolubility." (United States Federal Energy Regulatory Commission 1997, 13-4).

Decay in Storage (DIS): An idea whose time has not yet come: instead of dumping virtually uncontained ^{137}Cs (1/2 T = 30 yr) and other intermediate-level wastes into near surface landfills, these wastes would be stored onsite at their point of generation for periods of 50 to 300 years. Now widely accepted by the European community as the only viable waste storage option for intermediate wastes, DIS is only coming to the US as a result of failure of the US government to develop a viable waste disposal policy for high-level wastes. As ISFSIs are constructed for spent fuel, it's only a small step to expand these facilities to add intermediate-level waste storage including GTCC wastes.

Decay (Radioactive): "The change of one radioactive nuclide into a different nuclide by the spontaneous emission of radiation such as alpha, beta, or gamma rays, or by electron capture. The end product is a less energetic, more stable nucleus. Each decay process has a definite half-life."
(http://www.lbl.gov/abc/wallchart/glossary/glossary.html).

Delayed Pathway Transport: Exposure to accident-derived radioisotopes does not terminate when ambient radiation levels return to normal. After the Chernobyl accident, biologically significant quantities of Cs-137 (+370 Bq/kg) were transported to the United States in imported foods. Peak concentration levels of contamination were documented in imported foods brought to the US twelve to sixteen months after the accident ended. See Section

Derived Air Concentration (DAC): It is the concentration of a given radionuclide in the air, which, if breathed by the reference man for one working year (2,000 hours) under conditions of light work, results in an intake of one ALI (United States Department of Health & Human Services 1997, 308).

Derived Concentration Guideline Level (DCGL): A derived, radionuclide-specific activity concentration within a *survey unit* corresponding to the *release criterion*. The DCGL is based on the spatial distribution of the contaminant and hence is derived differently for the *nonparametric* statistical test (DCGL$_w$) and the Elevated Measurement Comparison (DCGL$_{EMC}$). DCGLs are derived from activity/dose relationships through various exposure pathway scenarios (MARSSIM 1996, GL-5).

Dose Assessment: An estimate of the radiation dose to an individual or a population group usually by means of predictive modeling techniques, often supplemented by the results of measurement (United States Department of Health & Human Services 1997, 309).

Dose Commitment: The dose that an organ or tissue would receive during a specified period of time (e.g. 50 or 70 years) as a result of intake (as by ingestion or inhalation) of one or more radionuclides from a given release (MARSSIM 1996, GL-4).

Dose Conversion Factor: A factor (Sv/Bq or rem/Ci) that is multiplied by the intake quantity of a radionuclide (Bq or Ci) to estimate the committed dose equivalent from radiation (Sv or rem). The dose conversion factor depends on the route of entry (inhalation or ingestion), the lung clearance class (D, W or Y) for inhalation, the fractional uptake from the small intestine to blood (f1) for ingestion, and the organ of interest. EPA provides separate dose conversion factor tables for inhalation and ingestion, and each provides factors for the gonads, breast, lung, red marrow, bone surface, thyroid, remainder, and effective whole body (United States Department of Health & Human Services 1997, 309).

Dose Equivalent (DE): A quantity used in radiation protection that expresses all radiation on a common scale for calculating the dose for purposes of radiation safety. It is defined as the product of the absorbed dose in rads and certain modifying factors. (The unit of dose equivalent is the rem. In SI units, the dose equivalent is the sievert, which equals 100 rem.) (United States Department of Health & Human Services 1997, 309).

Dose Reconstruction: A study process in which historical information is used to estimate the amounts of toxic materials released from a facility, how the materials could have moved offsite, and the exposure of the public to those materials. Dose reconstruction involves past releases, not present or future releases (Centers for Disease Control 1999).

Effective Action Level (FDA): Following the Chernobyl accident, the Food and Drug Administration implemented an unofficial protection action guideline when it observed high levels of Chernobyl-derived radiocesium contaminating imported foods approximately one year after the accident. The FDA seized and destroyed foods contaminated in excess of 10,000 pCi/kg (370 Bq/kg) thereby setting an *EFFECTIVE ACTION LEVEL,* which was significantly more conservative (lower) than the protection action guidelines promulgated by various US government agencies before and after the Chernobyl accident. See *Section VII. Radiation Protection Guidelines* for a more complete description of the wide variety of protection action guidelines.

Exposure Rate: The amount of ionization produced per unit time in air by X-rays or gamma rays. The unit of exposure rate is roentgens/hour (R/h); for decommissioning activities the typical units are microroentgens per hour (μR/h), i.e. 10^{-6}R/h (MARSSIM 1996, GL-7).

Fissile Material: A material such as an actinide with an odd neutron number (e.g. Pu-239), which, upon exposure to bombardment by thermal neutrons, is capable of sustaining a nuclear fission chain reaction.

French Drains: Chemical disposal wells utilized between 1945 and the late 1960s at most US weapons productions facilities involving fuel fabrication and spent fuel reprocessing. These wells were utilized for the quick and convenient disposal of highly toxic mixed waste streams, which included large quantities of radioactive effluents mixed with volatile organic compounds (VOCs). For national security reasons no unclassified data is available about the radioactivity of the mixed wastes disposed of by this method. The resulting plumes, many in either perched aquifers or in underlying aquifers, are frequently referenced in the DOE BEMR as well as in the extensive groundwater surveys of the USGS.

Fuel Cladding Failure: The most probable kind of nuclear accident (as in "probabilistic risk assessment") and one that characterizes the operations of most nuclear reactors. Cladding failure begins with pin hole leaks that release some of the gasses within the fuel rods (called gap release; ^{3}H, noble gasses, gaseous ^{131}I are the most common stack effluents). The most common cause of fuel failure is (fuel assembly) grid-to-(fuel)-rod fretting that results from (a) vibrations within the reactor containment (b) differences in pressure caused by use of different types of fuel and (c) deformations in or damage to fuel assembly grids that then result in fuel failure, which allows the spread of spent fuel pellets throughout the reactor containment. As fuel ages and undergoes long periods of burnup, it becomes much more vulnerable to fuel failure. Fuel failure can also result from defects in the manufacture of fuel rods; in the early days of the nuclear industry, aluminum cladding resulted in frequent fuel cladding failure in both DOE weapons production reactors and facilities such as the notorious Connecticut Yankee reactor at Haddam Neck. The cumulative effect of fuel cladding failure at CT Yankee constitutes the largest known accident since the Three Mile Island LORCA. The NRC has made every effort to cover up the ubiquitous nature of fuel cladding failure by labeling all failure as "leakage" and then asserting that this is a normal part of reactor operations. A more truthful way of stating the matter is that fuel failure is one of the most common forms of nuclear accident and can range from a few failed rods with small openings to large numbers of failed rods including those which split open and spill their entire contents in the reactor containment. What happens to the spilled fuel pellets after their release from the failed fuel rods is one of the most important issues facing the nuclear industry as it decommissions its aging reactors. For more information on the relatively small fuel failure accident at MYAPC see *Section IX. Nuclear Dada: The Maine Yankee Atomic Power Plant.*

Gamma Camera: Remotely operated gamma ray imaging system that generates photos showing radiation areas within the hot side of a nuclear power plant, fuel reprocessing facility or other nuclear installation. The gamma camera is particularly useful during decommissioning and remediation activities for identifying major hot spots in equipment such as reactor vessels, steam generators, or reactor water systems which may contribute to worker exposure.

Highly enriched uranium: Uranium with more than 20 percent of the ^{235}U isotope, used for making nuclear weapons and also as fuel for some isotope production, research and power reactors. Weapons grade uranium is a subset of this group (United States Department of Energy 1996, GL-4).

Historical Site Assessment (HSA): A detailed investigation to collect existing information, primarily historical in nature, on a site and its surroundings (MARSSIM 1996, GL-8).

High-LET: The characteristic ionization patterns by alpha particles, protons, or fast neutrons having a high relative specific ionization per unit path length (United States Department of Health & Human Services 1997, 311).

"Iron Fence": The most restricted alternative case for land use. It is characterized by containing, rather than actively remediating, contaminated sites. This means that soil and buried waste sites would be capped, ground-water contamination would be partially controlled from spreading by hydraulic controls and barriers, and facilities would be entombed (United States Department of Energy 1996, GL-5). This is the almost certain fate of the Fukushima Daiichi complex.

Independent Spent Fuel Storage Installation (ISFSI): These are onsite dry cask storage facilities built at nuclear reactors to hold spent fuel when the spent fuel pool gets too full. ISFSIs are one of the expensive and unfortunate consequences of the failure of the federal government to develop a safe and practical way to dispose of spent fuel wastes. Another term for ISFSI could be "monitored retrievable storage," the only probable future solution to storing intermediate and high-level wastes that is safe, economical, politically viable, practical, and likely to provide jobs for the next 10,000 years to former employees of nuclear power reactors and their descendants.

Isotope: Isotopes are forms of the same chemical element; they have the same atomic number (same number of protons in their nuclei) but different mass numbers (different number of neutrons n their nuclei). Most elements have more than one naturally occurring isotope. Many isotopes have been produced in reactors and scientific laboratories (http://www.lbl.gov/abc/wallchart/glossary/glossary.html).

Life Cycle Cost Estimate: A term used by the Department of Energy to designate the cost of complete remediation of weapons production facilities within the Environmental Management program. This term also applies to the decommissioning of nuclear power facilities. It may also be used in reference to the life cycle disposal costs of specific components in a contaminated site, e.g. spent fuel from a nuclear power plant, GTCC reactor vessel wastes, etc.

Linear Energy Transfer (LET): Another key concept in determining biological effectiveness and significance, LET expresses the combination of charge and speed in effecting the efficiency of ionizing radiation. LET describes "the amount of energy transferred per unit of path traveled by the ionizing particle (electron, alpha particle or other)" (Gofman 1981, 28). Alpha particles have twice the charge of a beta particle and, therefore, four times the efficiency of ionizing radiation per collision. **Alpha radiation is much slower than beta or gamma radiation; therefore, it is much more efficient** than the faster radiation, causing more ionizations per millimeter of distance traveled (Gofman 1981, 26-8). The efficient LET of alpha isotopes such as ^{239}Pu combines with their high decay energies to form the basis of their biological effectiveness. High radiotoxicity and great biological significance accompany these long-lived anthropogenic radionuclides in the environment.

Multi-Agency Radiation Survey and Site Investigation Manual (MARSSIM): A controversial publication issued by the EPA, NRC, and DOE that delineates the release criterion pertaining to the annual radiation dose that maximally exposed members of the public can receive, as a condition for decommissioning or remediating nuclear power plants or other NRC or DOE facilities. The MARSSIM (1996) is of particular importance now that the NRC has set 25 mrem/yr TEDE as the release criteria for decommissioning nuclear power plants under its jurisdiction. For an annotated description of this manual see *Section VII. Radiation Protection Guidelines.*

National Imagery and Mapping Agency (NIMA): Established within the Department of Defense on October 1, 1996 as a component of the US intelligence community, NIMA represents a consolidation of previously existing intelligence agencies into one centralized agency. NIMA incorporates the Defense Mapping Agency, the Central

Imagery Office (CIO), and other agencies as well as the functions of the CIA's National Photographic Interpretation Center. Also incorporated in NIMA is the imagery processing elements of the Defense Intelligence Agency, the National Reconnaissance Office (NRO), and the Defense Airborne Reconnaissance. All these agencies are of particular interest with respect to radiological monitoring because they incorporate sophisticated **remote sensing technologies**, including remote sensing of spectral data that document the plume pulse movement of nuclear accidents-in-progress. Up to October 1, 1996, much of the remote sensing data of interest to RADNET had been collected by the NRO and collated by the CIO (until recently the existence of both of these offices had been classified information). The remote sensing data which documents the plume pulse movement of nuclear accidents-in-progress (and the presence of any above ground source point) is still classified information. Most remote sensing data in the electromagnetic energy spectrum between ultraviolet and microwave regions is becoming available to the general public through electronic means; only spectral data pertaining to national security concerns (nuclear data) remains classified (+155 MeV). How long can the intelligence community keep their fingers in the dike?

Naturally occurring radiation (NOR): Cosmogenic (extraterrestrial) and terrestrial radiation usually but not always with an activity range of 5-10 micro roentgens per hour ($\mu R/hr$), and having the same biological consequences as artificial radiation (fission and activation products). Radon ($1/2T = 3.82$ d), one of many naturally occurring radionuclides, has been recently recognized as a significant source of exposure, particularly in well insulated homes overlying geological formations which produce large quantities of radon within the uranium decay series. Radon achieves biological significance when inhaled if, instead of being exhaled, it decays into four short-lived nuclides followed by the long-lived ^{210}Pb, all of which are surface-seeking particulates which become lodged in the lung. The daughter products in the uranium-radon decay series then become the source of the radiation dose from radon.

The term "naturally occurring" needs to be differentiated from "background radiation," which now includes the impact of the cumulative deposition from stratospheric fallout and nuclear accidents such as Chernobyl. In some contaminated areas, such accumulations of long-lived artificially produced radionuclides exceed natural background radiation levels. The term "background radiation," particularly when used by spokespersons for nuclear industries, can no longer be equated with the natural radiation background as it was before the advent of the nuclear age.

Negative void coefficient: The decline in the neutron and thermal heat output of a reactor due to the increase in the proportion of steam to liquid water inside the reactor.

Neutron source: The catalyst needed to begin a chain reaction at a nuclear power plant. After fissile activity is initiated in weapons grade U-235, the neutrons emitted by irradiated fuel, such as the highly energetic Pu-239 (5.155 MeV), continue the chain reaction. Neutron sources eventually are removed from the core and end up in the spent fuel pool and constitute one more component of "orphan" high-level waste (not spent fuel and not high-level waste either).

Noble gases: A significant percentage of the radioactive emissions in any accident consist of noble gases, which provides a percentage of dose exposure as measured in μSv per hour. Noble gases are, however, chemically inert and therefore are not biologically significant in contrast with radioiodine, radiocesium, or radiostrontium. As short-lived radioisotopes, they may provide some dose exposure from plume shine, or from inhalation, but provide no exposure via food web ingestion.

Nuclear fuel cycle: The primary source of the anthropogenic radionuclides documented in RADNET. The nuclear fuel cycle, which includes the **weapons production cycle**, has eight primary components, all of which result in the accumulation or release of significant quantities of radioactivity. The cycle begins with the mining of a naturally occurring radionuclide, uranium-238. It is later refined into weapons grade U-235, which in nuclear reactors becomes the target of neutron bombardment, producing energy and the highly energetic (heat-emitting) alpha emitting plutonium-239. It is the uranium-plutonium fission process that produces all other fission daughter products. The contamination that results from the processing or uranium in the nuclear fuel cycle is "anthropogenic" in that it derives from **human activities**. The nuclear fuel cycle involves the following steps:

1. **Uranium mining and milling:** Approximately 60 million tons of uranium ore were mined and milled in the United States for nuclear weapons production, resulting in the dispersion of toxic heavy metals as well as the natural radionuclides **radium** and **thorium.**
2. **Uranium enrichment:** ^{238}U was enriched and separated to produce weapons grade ^{235}U in the form of uranium hexafluoride gas, producing radioactive and hazardous wastes in locations such as Ohio, Kentucky, and Tennessee.
3. **Fuel and target fabrication:** Uranium hexafluoride gas was converted into metal (uranium targets) at fuel fabrication facilities in locations such as South Carolina and Washington.
4. **Reactor irradiation:** The uranium targets were irradiated in 14 production reactors to produce plutonium such as South Carolina and Washington.
5. **Chemical separation:** The resulting spent fuel was reprocessed at chemical separation facilities to produce fission products as well as weapons grade

uranium and plutonium such as South Carolina, Idaho, and Washington. This stage in the weapons production cycle produced the greatest amount of highly radioactive and hazardous chemical waste (USA: 100 million gallons).

6. **Fabrication of weapons components:** The machining of plutonium into warhead components such as Colorado, Washington, and Tennessee.

7. **Weapons assembly, disassembly, and maintenance:** Final assembling of nuclear warheads as well as dismantling and research occurs in 46 private sites in 14 states.

8. **Research, development, and testing:** Over 1,000 nuclear devices were exploded between 1945 and 1992 (NV, NM, Alaska as well as Pacific and South Atlantic Ocean sites) (United States Department of Energy Vol. 1, Appendix B).

Nuclear Reactor: "A device in which a fission chain reaction can be initiated, maintained, and controlled. Its essential components are fissionable fuel, moderator, shielding, control rods, and coolant."
(http://www.lbl.gov/abc/wallchart/glossary/glossary.html).

Nuclear Weapons Complex: The chain of foundries, uranium enrichment plants, reactors, chemical separation plants, factories, laboratories, assembly plants, and test sites that produced nuclear weapons. Sixteen major US facilities in 12 states formed the nuclear weapons complex (United States Department of Energy 1996, GL-6).

Photon: "A packet of electromagnetic energy. Photons have momentum and energy, but no rest mass or electrical charge."
(http://www.lbl.gov/abc/wallchart/glossary/glossary.html).

Plume: The concentration profile of an airborne or waterborne release of material as it spreads from its source (Centers for Disease Control 1999).

Quality Factor (Q): The linear-energy-transfer-dependent factor by which absorbed doses are multiplied to obtain (for radiation protection purposes) a quantity that expresses the biological effectiveness of the absorbed dose on a common scale for all ionizing radiation (United States Department of Health & Human Services 1997, 315).

Quick Release Accident: A quick release accident occurs when a sudden and total release of the inventory of radioactivity takes place at an operational (hot) nuclear power plant. The worst case scenario involves the vaporization of both the reactor vessel and associated spent fuel pool during a nuclear attack. Certain types of loss of reactor coolant and reactor vessel embrittlement accidents could result in a quick release accident, especially after pressure buildup in the reactor vessel, which might originate from a series of minor mishaps. The hydrogen explosions at the Fukushima Daiichi complex are examples of quick release accidents.

Radiocesium: The term radiocesium refers to the combined presence of Cs-134 (1/2 T = 2.1 yr) and Cs-137 (1/2 T = 30.1 yr). The long-lived isotope Cs-137 is often used as an indicator isotope tracking accident source terms, deposition levels, concentrations in biotic media, and pathway analysis. Estimates of the Chernobyl source term release fraction of Cs-134 and Cs-137 vary from 100 vs. 50 (Aarkrog 1994) to ±85 vs. ±54 (OECD 1995; UNSCEAR 2000; Fairlie 2006).

Radiolysis: A process by which radioactivity breaks down and hence changes chemical compounds. It is a principal cause of certain kinds of waste management problems, notably in relation to liquid radioactive wastes and wastes containing mixtures of radioactive materials and non-radioactive chemicals. Chemicals present in the waste break down over time due to the action of radiation unless they are in very stable forms. The breakdown products in turn create new chemical reactions with each other and with pre-existing chemicals. These processes make estimation of the chemical make-up of the waste very difficult. They also frequently result in the generation of hydrogen gas (Makhijani 1999, 21).

Radionuclide: "A radioactive nuclide. An unstable isotope of an element that decays or disintegrates spontaneously, emitting radiation." (http://www.lbl.gov/abc/wallchart/glossary/glossary.html).

Radon (Rn-222): A ubiquitous, tasteless, odorless, naturally occurring, inert radioactive gas formed by the decay of radium and associated with certain geological formations; exposure is usually by inhalation. Radon accumulates in domestic environments such as basements and attics and provides more background radiation exposure than any other naturally occurring radionuclide. An intensely energetic radioisotope, Rn-222 (5,590 MeV) has a short half-life of only 3.8235 days and quickly decomposes into polonium-218 (1/2 T = 3.10 min), lead-214 (1/2 T = 26.8 min), bismuth-214 (1/2 T 19.9 min), and polonium-214 (1/2 T = 0.164 min), which then decays to the highly toxic radioactive lead (Pb-210; 1/2T = 22.3 yr). When radon is inhaled, it is this nonvolatile, intensely radioactive, naturally occurring, alpha-emitting daughter product that is the primary cause of the tens of thousands of cases of lung cancer deaths in the US attributed to radon.

Residual Contamination Standards: The amount and concentrations of contaminants in soil, water, and other media that will remain following environmental management activities (United States Department of Energy 1996, GL-8).

Steam Generator Tubes: Also called fuel rods, these tubes are part of pressurized water reactors and are the essential component of fuel assemblies, containing the nuclear fuel (often U-235) that will undergo fission to produce the plutonium that drives a nuclear chain reaction. The heat that is thus produced creates the steam that generates nuclear electricity. The following subset of definitions is used by the NRC in Draft Regulatory Guide DG-1074 for inspecting the integrity of these tubes (United States Nuclear Regulatory Commission 1998).

- **Accident Leakage Rate:** The primary-to-secondary leakage rate occurring during postulated accidents other than a steam generator tube rupture.

- **Active Degradation Mechanisms:** New indications associated with defect types that have been identified during in-service inspection.

- **Buffer Zone:** A zone extending radially from the critical region.

- **Burst:** Gross structural failure of the tube wall.

- **Critical Region:** A region of the tube bundle that can be demonstrated to bound the region where a specific defect type is active.

- **Defective Tube:** A tube that exhibits an indication exceeding the applicable tube repair criteria.

- **Degradation Mechanism:** The general defect morphology and its associated causes, e.g., wear-induced thinning of the tube wall caused by adjacent support structures, high cycle fatigue cracking caused by flow-induced vibration of the tube, intergranular stress corrosion cracking caused by stress, material susceptibility, and environment.

- **Rupture:** Perforation of the tube wall such that the primary-to-secondary leak rate exceeds the normal charging pump capacity of the primary coolant system.

Stochastic and Non-Stochastic Effects: Stochastic health effects occur randomly, with the effect being independent of the size of the dose and for which the probability of the effect occurring, rather than its severity, is assumed to be a linear function of dose without threshold (e.g. hereditary effects, cancer, etc.) (*Federal Register* 10 CFR 20.1003). Non-stochastic effects are, in contrast, those that can be related directly to the dose received.

Technologically-Enhanced Naturally Occurring Radioactive Material (TENORM): "Large-volume, low-activity waste streams produced by industries such as mineral mining, ore benefaction, production of phosphate fertilizers, water treatment and purification, and oil and gas production. The majority of radionuclides in TENORM are found in the uranium and thorium decay chains. Radium and its subsequent decay products (radon) are the principal radionuclides used in characterizing the redistribution of TENORM in the environment by human activity ...TENORM is found in many waste streams; for example, scrap metal, sludges, slags, fluids, and is being discovered in industries traditionally not thought of as affected by radionuclide contamination. Not only the forms and volumes, but the levels of radioactivity in TENORM vary." (http://www.tenorm.com/intro.htm).

Thermal Neutrons: Slow, low energy neutrons, originally emitted by nuclear chain reactions as fast neutrons that have been slowed down by a collision with the atomic nuclei of a moderator, such as beryllium metal, that then induce and maintain continued fissile activity. Also called neutron flux, the resulting fissile activity can be controlled by the insertion of control rods into the central core of a nuclear reactor, which then blocks fissionable material such as Pu-239 from further bombardment.

Total Effective Dose Equivalent (TEDE): The sum of the deep dose equivalent (from external exposures) and the committed effective dose equivalent (from internal exposures) (United States Department of Health & Human Services 1997, 318).

Transuranic Elements: All elements beyond uranium on the periodic table, that is, all elements with a number greater than 92. All transuranic elements are manmade; the heat that they produce during fission chain reactions is the basis for nuclear energy production. Their high radiotoxicity is one of the external costs of nuclear power production. They include neptunium, plutonium, americium, and curium (United States Department of Energy 1996, GL-9).

Transuranic Waste: Waste material contaminated with ^{233}U and its daughter products, certain isotopes of plutonium, and nuclides with an atomic number greater than 92 (uranium); each with half-lives greater than 20 years and in concentrations of more than one ten-millionth of a curie per gram of waste. It is produced primarily by reprocessing spent fuel, by using plutonium to fabricate nuclear weapons, and during commercial nuclear electricity production (United States Department of Energy 1996, GL-9).

Transuranium Nuclides: Elements of a higher atomic number than uranium (92), most transuranic isotopes are highly toxic alpha-emitting radionuclides with great biological significance which do not occur naturally in any significant quantities, but which are an artificial product of the fission process and emit radiation having much higher energy than other radionuclides which are also produced in the fission process; e.g. tritium, carbon-14 and strontium-90. The transuranic nuclides of the greatest significance are neptunium-237, plutonium-238, 239, 241, americium-241, and curium-242, 244. (See the *Checklist of Biologically Significant Radionuclides*.)

Tritium: The heaviest isotope of the element hydrogen (H-3). It is three times heavier than hydrogen. Tritium gas is used to boost the explosive power of most modern nuclear weapons and has a half-life of over 12 years.

Uranium: It is a slightly radioactive, naturally occurring, heavy metal that is more dense than lead. Its principal naturally occurring isotope is U-238, which is fissionable by fast neutrons and is often transmuted to Pu-239 in nuclear reactors. U-235, which is much less common than U-238, is the only naturally occurring fissile isotope. When mined, milled, and concentrated in nuclear reactor fuels in sufficient concentration, it can maintain a sustained nuclear reaction. Its second daughter product, neptunian-239 decays into plutonium-239; U-238, U-235, and Pu-239 are the principal sources of most nuclear energy, either for weapons production or the generation of nuclear power. The half-life of U-238 is almost 4.5 billion years; it is 40 times more common than silver and is often extracted from uraninite, a uranium-bearing mineral.

Uranium Mill Tailings Radiation Control Act of 1978 (UMTRA): This act directed the Dept. of Energy to stabilize and control 24 designated inactive uranium processing sites and an estimated 5,048 vicinity properties. These sites were and are the source point of significant quantities of radioactive contamination in the form of windblown sand-like ore tailings. An associated environmental remediation program is the Formerly Utilized Sites Remediation Action Program of 1974 (**FUSRAP)**, the purpose of which was to remediate sites associated with research, development, processing, and production of uranium and thorium. All of these important source points of anthropogenic radioactivity, which result from the first stage of the **weapons production cycle**, are listed in the DOE Baseline Environmental Management Report (United States Department of Energy 1996).

Vitrification: The process by which waste is transformed from a liquid or sludge into an immobile solid that traps radionuclides and prevents waste from contaminating soil, ground water, and surface water (United States Department of Energy 1996, GL-10).

Volatile organic compounds (VOCs): The generic name for a variety of toxic chemicals utilized in the reprocessing of spent nuclear fuel as well as in other industrial applications pertaining to weapons production. The principal chemicals of interest include: **trichloroethylene, carbon tetrachloride, benzene, acetone, toluene, methylene chloride, xylenes, chlorobenzene, naphthalene, etc.** These and other chemicals characterize the hydrologic plumes of weapons production-derived wastes now being monitored in confined and unconfined aquifers at many US weapons production facilities. The large quantities of VOCs released to surface water supplies and evaporating ponds and pits were subject to rapid evaporation and redistribution as chemical fallout; only those VOCs which seeped into the soil and underlying aquifers, or which were deliberately released as shallow well or deep well injections, remain in the underground plumes now a component of DOE environmental remediation efforts and USGS ground water monitoring programs. See *Section V. Plume Source Points*, especially the citations documenting the Hanford and Savannah River plumes.

Waste Isolation Pilot Plant (WIPP): A geologic repository intended to provide permanent deep underground disposal for transuranic waste. "On Saturday, March 26, 2011, the Department of Energy's Waste Isolation Pilot Plant marked another anniversary. It has now been 12 years since WIPP received its first shipment of transuranic waste." (http://www.wipp.energy.gov).

Wet/Dry: A reference to characterization of contamination within a sample of any media during laboratory analysis, especially spectroanalysis. Specimens being analyzed are either wet weight (ww) or dry weight (dw), meaning the wet samples have been ashed to remove all water. As a result, dry samples have much more contamination per unit weight than wet samples: a convenient conversion factor is 8; that is, a dry sample will generally have 8 times the contamination per kg than a wet sample.

Table of Prefixes

multiple	prefix	symbol	fraction	prefix	symbol
10^{18}	exo	E	10^{-1}	deci	d
10^{15}	peta	P	10^{-2}	centi	c
10^{12}	tera	T	10^{-3}	milli	m
10^{9}	giga	G	10^{-6}	micro	μ
10^{6}	mega	M	10^{-9}	nano	n
10^{3}	kilo	k	10^{-12}	pico	p
10^{2}	hecto	h	10^{-15}	femto	f
10	deka	da	10^{-18}	atto	a

IV. Baseline Information

Historic Baseline Contamination Levels

The baseline contamination levels reprinted below provide an important guide for evaluating the source term (release inventories) of any nuclear accident, including the ongoing multiple interlocking meltdown event (MIME), which began at the Fukushima Daiichi complex on March 11, 2011.

Nuclear weapons fallout levels

- Denmark: maximum fallout deposition in1963; 988 Bq/m^2 Cs-137 (Aarkrog 1992)

- US: maximum fallout deposition in 1963; ±2,500 Bq/m^2 Cs-137 (in downwind locations. Weapons testing location depositions were much higher, but deposition levels are classified information.)

The RISO National Laboratory cumulative fallout index reprinted below is the only publically available comprehensive survey of atmospheric weapons-test-derived fallout levels.

Chernobyl-derived fallout levels

See *Section VIII. Chernobyl Fallout Data* for a more detailed survey of Chernobyl fallout patterns.

- Denmark: Denmark was considered an "unaffected" area; maximum deposition was 1,210 Bq/m^2 Cs-137 in1986

- Northern England, Sweden, Finland: ±100,000 Bq/m^2 Cs-137 in 1986

- Areas of maximum fallout in Russia: ±1,000,000 Bq/m^2 Cs-137

Chernobyl Source Terms

Source term refers to the quantity of isotopes released during an accident. The following Chernobyl accident release inventories are contained in the RISO National Laboratory's 1994 report by A. Aarkrog.

Aarkrog, A. (1994). *Source terms and inventories of anthropogenic radionuclides.* Riso National Laboratory, Roskilde, Denmark.

102

Radionuclide	Total released radioactivity (Curies)
^{137}Cs	2,700,000
^{134}Cs	1,350,000
^{90}Sr	216,000
^{106}Ru	948,000
^{144}Ce	2,430,000
^{110m}Ag	40,500
^{125}Sb	81,000
^{239,240}Pu	1,480
^{238}Pu	700
^{241}Pu	135,000
^{241}Am	162
^{242}Cm	16,200
^{243,244}Cm	162

Chernobyl Fallout in the US

US Nuclear Regulatory Commission. (1987). *Report on the accident at the Chernobyl nuclear power station*. Report No. NUREG-1250, Rev. 1. Government Printing Office, Washington, D.C. This report contains very little media-specific data on Chernobyl fallout. It does, however, contain some important specific data for fallout at one US site in Chester, NJ, (5/6/86-6/2/86) reported in picocuries per square meter, which expresses the hemispheric extent of Chernobyl fallout.

- I-131: 2,380 pCi/m^2; 88 Bq/m^2
- Cs-137: 650 pCi/m^2; 24 Bq/m^2
- Cs-134: 290 pCi/m^2; 11 Bq/m^2
- Ru-103: 720 pCi/m^2; 27 Bq/m^2 (pg. 8-3)

Chernobyl-Derived Radiocesium in US Imported Foods

US Food and Drug Administration
1986-1988 Imported Food Survey Data on Elevated Levels of Chernobyl-derived radiocesium Cs-137 Pulse Peak Analysis

Between May 5, 1986 and December 22, 1988, the FDA tested 1,749 samples of imported foods for levels of Chernobyl-derived radioactive contamination. From analysis of the results, it appears that the peak levels of Chernobyl-derived radiocesium in imported foods occurred approximately one year after the Chernobyl accident. Peak values are summarized below for the eight month period between February 1 and October 4, 1987, when contamination levels were highest.

Feb. 1 – Oct. 4, 1987

Total number of FDA Samples of imported foods: 411

Number of samples with elevated levels of Cs-137:

> 183 samples with levels over 100 pCi/kg (44% of samples)
> 85 samples with levels over 1000 pCi/kg (20%)
> 19 samples with levels over 5000 pCi/kg (approx. 5%)

Feb 5. – June 25, 1987

Total number of FDA Samples of imported foods: 161

Number of samples with elevated levels of Cs-137:

> 83 samples with levels over 100 pCi/kg (more than 50% of samples)
> 48 samples with levels over 1000 pCi/kg (almost 30%)
> 11 samples with levels over 5000 pCi/kg (approx. 7%)

In 1998, FDA survey data was still showing elevated levels of Cs-137 in imported foods. An Oct. 27, 1988 sample of macaroni from Greece contained Cs-137 levels of 11,100 pCi/kg (410 Bq/kg).

Chernobyl Radioactivity in Turkish Tea

During and following the Chernobyl accident, spiraling plumes of reactor-derived radioactive contamination were distributed over wide areas of the northern hemisphere,

104

including locations thousands of kilometers from the accident site. The eastern Black Sea region of Turkey was the location of a rainfall event that brought extensive contamination to farmlands in which various agricultural crops, including hazelnuts and tea, were grown. Accident-derived cesium-137 was detected in hazelnuts in a range of 2000-2500 Bq/kg. Even larger quantities of radiocesium followed the potassium cycle and were reported in Turkish tea in quantities ranging from 1,064 to 44,000 Bq/kg (Gedikoglu 1989). The ratio of radioactivity transferred to brewed tea was 65%. The documentation of Chernobyl radioactivity in Turkish tea provides a model for what could occur in areas downwind from Japan in a worst case scenario. The challenge for the general public in the age of information technology, in the event of a catastrophic accident in any location, is to obtain media-specific data recorded in now-standard reporting units of becquerels per liter or per kilogram, and to have access to information databases that allow rational evaluation of this contamination. Political and corporate interests may make data-specific information difficult to obtain through most mass media outlets.

Nuclear Weapons plutonium baseline

As a footnote to the Fukushima Daiichi disaster, the Japanese government is reporting background levels of plutonium from weapons testing fallout as 0.61 Bq per kilogram of soil. Any reports of plutonium in soil above 1 Bq/kg would almost certainly be accident-derived plutonium from MOX fuel or spent fuel dispersed after being absorbed by particulate matter, such as smoke.

Chernobyl Cs-137 Deposition

The following tables are from *The Other Report on Chernobyl (TORCH)* (Fairlie 2006, 40). Note 1 PBq = 27,000 Ci.

Table 3.4 Cs-137 deposition ranked by country

Country	PBq	Country	PBq	Country	PBq
Russia (Europe part)	29	Italy	0.93	Ireland	0.35
Belarus	15	France	0.93	Slovak Rep	0.32
Ukraine	13	United Kingdom	0.88	Latvia	0.25
Finland	3.8	Czech Rep	0.6	Estonia	0.18
Sweden	3.5	Lithuania	0.44	Turkey (Europe part)	0.16
Norway	2.5	Moldova	0.4	Denmark	0.087
Rumania	2.1	Slovenia	0.39	Netherlands	0.062
Germany	1.9	Spain	0.38	Belgium	0.053

Country	PBq	Country	PBq	Country	PBq
Austria	1.8	Croatia	0.37	Luxembourg	0.008
Poland	1.2	Switzerland	0.36	**Total**	**85**
Greece	0.95	Hungary	0.35		

data reproduced from table III.1 in European Commission, 1998

Table 3.5 Cs-137 deposition ranked by country

Country	PBq
Yugoslavia	5.4
Bulgaria	2.7
Albania	0.4
TOTAL	**8.5**

data reproduced from Goldman, US DoE, 1987
[Yugoslavia reduced by 0.76 PBq to avoid double-counting Slovenia and Croatia in table 3.4]

Cumulative Fallout Index

The most important resource for information about the historic accumulation of radioactive contamination from weapons testing and the Chernobyl accident is the database compiled by the RISO National Laboratory in Denmark. Many of the extensive databases compiled by the United States government, especially those compiled by the US Department of Defense via the National Imagery and Mapping Agency (NIMA), are classified information and not available to the general public. The following data compiled by the RISO National Laboratory provides important baseline information for evaluating the significance of fallout derived from the Japan disaster. Atomic weapons testing began in the early 1950s; the RISO National Laboratory database provides summaries of the annual and cumulative deposition of both strontium-90 and radiocesium. The accidents at Chernobyl and in Japan released much more radiocesium and much less radiostrontium than weapons testing, therefore, the database of Cs-137 deposition is referenced in this accident guideline synopsis. The RISO National Laboratory measured fallout in three locations, Denmark, Jutland, and the Faroe Islands. The cumulative fallout for selective years is reproduced below. The peak year for weapons testing fallout was 1963 with rapidly declining fallout levels after 1965 until the Chernobyl accident in 1986. The annual (DI) and the cumulative (AI) fallout are reported in **becquerels/m^2**.

FALLOUT RATES AND CUMULATIVE FALLOUT ($Bq\ Cs_{137}\ M\text{-}2$) IN DENMARK, JUTLAND, AND THE FAROE ISLANDS 1950-1991 DI = ANNUAL DEPOSITION, AI = CUMULATIVE RADIATION						
	DENMARK		JUTLAND		ISLANDS	
YEAR	DI	AI(30.2)	DI	AI(30.2)	DI	AI(30.2)
1950	1.243	1.215	1.302	1.273	1.184	1.157
1951	5.979	7.030	6.749	7.838	5.210	6.221
1952	11.722	18.323	13.261	20.618	10.182	16.029
1953	29.600	46.830	33.507	52.889	25.693	40.770
1954	112.539	155.731	127.398	176.173	97.680	135.290
1955	148.059	296.857	167.595	335.922	128.523	257.792
1956	183.579	469.471	207.792	531.304	159.366	407.637
1957	183.579	638.145	207.792	722.227	159.366	554.062
1958	254.678	872.445	288.245	987.409	221.053	757.424
1959	361.238	1205.526	408.954	1364.492	313.582	1046.561
1960	67.488	1243.959	76.427	1408.032	58.608	1079.940
1961	87.675	1301.241	99.219	1472.849	76.072	1129.632
1962	439.738	1701.242	472.179	1900.635	407.296	1501.849
1963	**988.344**	**2628.199**	**1092.418**	**2924.739**	**884.270**	**1331.659**
1964	616.390	3170.535	691.752	3533.949	541.029	2807.121
1965	234.077	3326.905	248.877	3696.486	219.277	2957.324
1966	126.984	3375.057	128.227	3737.418	125.741	3012.697
1967	61.982	3358.593	69.619	3720.145	54.346	2997.040
1968	83.058	3363.098	92.826	3725.944	73.230	3000.195
1969	61.272	3346.212	73.467	3712.693	49.077	2979.675
1970	97.502	3365.115	117.986	3743.247	77.019	2986.928
1971	89.155	3375.430	102.179	3757.659	76.131	2993.148
1972	25.752	3323.554	27.054	3698.331	24.450	2948.724
1973	11.366	3258.804	12.728	3626.358	9.946	2891.141

FALLOUT RATES AND CUMULATIVE FALLOUT (BQ CS$_{137}$ M-2) IN DENMARK, JUTLAND, AND THE FAROE ISLANDS 1950-1991 DI = ANNUAL DEPOSITION, AI = CUMULATIVE RADIATION						
	DENMARK		JUTLAND		ISLANDS	
YEAR	DI	AI(30.2)	DI	AI(30.2)	DI	AI(30.2)
1974	42.032	3225.498	46.117	3588.654	38.066	2862.350
1975	24.509	3175.828	26.758	3532.894	22.259	2818.771
1976	6.098	3109.302	6.867	3458.970	5.328	2759.642
1977	22.733	3060.549	23.976	3403.451	21.430	2717.597
1978	27.410	3017.479	31.850	3356.893	22.970	2678.016
1979	9.827	2958.211	10.301	3290.341	9.235	2625.917
1980	5.606	2896.171	6.766	3221.854	4.591	2570.470
1981	17.059	2846.738	18.316	3166.216	15.948	2527.385
1982	2.706	2784.409	2.851	3096.736	2.561	2472.203
1983	2.151	2722.959	2.126	3028.134	2.175	2417.902
1984	1.751	2662.521	1.935	2960.911	1.567	2364.247
1985	1.290	2603.012	1.191	2894.495	1.388	2311.642
1986	**1210.000**	**3725.984**	**1340.000**	**4137.847**	**1080.000**	**3314.232**
1987	29.000	3669.280	32.000	4074.674	26.000	3263.994
1988	11.900	3597.161	13.400	3994.768	10.300	3199.562
1989	3.500	3518.480	4.510	3907.998	2.530	3129.007
1990	2.63	3440.744	3.85	3822.564	1.41	3058.968
1991	1.63	3363.805	1.92	3737.194	1.36	2990.480

US DOE Spent Fuel and Radioactive Waste Inventories

The *US DOE Integrated Database for 1992: U. S. Spent Fuel and Radioactive Waste Inventories, Projections, and Characteristics* provides a now classified database with information that can be easily interpreted by a layperson to evaluate potential releases at the Japan facilities. As with the RISO National Laboratory reporting system, the DOE tabulated both the annual and cumulative radioactivity produced at America's 104 light water reactors. 1992 was the most recent year that data was available pertaining to inventories of biologically significant radioisotopes in US nuclear reactors. The following summary is extracted from Vol. 3 of *Biocatastrophe the Legacy of Human Ecology*, a publication sponsored by the Davistown Museum, and is available from amazon.com in paperback and eBook. In turn, this summary was extracted from Davistown Museum Special Publication 47 (3/21/2007) which is a reprint of the Oak Ridge National Laboratory database (October 1992) (DOE/RW-0006, Rev 8).

Atomic #	Element	Mass # of nuclide	Half-life	Mass, g		Radioactivity, Ci	
				Annual	Cumulative	Annual	Cumulative
27	Cobalt	60	5.2714 y	1.29E+04	7.00E+04	1.46E+07	7.92E+07
38	Strontium	90	28.90 y	9.93E+05	8.84E+06	1.35E+08	1.21E+09
53	Iodine	129	15.7×10^6 y	3.40E+05	3.53E+06	6.00E+01	6.23E+02
55	Cesium	137	30.07 y	2.24E+06	2.01E+07	1.95E+08	1.75E+09
94	Plutonium	238	88 y	2.65E+05	2.39E+06	4.53E+06	4.10E+07
94	Plutonium	239	2.41×10^4 y	9.52E+06	1.14E+08	5.92E+05	7.08E+06
94	Plutonium	241	14 y	2.32E+06	1.92E+07	2.39E+07	1.98E+09

Source: US Department of Energy. (October 1992). *Integrated Data Base for 1992: U.S. Spent Fuel and Radioactive Waste Inventories, Projections, and Characteristics.* Oak Ridge National Laboratory, Oak Ridge, Tennessee.

The cumulative inventory of radioactivity at US nuclear reactors is summarized in the far right hand column for the most biologically significant isotopes in a nuclear reactor inventory. These inventories express the contents of the spent fuel pools accumulated over the lifetime of reactor operations; they do not include the inventories within operating nuclear reactor vessels. A thumbnail estimate of reactor vessel inventories would be about 10% of the cumulative spent fuel pool inventory. Taking Cs-137 as an example, the 1992 total spent fuel pool inventory at US reactors was 1.75E +09, or in plain English 1,750,000,000 curies. In 1992, there were 104 operating reactors in the US; a thumbnail sketch of current inventories of radiocesium at US reactors can be obtained by dividing this number by 100 providing an approximate estimate of 17,500,000 curies of radiocesium at each US reactor site. During the interim between 1992 and 2011, most US reactors will have accumulated much more than this baseline figure considering that the radioactive decay rate of radiocesium (1/2T= 30.07 years) is much less than the annual productivity of any given reactor. The same observation applies to the cumulative inventory of other biologically significant isotopes noted above.

Fuel Assembly Inventories in the United States

The following 2002 spent nuclear fuel storage inventory, now almost ancient history, of fuel assemblies accumulated by both nuclear weapons production reactors and nuclear power reactors provides an insight into the legacy costs of nuclear industries. The combined total of 383,653 fuel assemblies breaks down into two principal categories.

- 133,983 fuel assemblies derived from weapons production, including 4,834 assemblies from university and industry sources

- 249,670 assemblies derived from the waste created by the generation of electricity

A sketch of the Cs-137 inventory in these fuel assemblies, only one isotope among many in spent fuel storage wastes (see *Figure 10* for a listing of all isotopes in spent fuel at 100 days cooling) can be made by multiplying 249,670 by 10,000 Ci, the average per assembly estimate of Cs-137 in fuel assemblies. The legacy of 2,496,700,000 Ci of radiocesium from the generation of nuclear power contrasts with the ±20,000,000 Ci of Cs-137 released from nuclear weapons tests. The safe storage and maintenance, of what is, by 2011, well over 3 billion Ci of Cs-137 and tens of billions of curies of other biologically significant isotopes derived from the operation of nuclear power plants in the United States is a huge legacy cost that must be borne by future generations. Currently, because the cancellation of the Yucca Mountain repository for these wastes due to environmental opposition (the alleged dangers of transport) and the objection of Nevada residents to the construction of the Yucca Mountain facility, no safe practical storage solution for these wastes has been formulated. Rather, other than the 17,826 assemblies in independent spent fuel storage installations (ISFSIs), the majority of America's nuclear wastes are tightly packed in spent fuel pools that were designed as temporary storage facilities. At many nuclear power plants, due to our inability to safely store these wastes at Yucca Mountain, fuel assemblies are packed in spent fuel pools so tightly, often at two to five times their waste storage capacity, that fuel assembly storage may be aptly described as:

Least safe storage of nuclear wastes (LSSNW)

This acronym should join multiple interlocking meltdown events (MIMEs) as symbols of our collective incompetence in designing safe nuclear power plants and waste storage facilities. Many other nations follow our example in the risky design and operation of nuclear power facilities and ad hoc fly-by-night waste disposal strategies. Avoiding the controversial subject of reprocessing spent fuel, only France can be complimented as having relatively safe uniform nuclear power plant designs with onsite underground storage of vitrified spent fuel wastes. Such vitrified fuel assemblies will never be subject to terrorist attacks or loss of coolant accidents.

Facility			Assemblies	MT	Facility			Assemblies	MT
1. Arkansas Nuclear One	AK	P	1,517	666.7	46. Shearon Harris Nuc Pwr Plnt	NC	P	3,814	964.5
		I	552	241.4	47. Cooper Nuclear Station	NE	P	1,537	278.6
2. Browns Ferry Nuclear Plant	AL	P	6,696	1,230.2	48. Fort Calhoun Station	NE	P	839	305.0
3. J M Farley Nuclear Plant	AL	P	2,011	903.8	49. Seabrook Nuclear Station	NH	P	624	287.2
4. Palo Verde Nuc Gen Station	AZ	P	2,747	1,157.8	50. Hope Creek Gen Station	NJ	P	2,376	431.5
5. Diablo Canyon Power Plant	CA	P	1,736	760.9	51. Oyster Creek Generating Sta	NJ	P	2,556	455.9
6. GE Vallecitos Nuc Center	CA	I	fragments	0.2			I	244	47.6
7. Humboldt Bay Power Plant	CA	P	390	28.9	52. Salem Nuc Generating Sta	NJ	P	1,804	832.7
8. Rancho Seco Nuc Gen Station	CA	I	493	228.4	53. Sandia National Laboratory	NM	F	503	0.3
9. San Onofre Nuc Gen Station	CA	P	2,490	1,013.3	54. Brookhaven National Lab	NY	F	40	<0.1
10. Fort St. Vrain Power Station	CO	F	1,464	14.7	55. JA Fitzpatrick Nuc Pwr Plant	NY	P	2,460	446.5
11. Connecticut Yankee Atom Pwr	CT	P	1,019	412.3			I	204	37.2
12. Millstone Nuc Power Station	CT	P	4,558	1,227.9	56. Indian Point Energy Center	NY	P	2,073	903.6
13. Crystal River Nuc Power Plant	FL	P	824	382.3	57. Nine Mile Point Nuclear Station	NY	P	4,456	801.6
14. St. Lucie Nuc Power Plant	FL	P	2,278	870.7	58. R E Ginna Nuclear Power Plant	NY	P	967	357.4
15. Turkey Point Station	FL	P	1,862	851.7	59. Davis-Besse Nuclear Pwr Sta	OH	P	749	351.3
16. AW Vogtle Electric Gen Plant	GA	P	1,639	720.8			I	72	33.9
17. EL Hatch Nuclear Plant	GA	P	5,019	909.3	60. Perry Nuclear Power Plant	OH	P	2,088	378.4
		I	816	151.2	61. Trojan Nuclear Power Plant	OR	P	780	358.9
18. D Arnold Energy Center	IA	P	1,912	347.9	62. Beaver Valley Power Station	PA	P	1,456	672.9
19. Idaho National Eng & Env Lab	ID	F	93522	299.3	63. Limerick Generating Station	PA	P	4,601	824.0
20. Argonne National Lab East	IL	F	78	0.1	64. Peach Bottom Atm Pwr Sta	PA	P	5,905	1,062.7
21. Braidwood Generating Sta	IL	P	1,485	628.7			I	1,020	190.3
22. Byron Generating Station	IL	P	1,786	756.4	65. Susquehanna Steam Elec Sta	PA	P	4,240	738.4
23. Clinton Power Station	IL	P	1,580	288.8			I	1,300	238.5
24. Dresden Generating Station	IL	P	5,698	1,009.2	66. Three Mile Island Nuc Station	PA	P	898	416.1
		I	1,155	146.9	67. Catawba Nuclear Station	SC	P	1,780	782.4
25. GE Morris Operation	IL	I	3,217	674.3	68. HB Robinson Steam Elec Plt	SC	P	344	147.9
26. LaSalle County Gen Sta	IL	P	4,106	744.6			I	56	24.1
27. Quad Cities Gen Station	IL	P	6,116	1,106.5	69. Oconee Nuclear Station	SC	P	1,419	665.8
28. Zion Generating Station	IL	P	2,226	1,019.4			I	1,726	800.4
29. Wolf Creek Gen Station	KS	P	925	427.3	70. Savannah River Defense Site	SC	F	9,657	28.9
30. River Bend Station	LA	P	2,148	383.9	71. VC Summer Nuclear Station	SC	P	812	353.9
31. Waterford Gen Sta	LA	P	960	396.4	72. Sequoyah Nuclear Power Plant	TN	P	1,699	782.6
32. Pilgrim Nuclear Station	MA	P	2,274	413.9	73. Watts Bar Nuclear Plant	TN	P	297	136.6
33. Yankee Rowe Nuc Power Sta	MA	I	533	127.1	74. Comanche Peak Steam Elec Sta	TX	P	1,273	540.7
34. Calvert Cliffs Nuc Pwr Plt	MD	P	1,348	518.0	75. South Texas Project	TX	P	1,254	677.8
		I	960	368.1	76. North Anna Power Station	VA	P	1,410	652.7
35. Maine Yankee Atomic Pwr Plt	ME	I	1,434	542.3			I	480	220.8
36. Big Rock Point Nuc Plt	MI	I	441	57.9	77. Surry Power Station	VA	P	794	365.4
37. D C Cook Nuclear Plant	MI	P	2,198	969.0			I	1,150	524.2
38. Enrico Fermi Atomic Pwr Plt	MI	P	1,708	304.6	78. Vermont Yankee Gen Station	VT	P	2,671	488.4
39. Palisades Nuclear Pwr Sta	MI	P	649	260.7	79. Columbia Generating Station	WA	P	1,904	333.7
		I	432	172.4			I	340	61.0
40. Monticello Nuclear Gen Plant	MN	P	1,342	236.1	80. Hanford Defense Site	WA	F	110,140	2,128.9
41. Prairie Isl. Nuc Gen Plt	MN	P	1,135	410.3	81. Kewaunee Nuclear Power Plant	WI	P	904	347.6
		I	680	262.3	82. La Crosse Nuclear Gen Station	WI	P	333	38.0
42. Callaway Nuclear Plant	MO	P	1,118	479.0	83. Point Beach Nuclear Plant	WI	P	1,353	507.4
43. Grand Gulf Nuclear Station	MS	P	3,160	560.2			I	360	144.1
44. Brunswick Stm Elec Plt	NC	P	2,227	477.4	Other: University & Industry		F	4,834	1.7
45. W B McGuire Nuc Sta	NC	P	2,232	1,001.1					
		I	160	68.6	Combined Total			383,653	49,401.2

			Assemblies	MT	
Reactor Pool		P	145,589	41,564.1	**Sources:** Energy Information Administration, and DOE National Spent Nuclear Fuels Program
ISFSI		I	17,826	5,363.2	
Federal and Other		F	220,238	2,473.9	MT: metric ton (1,000 kg)

Figure 9. 2002 Reactor storage pools, independent spent fuel storage installations, federal, and other sites (Andrews 2004, Table 1 pg. CRS-5).

Baseline Data: Riso National Laboratory

The Riso National Laboratory in Roskilde Denmark has compiled a comprehensive data base of unclassified information about the dietary intake of radionuclides. In providing the following summary of Riso reports, RADNET would like to acknowledge and thank Asker Aarkrog and fellow researchers at the Riso Laboratory who collected this data. The following summary of Riso tabulations selects sea vegetables, milk, fish, tea and meat/eggs as well as total annual dietary intake for the radionuclide cesium-137 as an indicator of radioactive contamination from man-made sources. The Chernobyl accident and the high levels of contamination that resulted in locations outside Denmark illustrate the importance of the Riso database. In reviewing the following summary, the reader is urged to remember that Denmark escaped most of the impact of the Chernobyl accident, having received 1,210 Bq^{137}Cs m^2. In comparison, Chernobyl fallout in England and Wales reached levels exceeding 60,000 Bq ^{137}Cs m^2, in Sweden 100,000 Bq ^{137}Cs m^2 in scattered locations, and, in some areas downwind from Chernobyl in Byelorussia, contamination levels were measured in the hundreds of millions of Bq ^{137}Cs m^2. In the following summary, all data are **averages** unless followed by **pv**, meaning the **peak value** (peak concentrations) among a number of samples.

RADIOCESIUM DIETARY INTAKE BASELINE
All data in Bq/yr. or Bq/kg and are mean values unless noted as peak values (pv).

Location/yr.	Dietary Intake/yr.	Sea* Vegetables	Milk and Cream	Fish	Coffee/Tea	Meat
Greenland/1981	531.75	0.61	0.134	0.46	2.21	9.13
Faroes Is./1981	2,520.7	1.29	3.86	0.34	2.21	48.6
Denmark/1981	160.06	1.79 pv	0.134	5.70	2.21	0.36
Denmark/1982	153.55	15.6 pv	0.105	4.2	6.50 pv	0.38
Greenland/1982	651.66	3.1	1.05	0.28	2.53	11.16
Faroes Is./1982	1,887.8		3.33	0.26	2.53	24.6
Denmark/1983	103.24	11.6 pv	0.076	4.0	2.53	0.28
Greenland/1983	593.88		0.076	0.28	2.53	11.4
Faroes Is./1983	1,749.7		3.24	0.25	2.53	3.33

RADIOCESIUM DIETARY INTAKE BASELINE

All data in Bq/yr. or Bq/kg and are mean values unless noted as peak values (pv).

Location/yr.	Dietary Intake/yr.	Sea* Vegetables	Milk and Cream	Fish	Coffee/Tea	Meat
Denmark/1984	84.90	12.9 pv	0.085	14.5 pv	1.53	0.18
Greenland/1984	625.32	1.16	0.085	0.28	1.53	12.3
Faroes Is./1984	1,560.7	1.16	3.10	0.29	1.53	19.57
Denmark/1985	83.58	22 pv	0.076	3.15	1.53	0.25
Greenland/1985	856.06	1.56 pv	0.076	0.39	1.53	17.00
Faroes Is./1985	985.9	0.78 pv	1.82	0.29	1.53	11.34
Denmark/1986	483.57	33 pv	1.062	8.1 pv	1.29	1.16
Denmark/1987	571.54	18.1 pv	0.60	4.7	1.29	2.53
Greenland/1987	826.89		0.60	0.55	1.23	5.6
Faroes Is./1987	3,508.01	2.16 pv	5.85	0.63	1.29	53.8
Denmark/1988	153.44	60 pv	0.28	3.6	0.82	0.72
Faroes Is./1988	2,120.5	0.78 pv	2.77	0.44	7.3	24.1
Denmark/1989	171.14	21.4 pv	1.78	6.6	0.819	0.728
Faroes Is./1989	1,792.8	0.84 pv	1.89	0.36	7.3	22.6
Denmark/1990	177.62	25.0 pv	1.25	5.8	0.44	1.28
Greenland/1990	329.02		0.125	0.22	0.44	3.9
Faroes Is./1990	817.6	0.56 pv	1.61	0.23	0.44	6.64
Denmark/1991	146.73	26.0 pv	0.092	7.7	0.44	0.47
Greenland/1991	479.10		0.092	0.27	0.44	5.9

RADIOCESIUM DIETARY INTAKE BASELINE
All data in Bq/yr. or Bq/kg and are mean values unless noted as peak values (pv).

Location/yr.	Dietary Intake/yr.	Sea* Vegetables	Milk and Cream	Fish	Coffee/Tea	Meat
Faroes Is./1991	1196.9	1.36 pv	1.28	0.18	0.44	16.6

*dry weight

Several observations can be made about the data collected by the Riso National Laboratory. The yearly dietary intake of ^{137}Cs and its presence in the sea vegetables and market basket items cited in this chart vary widely, both before and after the Chernobyl accident. The Riso data document a modest increase in the dietary intake of cesium-137 in and after 1986. The question arises as to whether an expanded survey including imported and processed foods in Denmark after 1986 would document an even larger increase of Chernobyl-derived radiocesium similar to that documented by the FDA in the US. (See FDA peak pulse analysis at the end of this section). A yearly intake of 571.54 Bq of cesium-137 in Denmark in 1987 supports the argument that Denmark escaped the impact of the Chernobyl accident. The consistently low contamination levels in Danish tea and coffee help illustrate the significance of reports in RADNET Section 10: Chernobyl Annotated Bibliography of Chernobyl-derived radiocesium in Turkish tea in the tens of thousands of becquerels per kilogram of brewed tea, to mention one example of the many citations of high levels of contamination in pathways to human consumption which resulted from the Chernobyl accident.

The following data has also been extrapolated from the Riso Reports and complements the dietary intake data cited above.

*The following data express peak values unless otherwise noted.

Date	Location	Media	Nuclide	Activity*
1969-70	Greenland	Fucus vesiculosis	^{137}Cs	35.6 Bq/kg
1969-70	Greenland	Fucus vesiculosis	^{239}Pu	8.40 Bq/kg

- Sample taken near the impact of a B-52 carrying nuclear weapons which crashed near Thule, Greenland.

114

Date	Location	Media	Nuclide	Activity*
1981	Greenland	Reindeer meat	^{137}Cs	102 Bq/kg
1981	Greenland	Lamb	^{137}Cs	118 Bq/kg
1981	Greenland	Moss	^{137}Cs	242 Bq/kg
1982	Denmark	Brown algae	^{239}Pu	0.176 Bq/kg

- The Riso report notes that 27% of cesium-137 in fish in Denmark originated from Sellafield, a nuclear fuel reprocessing facility located in Northern England on the Irish Sea and often noted as impacting dietary intake of radionuclides, particularly in the Faroes Islands.

Date	Location	Media	Nuclide	Activity*
1982	Sellafield facility vicinity	Fucus vesiculosis	^{137}Cs	3 km: 2,600 Bq/kg
1982	Sellafield facility vicinity	Fucus vesiculosis	^{137}Cs	23 km: 1,410 Bq/kg
1982	Sellafield facility vicinity	Fucus vesiculosis	^{137}Cs	1050 km: 20.5 Bq/kg
1982	Greenland	Moss	^{239}Pu	60 Bq/kg
1982	Greenland	Mytilus edulus	^{239}Pu	0.64 Bq/kg

- The Riso report also notes a peak value of cesium-137 in lichen carpet at 1,630 Bq/m^2, with an estimated total accumulation of 2,400 Bq/m^2 from stratospheric fallout.

Date	Location	Media	Nuclide	Activity*
1983	Denmark	Mytilus edulus	^{137}Cs	5.1 Bq/kg
1984	Thule, Greenland	Sediment	^{239}Pu	43.5 Bq/kg

- 13.3 km from the impact of the B-52 bomber crash.

Date	Location	Media	Nuclide	Activity*
1984	Greenland	Seals	^{137}Cs	300 Bq/kg
1984	Greenland	Reindeer	^{137}Cs	229 Bq/kg

- The above two citations are weapons-fallout radiocesium.
- 1985: The Riso report notes that 85% of the ^{137}Cs content in Danish fish was Sellafield-derived. Without the contribution of Sellafield radiocesium the total dietary intake for 1985 would be 107.25 Bq of ^{137}Cs instead of 160.06 (Report no. Riso R-469, pg. 63).
- 1986: The Riso report notes that in Denmark, an area relatively unaffected by Chernobyl fallout, the dietary intake of strontium 90 remained the same as the previous year, but levels of cesium-137 were 13.8 times higher than 1985 (Report no. Riso R-549, pg. 157).
- The report also notes a peak value of iodine-131 in stinging nettle at 670 Bq/kg and of cesium-137 at 200 Bq/kg. Mean concentration of cesium-137 in air increased by a factor of 2000 (Report no. Riso R-549, pg. 45).

Date	Location	Media	Nuclide	Activity*
1987	Greenland	Lichens	^{137}Cs	5,800 Bq/kg
1989	Denmark	Mushrooms	^{137}Cs	950 Bq/kg
1989	Denmark	Mushrooms	134Cs	76 Bq/kg

The Riso reports cited in this section are: Riso R--470, 471, 487, 488, 489, 509, 510, 527, 528, 540, 549, 563, 564, 570, 571(EN), R-621(EN).

These summaries of data represent a pre-Chernobyl baseline for the dietary intake of radiocesium and provide a reference for the evaluation of the Chernobyl accident and its impact outside of Denmark.

Fukushima Daiichi Waste Inventories

The reactors at the Japan facility were designed and produced by General Electric and have the same boiling water reactor design as used in 23 of 35 aging boiling water reactors operating in the United States. In this context, a conservative estimate of the maximum releases of other biologically significant isotopes at the four reactors at risk at Fukushima Daiichi can be obtained by multiplying the cumulative inventories listed in the Oak Ridge database by 0.04 (4%). Potential releases of plutonium-239, not including the extra inventories of plutonium in the number three reactor using MOX fuel, could be reasonably estimated at 4% of 7,080,000 curies, or 283,200 curies of the highly toxic alpha-emitting plutonium-239. The inventory of Pu-239 at the four reactors in Japan is actually much higher than this figure, since the ORNL database was last compiled in 1992. A conservative estimate of the Cs-137 inventories at the four reactors in Japan has already been noted in *Section I*. Comments on our preliminary estimates of the cumulative radioactivity available for release at the Fukushima Daiichi facilities are welcomed.

Radionuclide Content of Reactor Fuel

This inventory of the radionuclide content of spent fuel at 150 days cooling provides a guideline to the source term resulting from the dispersal of the contents of the fuel assemblies in any nuclear accident.

Radionuclide Content of Light-Water Reactor Fuel at 150 Days Cooling[1]

Radionuclide	Element	$T_{1/2}$		Curies per mwyr(e) [2]
[3]H	Hydrogen	12.5 yrs		23
[85]Kr	Krypton	9.4 yrs		375
[89]Sr	Strontium	50.5 days		3220
[90]Sr	Strontium	28.5 yrs		2570
[90]Y*	Yttrium	65 hours		2570
[91]Y	Yttrium	57.0 days		5330
[95]Zr	Zirconium	65.0 days		9250
[95]Nb	Niobium	35.0 days		17360
[95m]Nb*	Niobium	90.0 hrs		196
[103]Ru	Ruthenium	39.8 days		2986
[106]Ru	Ruthenium	1.0 yr		13740
[103m]Rh*	Rhodium	57 min		2990
[106]Rh*	Rhodium	30 sec		13740
[110m]Ag	Silver	270 days		8.7
[125]Sb	Antimony	2.7 yrs		272
[125m]Te	Tellurium	58 days		110
[127m]Te	Tellurium	113 days		207
[127]Te*	Tellurium	9.3 hrs		205
[129m]Te	Tellurium	34 days		224
[129]Te*	Tellurium	72 min		144
[129]I	Iodine	1.7×10^7 yrs		0.0013
[131]I	Iodine	8.04 days		0.073
[134]Cs	Cesium	2.3 yrs		7140
[137]Cs	Cesium	30.2 yrs		3550
[137m]Ba*	Barium	2.7 minutes		3340
[140]Ba	Barium	12.8 days		14.4
[140]La*	Lanthanum	40.0 hours		16.6
[141]Ce	Cerium	32.5 days		1900
[144]Ce	Cerium	290 days		25810
[143]Pr	Praseodymium	13.8 days		23.3
[144]Pr*	Praseodymium	17 minutes		25810
[147]Pm	Promethium	2.6 years		3330
[151]Sm	Samarium	70 yrs		38.5
[154]Eu	Europium	5.4 yrs		229
[155]Eu	Europium	1.7 yrs		214
[160]Tb	Terbium	71 days		10.1
[239]Np	Neptunium	2.33 days	β	0.6
[238]Pu	Plutonium	92 yrs	α	94.0
[239]Pu	Plutonium	24,400 yrs	α	11.0
[240]Pu	Plutonium	6,580 yrs	α	16.0
[241]Pu	Plutonium	14 yrs	β	3,850.0
[241]Am	Americium	475 yrs	α	6.7
[242]Cm	Curium	162 days	α	503.0
[244]Cm	Curium	19 yrs	α	84.0

Figure 10. Light water reactor fuel (Brack 1984, Appendix VI pg 65).

Cesium Baseline (to 1986)

Cesium-137 is one of two indicator nuclides frequently used to document the impact of nuclear accidents (the other is I-131). This *Handbook* uses it for the documentation of both stratospheric fallout contamination and Chernobyl-derived fallout. Strontium-90 is the nuclide most associated with weapons fallout, and it occurred in larger quantities in stratospheric fallout than radiocesium (Cs-134 and Cs-137). However, radiocesium is the largest constituent of spent fuel, the most prevalent long-lived component of the tropospheric plume which originated at Chernobyl, the most ubiquitous long-lived contaminant from Fukushima Daiichi, and an omnipresent fission product resulting from most nuclear industries and activities.

- Most of the following baseline data are pre-Chernobyl peak concentrations unless otherwise noted.
- Data is compiled from dry weight samples unless noted as wet weight, except milk.
- Baseline data is listed in order of the date of the research, beginning with data collected in the early days of nuclear testing.
- The activity levels noted do not include the shorter-lived Cs-134 (1/2 T = 2.1yr).

Booker, D.V. (1959). Cesium-137 in dried milk. *Nature*,183. pg. 921-924.

DATE	LOCATION	MEDIA	NUCLIDE	ACTIVITY
1957	England	Milk	^{137}Cs	870 pCi/kg

- Data collected near the Windscale accident.

Yamagata, N., Kodaira, K. and Hiroshi, H. (1962). Cesium-137 in Japanese people and diet. *Journal of Radiation Research*, 3, 3, 182-192.

DATE	LOCATION	MEDIA	NUCLIDE	ACTIVITY
Feb. 1962	Japan	Daily intake	^{137}Cs	50 pCi/day

- The impact from stratospheric fallout increased the daily intake of radiocesium well above the previous August 1960 average of 22 pCi/day. 40% of the dietary intake of radiocesium in Japan comes from polished rice.

Baxter, A.J. and Camplin, W.C. (1993). Radiocaesium in the seas of northern Europe: 1962-69. *Fish. Res. Data Rep.,* MAFF Direct. Fish. Res., Lowestoft. 31. pg. 1-69.

Wilson, A. R. and Spiers, F.W. (1967). Fallout cesium-137 and potassium in new-born infants. *Nature.* 215. pg. 470-474.

DATE	LOCATION	MEDIA	NUCLIDE	ACTIVITY
1966	Leeds, England	Mother	^{137}Cs	0.202 nCi/kg (202 pCi/kg)
1966	Leeds, England	Infant	^{137}Cs	0.196 nCi/kg (196 pCi/kg)

- "The biological half life for ^{137}Cs in an adult averages about 100 days for the principal component, comprising about 90 percent of the intake by ingestion. In infants the biological half life is 5 to 10 times lower." (p. 470).

Pelletier, C.A. and Voilleque, P.G. (1971). The behavior of 137Cs and other fallout radionuclides on a Michigan dairy farm. Health Physics, 21, 777- 792.

DATE	LOCATION	MEDIA	NUCLIDE	ACTIVITY
1964-65	Tecumseh, MI	Milk	^{137}Cs	140 pCi/l

- Reflects stratospheric fallout from weapons testing.

Gruter, H. (1970). Radioactive fission product Cs-137 in mushrooms in West Germany during 1963-1970. *Health Phys.* 20. pg. 655-656.

Brisbin, I.L., Geiger, R.A. and Smith, M.H. (1973). Accumulation and redistribution of radiocaesium by migratory waterfowl inhabiting a reactor cooling reservoir. IN: *Environmental Behavior of Radionuclides Released in the Nuclear Industry.* International Atomic Energy Agency, Vienna.

DATE	LOCATION	MEDIA	NUCLIDE	ACTIVITY
1971-72	Savannah River SC	(Waterfowl) Common Gallinules	^{137}Cs	1,500 pCi/kg live weight

- This is a highly contaminated ecosystem. See below for additional information on the Savannah River source point.

Bowen, V.T., Noshkin, V.E., Volchok, H.L., Livingston, H.D. and Wong, K.M. (1974). Cesium-137 to Strontium 90 ratios in the Atlantic Ocean 1966 through 1972. *Limnology and Oceanography, 19*, 4, 670-681.

Hawthorne, H.A., Zellmer, S.D., Eberhard, L.L. and Thomas, J.M. (1976). 137 Cesium cycling in a Utah *dairy* farm. *Health Physics, 30*, 447- 464.

DATE	LOCATION	MEDIA	NUCLIDE	ACTIVITY
1963-67	Utah Dairy Farm	Alfalfa	^{137}Cs	3,058 pCi/kg (112.4 Bq/kg)

- Peak value of ^{137}Cs in fallout collectors (funnels): 145.4 Bq/m^2; reflects contamination from local nuclear weapons tests.

Richie, J.C. and McHenry, J.R. (1978). Fallout cesium-137 in cultivated and noncultivated North Central United States watersheds. *J. Envrion. Qual.* 7(1). pg. 40-4.

- ^{137}Cs concentrations in watershed soils ranged from 56-149 nCi/m^2 (2,080 - 5,520 Bq/m^2) with peak values in reservoir sediments up to 1,280 nCi/m^2 and a mean value of 676 nCi/m^2 for cultivated watersheds. Both the cultivated soil and the reservoir sediments serve as a sink or a trap for cesium being washed out of the watershed soils.

Toonkel, L.E. (1980). *Environmental Measurements Laboratory: Environmental quarterly, October 1,* 1980. EML-381. Appendix. US Department of Energy, New York, NY.

- Record keeping of air concentrations for ^{137}Cs begin in 1963 and ended in 1979; peak concentrations are noted as 149 femtocuries/m^3 in April of 1963 at Miami, Florida, and gradually drop to 10 femtocuries/m^3 or less in the spring of 1966; no data is available between May, 1966, and June, 1968, after which time low concentrations of ^{137}Cs are consistently observed until November of 1979. Most concentrations in this latter time frame are less than 5 femtocuries/m^3 and in many cases are below 1 femtocurie/m^3.

- EML publications contain surprisingly little data about cumulative ground deposition of the most significant weapons testing-derived radionuclides and almost no data whatsoever pertaining to media-specific concentrations of these nuclides in the biotic environment. A comparison of the EML publications and the EPA's Environmental Radiation Data Reports with the publications of the Riso National Laboratory graphically illustrate that, while the United States is in the vanguard of creating radioactive contamination from anthropogenic source points, it is still in the late Stone Age when it comes to monitoring these effluents.

Camplin, W.C. and Steele, A.K. (1991). Radiocaesium in the seas of northern Europe: 1980-84. *Fish. Res. Data Rep.,* MAFF Direct. Fish. Res., Lowestoft. 25. pg. 1-174.

National Research Council of Canada. (1983). *Radioactivity in the Canadian aquatic environment*. NRCC Report no. 19250.

Holm, E., Persson, B.R.R., Hallstadius, L., Aarkrog, A. and Dahlgaard, H. (1983). Radiocesium and transuranium elements in the Greenland and Barents Seas. *Oceanol. Acta.* 6(4). pg. 457-62.

DATE	LOCATION	MEDIA	NUCLIDE	ACTIVITY
June-Oct. 1980	Barents Sea	Sediment	^{137}Cs	18.1 Bq/kg
June-Oct. 1980	Barents Sea	Sediment	^{239}Pu	1.10 Bq/kg

Davis, R.B., Hess, T.C., Norton, S.A., Hanson, D.W., Hoagland, K.D. and Anderson, D.S. (1984). 137Cs and ^{210}Pb dating of sediments from soft-water lakes in New England (U.S.A.) and Scandinavia, a failure of ^{137}Cs dating. *Chemical Geology*. 44. pg. 151-185.

DATE	LOCATION	MEDIA	NUCLIDE	ACTIVITY
1978	Norway / pond	Sediment	^{137}Cs	39,900 pCi/kg
1979	Maine / pond	Sediment	^{137}Cs	21,400 pCi/kg
1979	Norway / pond	Sediment	^{137}Cs	32,100 pCi/kg

- Natural processes also mixed ^{137}Cs with pre-fallout sediments.

Casso, S.A. and Livingston, H.D. (1984). *Radiocesium and other nuclides in the Norwegian-Greenland seas*. WHOI-84- 40. Woods Hole Oceanographic Institute, Woods Hole, MA.

DATE	LOCATION	MEDIA	NUCLIDE	ACTIVITY
1981	Norwegian Sea	Seawater	^{137}Cs	570 d.p.m./100 kg

- Median values were less than 100 d.p.m./100 kg with the greatest concentration ratios of radiocesium near or at the ocean surface.

Baxter, A.J. and Camplin, W.C. (1993). Radiocaesium in the seas of northern Europe: 1985-89. *Fish. Res. Data Rep*. MAFF Direct. Fish. Res., Lowestoft. 32. pg. 1-179.

Finnish Centre for Radiation and Nuclear Safety. (1986). *Studies on environmental radioactivity in Finland in 1986: Annual Report*. Report No. STUK-A55. Finnish Centre for Radiation and Nuclear Safety, Helsinki, Finland.

DATE	LOCATION	MEDIA	NUCLIDE	ACTIVITY
Dec.-April 1986	Finland	Air concentration	^{131}I	4.3 µBq/m^3
Dec.-April 1986	Finland	Air concentration	^{134}Cs	0.25 µBq/m^3
Dec.-April 1986	Finland	Air concentration	^{137}Cs	1.25 µBq/m^3
Dec.-April 1986	Finland	Milk	^{137}Cs	0.23 Bq/l
Dec.-April 1986	Finland	Beef	^{137}Cs	0.9 Bq/kg
1980-1985	Finland	Pike	^{137}Cs	42 Bq/kg mean concentration

- This report contains detailed information on pre-Chernobyl dietary intake of radiocesium in many media as well as extensive data on the body burdens of radiocesium in reindeer herders and fisher folk.
- The air concentrations listed above are very significant for two reasons: they give expression to the very sophisticated radiological surveillance technology in place in Finland prior to the Chernobyl accident whereby very small amounts of air

concentrations of radioactive contamination can be measured by air sampling equipment which is far superior in its sensitivity to any equipment in the United States available a decade later. The antiquated equipment still in use in the United States has a limit of detection (LOD) two orders of magnitude higher than the Finnish air filters which routinely measure air contamination in millionths of a becquerel per cubic meter of air.

- During the passage of the Chernobyl accident plume, air concentrations of dozens of volatile and/or vaporized radionuclides reached the millions, tens of millions and hundreds of millions of $\mu Bq/m^3$, not only in Finland, but in far-field locations such as Canada.

- These seemingly insignificant pre-Chernobyl air concentrations provide an important baseline for evaluating the significance of the Chernobyl plume, which would soon pass over Finland, and airborne contamination from the Fukushima Daiichi accident.

Hunt, G.J. (1986). Radioactivity in surface and coastal waters of the British Isles, 1985. *Aquatic Environment Monitoring Report, No. 14*. Ministry of Agriculture, Fisheries and Food, Directorate of Fisheries Research, Lowestoft, Great Britain.

Scottish Development Department. (1987). *Statistical Bulletin, Number 1(E), 1987*. Edinburgh: Government Statistical Service.

- Annual mean concentrations of cesium-137 in milk 1981-1985: all readings at or below 0.2 Bq/l.
- Annual mean concentrations of cesium-137 in salmon 1981 2.6 (Bq/kg wet weight); 1982 4.1; 1983 1.3; 1984 1.5; 1985 3.1.
- This publication contains extremely detailed data about pre-Chernobyl activity levels in many media; much of the data was collected in the vicinity of Scottish nuclear power stations, naval installations, and other source points of radioactivity and constitutes a comprehensive baseline of environmental radioactivity levels prior to the Chernobyl accident. This report does not differentiate activity from weapons test fallout from local source point activity; at this point in time, background artificial radiation levels have many origins.

Mitchell, P.I., Vidal-Quadras, A., Font, J.L. and Oliva, M. (1988). Gamma radioactivity in the Iberian marine environment closest to the NEA dumping site. *J. Environ. Radioactivity*. 6. pg. 77-89.

DATE	LOCATION	MEDIA	NUCLIDE	ACTIVITY
1984	Iberian Coastline	Fish	^{137}Cs	Range 1.1-6.2 Bq/kg

Cunningham, J.D., O'Grady, J. and Rush, T. (1988). *Radioactivity monitoring of the Irish marine environment, 1985-86*. Nuclear Energy Board, Dublin.

DATE	LOCATION	MEDIA	NUCLIDE	ACTIVITY
1985-86	Irish Sea	Seaweed	^{137}Cs	0.5 Bq/kg pre-Chernobyl mean
1985	Irish Sea	Whiting (fish)	^{137}Cs	pv 91 Bq/kg pre-Chernobyl

- Reflects Sellafield contamination. Mean values of all fish and shellfish 1985: 8.4 - 35.2 Bq/kg.

Potter, C.M., Brisbin, I.L., McDowell, S.G. and Whicker, F.W. (1989). Distribution of ^{137}Cs in the American Coot (Fulica *americana*). *J. Environ. Radioactivity*. 9. pg. 105-115.

DATE	LOCATION	MEDIA	NUCLIDE	ACTIVITY
Dec. 1986	S. Carolina: Savanna River Plant	American Coot-skeletal muscle	^{137}Cs	3,950 Bq/kg
Dec. 1986	S. Carolina: Savanna River Plant	Am. Coot - GI contents	^{137}Cs	3,940 Bq/kg

- These pond "B" peak values are 26 times higher than the relatively uncontaminated north arm of the nearby Par Pond.

Ancient History: Fallout Contamination in Milk

In 1976, a Chinese atmospheric weapons test produced a significant pulse of radiostrontium that was picked up by the US EPA/ERAMS (Environmental Protection Agency/Environmental Radiation Ambient Monitoring System). Peak values of 41 becquerels per liter (1,120 picocuries) were recorded in raw milk from dairies in

Amherst and Montague, MA (10/10/76). Mixed milk from a variety of sources being sold in Boston area grocery stores had radiostrontium contamination levels of less than 20% of the raw milk values from western MA. The passing plume of weapons-derived contamination had its maximum impact on October 10-11. Within 15 days, contamination levels at the dairies had dropped to 8 Bq/liter and continued to fall after that. At the time of the plume passage, MA State agencies ordered dairies to switch to stored feed to allow time for the dissipation of the radiostrontium. Needless to say, this information was not made public until later publication (Simpson 1980).

Plutonium and Americium Baseline

All data are peak values unless noted as mean or median values.

Romney, E.M., Mork, and Larson, K.H. (1970). Persistence of plutonium in soil, plants and small mammals. *Health Physics*. 19. pg. 487-491.

DATE	LOCATION	MEDIA	NUCLIDE	ACTIVITY
1955-66	Nevada Test Site	Soil	^{239}Pu	208,333.00 Bq/kg
1955-66	Nevada Test Site	Clover	^{239}Pu	375 Bq/kg
1955-66	Nevada Test Site	Jackrabbit Bone	^{239}Pu	11,900 Bq/kg

- Reflects extensive contamination from local nuclear weapons tests.

Bowen, V.T., Wong, K.M. and Noshkin, V.E. (1971). ^{239}Plutonium in and over the Atlantic Ocean. *Journal of Marine Research*. 29. pg. 1-10.

DATE	LOCATION	MEDIA	NUCLIDE	ACTIVITY
1969	Atlantic Ocean 37^{01} N 00^{03} E 37^{01} N 00^{03} E	Seawater	^{239}Pu	7.7 d.p.m./1000 kg

- A classic in radiological surveillance of the abiotic environment.

Wong, K.M., Burke, J.C. and Bowen, V.T. (1971). Plutonium concentration in organisms of the Atlantic Ocean. *Proceedings, Fifth Annual Health Physics Society*

Midyear Topical Symposium: Health Physics Aspects of Nuclear Facility Siting. 2. pg. 529-539.

Noshkin, V.E. (1972). Ecological aspects of plutonium dissemination in aquatic environments. *Health Physics.* 22. pg. 537-549.

DATE	LOCATION	MEDIA	NUCLIDE	ACTIVITY
1970	Cape Cod Bay	Sediment	^{239}Pu	3 samples: range 20-60 pCi/kg

Hardy, E.P., Krey, P.W. and Volchok, H.L. (February 16, 1973). Global inventory and distribution of fallout plutonium. *Nature.* 241. pg. 444-445.

- This is a landmark document in the literature of radiological surveillance and is not only one of the first summaries of weapons testing ^{239}Pu fallout (worldwide inventory: 325 ±36 kCi as of January 1971, pg. 445) but also includes the inventory of ^{238}Pu dispersed by the April 21, 1964 SNAP-9A navigational satellite failure, which released 17,000 curies of radioactivity into the biosphere over the Southern Indian Ocean.
- A relatively uniform pattern of fallout plutonium distribution characterizes soil deposition inventories of ^{239}Pu; the highest accumulations are located in the northern hemisphere in the latitudinal band of 50 to 40 degrees, with an average deposition of 2.2 mCi per km^2, dropping off to 0.10+ mCi per km in the arctic regions (90-80 degrees) and 0.13 to 0.24 in the tropical regions of the northern hemisphere. Fallout deposition levels in the southern atmosphere are approximately an order of magnitude less.
- Hardy's data thus indicate that 5.2 metric tons of plutonium-239/240 has been globally distributed between the beginning of weapons testing and January 1970.

United States Environmental Protection Agency. (1974). *Radiation data and reports, volume 15, number 23.* US EPA, Washington, D.C.

DATE	LOCATION	MEDIA	NUCLIDE	ACTIVITY
Jan.-Dec. 1972	Mound Laboratory, OH	Avg. air concentration	^{238}Pu	400 aCi/m3
Jan.-Dec. 1972	Mound Laboratory, OH	Pond silt	^{238}Pu	0.582 pCi/kg pv

- ^{238}Pu emissions are frequently associated with reprocessing spent fuel.

Wrenn, M.E. (1974). Environmental levels of plutonium and transplutonium elements, *WASI, 1359*, 89.

Bennet, B.G. (1976). Fallout *in ^{239,240}Pu in Diet*. Report No. HASL-306. Washington, D.C.

DATE	LOCATION	MEDIA	NUCLIDE	ACTIVITY
1972-74	New York City	Shellfish	^{239}Pu	0.012 pCi/kg
1972-74	New York City	Milk	^{239}Pu	0.00051 pCi/kg

- Mean concentrations in annual dietary intake. Most other food products had a range of mean concentrations from 0.001 to 0.01 pCi/kg.
- Uptake of ingested plutonium from the gastro-intestinal tract is estimated at 3×10^{-5} to 3×10^{-6} as a percent of ingested plutonium. This illustrates that plutonium oxide is not readily absorbed when ingested in this chemical form.

Bennet, B.G. (1976). Transuranic element pathways to man. *Transuranium nuclides in the environment*. Vienna: International Atomic Energy Agency.

- The cumulative deposit of fallout of ^{239,240}Pu, which was first measured in 1954 at 0.07 mCi/km^2 had reached 2.68 mCi/km^2, by 1968.

Bowen, V.T., Livingston, H.D. and Burke, J.C. (1976). Distribution of transuranium nuclides in sediment and biota of the North Atlantic Ocean. IN: *Transuranium nuclides in the environment*. Vienna: International Atomic Energy Agency.

DATE	LOCATION	MEDIA	NUCLIDE	ACTIVITY
1968	Buzzards Bay, MA	Sediment Cone	^{239}Pu	4.2 mCi/km^2 integrated mean sediment load
1974	N. Atlantic Ocean	Menhaden guts, pooled sample	^{239}Pu	4.49 d.p.m./kg wet weight

128

Livingston, H.D. and Bowen, V.T. (1976). Americium in the marine environment: Relationships to plutonium. *Environmental toxicity of aquatic radionuclides: models and mechanisms*. Woods Hole, MA: Woods Hole Oceanographic Institute.

- All production of ^{241}Am comes solely from decay of its parent, ^{241}Pu.
- Up until this time, most ^{241}Pu came from the 1961-62 US and USSR tests.
- Both plutonium and americium are rapidly transferred from the water to the sediment with near-shore sediment inventories showing some transfer to deep water sediments.

Bennett, B.G. (1978). *Environmental Measurements Laboratory: Environmental aspects of americium, December 1978*. EML-348. Department of Energy, New York, NY.

- Extensive data on air concentrations and ground deposition of ^{241}Am from 1950-1984 as well as its comparison with plutonium deposition levels.
- The majority of ^{241}Am contamination in the environment derives from its growth out of the decay of ^{241}Pu. Table 4, page 58 indicates that 4,500,000 curies of ^{241}Pu were produced by weapons tests through 1978 and will decay into 150,000 curies of ^{241}Am.
- Table 7 on page 67 provides a cumulative deposition record for both plutonium and americium. For the year 2000, cumulative deposition of ^{241}Am in the New York region is estimated at .80 mCi/km^2 (29.4 Bq/m^2); the cumulative deposit of ^{239,240}Pu in the New York region reached 2.2 mCi/km^2 in 1979 (81.6 Bq/m^2, slightly higher than levels recorded by Aarkrog and other researchers for mid-latitude locations).

Goldberg, E.D., Bowen, V.T., Farrington, J.W., Harvey, G., Martin, J.H. (1978). The mussel watch. *Environmental Conservation*. 5. pg. 101-125.

DATE	LOCATION	MEDIA	NUCLIDE	ACTIVITY
1976	Bodega Bay, CA	Mussel	^{239,240}Pu	4.4 d.p.m./kg
1976	Fallaron IsCA	Mussel	^{239,240}Pu	7.5 d.p.m./kg
1976	Fallaron Is., CA	Mussel	^{241}Am	19.6 d.p.m./kg

Cutter, G.A., Bruland, K.W. and Risebrough, R.W. (1979). Deposition and accumulation of plutonium isotopes in Antarctica. *Nature*. 279. pg. 628-9.

DATE	LOCATION	MEDIA	NUCLIDE	ACTIVITY
Jan. 1977	Antarctica	Snow Meltwater	^{239}Pu	5.62 d.p.m./100 l

Livingston, H.D. and Bowen, V.T. (1979). Pu and ^{137}Cs in coastal sediments. *Earth and Planetary Science Letters*. 543. pg. 29-45.

DATE	LOCATION	MEDIA	NUCLIDE	ACTIVITY
1973	Buzzards Bay	Dry sediment	^{239}Pu	160 d.p.m./kg

- The range of surface sediment fallout plutonium in eight Buzzards Bay samples was 117-160 d.p.m./kg, with plutonium concentrations falling off sharply below 10 cm in depth.

Holm, E., Ballestra, S., Fukai, R. and Beasley, T.M. (1980). Particulate plutonium and americium in Mediterranean surface waters. *Oceanolgica Acta*. 3(2). pg. 157-160.

- ^{241}Am / ^{239}Pu activity ratios are enriched in biogenic particulates by a factor of three (9:1) versus the activity ratio in surface seawater (3:1).

Santschi, P.H., Li, Y.H., Bell, J.J., Trier, R.M. and Kawtaluk, K. (1980). Pu in coastal marine environments. *Earth and Planetary Science Letters*. 51. pg. 248-265.

- The Pu inventories in the sediment are subject to resuspension and transfer back into the water column as well as extensive bioturbation and mixing with underlying sediments.

Benniger, L.K. and Krishnaswami, S. (1981). Sedimentary processes in the inner New York Bight: Evidence from excess ^{210}Pb and ^{239,240}Pu. *Earth and Planetary Science Letters*. 53. pg. 158-174.

DATE	LOCATION	MEDIA	NUCLIDE	ACTIVITY
Oct. 1975	NY Bight	Sediment	^{239}Pu	294 d.p.m./kg (5 Bq/kg)

- "The measured inventories are greatly in excess of quantities supportable from atmospheric deposition alone; most of the excess... ^{239,240}Pu stored in sediments at these stations was introduced by lateral transport. Whether these inventories record mainly a terrigenous input or the focusing of atmospheric deposition over coastal water cannot be determined from these data." (p. 171).

Beasley, T.M., Carpenter, R. and Jennings, C.D. (1982). Plutonium, ^{241}Am and ^{137}Cs ratios, inventories and vertical profiles in Washington and Oregon continental shelf sediments. *Geochimica et Cosmochimica Acta*. 46. pg. 1931- 1946.

DATE	LOCATION	MEDIA	NUCLIDE	ACTIVITY
1975-76	Washington/Oregon Continental Shelf	Sediment	^{239}Pu	311 d.p.m./kg
1975-76	W/O Cont. Shelf	Sediment	^{241}Am	86 d.p.m./kg
1975-76	W/O Cont. Shelf	Sediment	^{137}Cs	3,176 d.p.m./kg

- Reflects contamination from the Hanford, WA reservation which has been deposited on the Continental Shelf via the Columbia River.

Hotzl, H., Rosner, G. and Winkler, R. (1983). *Radionuclide concentrations in ground level air and precipitation in South Germany from 1976 to 1982*. GSF- Report S-956. Gesellschaft fur Strahlen-und Umweltforschung MBH, Munchen.

DATE	LOCATION	MEDIA	NUCLIDE	ACTIVITY
1976-82	South Germany	Annual Deposition	^{239}Pu	Range: 4.3 to 14.2 E-3 nCi/m^2
1976-82	South Germany	Annual Deposition	^{137}Cs	Range: 180 to 1100 pCi/m^2

- Peak value for both ^{238}Pu and 137Cs deposition was in 1978.

Palmieri, J., Livingston, H. and Farrington, J.W. (1984). *U.S. "mussel watch" program: Transuranic element data from Woods Hole Oceanographic Institution 1976-1983*. Woods Hole Oceanog. Inst. Tech. Rept. WHOI-84-28. Woods Hole Oceanographic Institute, Woods Hole, MA.

DATE	LOCATION	MEDIA	NUCLIDE	ACTIVITY
1979	Bodega Bay, CA	Mussel	^{239}Pu	7.46 d.p.m./kg dry weight

- Data comparison shows no significant differences in ^{239,240}Pu content for West Coast and East Coast mussels except central California coastal stations with had slightly elevated levels due to upwelling.

Jennings, C.D., Delfanti, R. and Papucci, C. (1985). The distributions and inventory of fallout plutonium in sediments of the Ligurian Sea near La Spezia, Italy. *J. Environ. Radioactivity.* 2. pg. 293-310.

- "The Integrated inventory of ^{239,240}Pu in a sediment core is calculated to be 3-5 mCi/km^2, nearly twice the average input from fallout at these latitudes, apparently because Pu is removed from seawater by particle scavenging." (p. 293).

Livingston, H.D. (1985). Anthropogenic radiotracer evolution in the Central Greenland Sea. *Rit Fiskideildar.* 9. pg. 43-54.

DATE	LOCATION	MEDIA	NUCLIDE	ACTIVITY
1979	Greenland	Total Water Column Inventory	^{239,240}Pu	0.00190 mCi/m^2

Livingston, H.D., Bowen, V.T., Casso, S.A., Volchok, H.L., Noshkin, V.E., Wong, K.M. and Beasley, T.M. (1985) *Fallout nuclides in Atlantic and Pacific Water Columns: GEOSECS data.* WHOI-85-19. Woods Hole, MA: Woods Hole Oceanographic Institute.

DATE	LOCATION	MEDIA	NUCLIDE	ACTIVITY
1972	North Atlantic	Ocean Water Column	^{239}Pu	3.06 mCi/km^2
1973	Pacific	Ocean Water Column	^{239}Pu	4.04 mCi/km^2

- A very detailed survey of plutonium deposition in ocean water.

Hallastadius, L, Aarkrog, A., Dahlgaard, H., Holm, E., Boelskifte, S., Duniec, S. and Persson, B. (1986). Plutonium and americium in Arctic waters, the North Sea and Scottish and Irish coastal zones. *J. Environ. Radioactivity*. 4. pg. 11- 30.

DATE	LOCATION	MEDIA	NUCLIDE	ACTIVITY
1982	Irish Sea	Fucus vesiculosis	^{239}Pu	270 Bq/kg near Sellafield
1982	Irish Sea	Fucus vesiculosis	^{239}Pu	0.6 to 2.0 Bq/kg uncontaminated areas
1983	Irish Sea	Fucus vesiculosis	^{241}Am	84 Bq/kg near Sellafield

American Nuclear Society. (January 1999). Antinuclear story continues, with moves toward phaseout. *Nuclear News*. pg. 47.

- "GSF has pointed out that the whole of Germany -- and, for that matter, most of the world -- is covered with some 50 becquerels per square meter (Bq/m^2) of Pu-239 and Pu-242 from weapons tests. When atmospheric tests ceased in 1980, there was an additional 300 Bq/m^2 of Pu-241 in fallout, and even today this has decayed only to between 20 and 30 Bq/m^2." (pg. 47).

Albright, D. and Barbour, L. (May 1999). *Separated inventories of civil plutonium continue to grow*. Institute for Science and International Security, Washington, DC.

Lobsenz, George. (May 26, 1999). Civilian plutonium stockpile growing. *The Energy Daily*.

- "Civilian stockpiles of plutonium worldwide increased 65 metric tons in 1998 to 1,115 metric tons-including an estimated 13 metric-ton rise in the amount of proliferation-sensitive separated plutonium, according to a new report."
- "At the end of 1998, there were 920 metric tons of plutonium in spent nuclear fuel, compared to 868 metric tons at the end of 1997, according to the ISIS [Institute for Science and International Security] report."
- "The report said separated plutonium stocks reached 195 metric tons at the end of 1998, compared to 182 metric tons at the end of 1997."
- "Most of the separated plutonium is in France and Britain, both of which have large state-backed commercial companies that reprocess spent reactor fuel to recover plutonium and then refabricate it into MOX fuel for re-use in civilian

reactors. Major customers have included Japanese, German and other European reactor operators."

- "In total, the civilian stockpile of 1,115 metric tons is nearly three times larger than the roughly 250 metric tons held by the military worldwide, ISIS said. However, it said 90 percent of the military stockpile was in separated form."

Kudo, A., Ed. (2001). *Radioactivity in the environment: Plutonium in the environment: Edited proceedings of the Second Invited International Symposium.* Volume 1. Elsevier, NY, NY.

Radioiodine, Strontium, and Other Nuclides

All data are peak values unless noted as mean or median values.

Book, S.A., Garner, R.J., Soldat, J.K. and Bustad, L.K. (1977). Thyroidal burdens of ^{129}I from various dietary sources. *Health Physics.* 32. pg. 143-148.

- This study uses modeled ^{129}I intake rather than actual intake and is one of the first studies to explore the potential pathways of the very long-lived ^{129}I which will become an important source of exposure thousands of years in the future.

Schink, D.R., Santschi, P.H., Corapcioglu, O., Sharma, P. and Fehn, U. (October 1995). ^{129}I in Gulf of Mexico waters. *Earth and Planetary Science Letters.* 135(1-4). pg. 131-138.

Simpson, R.E., Shuman, F.G.D., Baratta, E.J. and Tanner, J.T. (1981). Projected dose commitment from fallout contamination in milk resulting from the 1976 Chinese atmospheric nuclear weapons test. *Health Physics.* 40. pg. 741-744.

Date	Location	Media	Nuclide	Activity
Oct. 1976	Amherst, MA	Milk	^{131}I	1,150 pCi/l (42.2 Bq/l)

- Reflects a very strong pulse of radioiodine in the cattle-forage pathway after a particularly dirty Chinese test explosion.
- These contamination levels reached range III of the 1961 FRC protective action guidelines.

- For more comments including a scan of this article, see our special the online RADNET appendix: Contaminated milk: A paradigm (http://www.davistownmuseum.org/cbm/RadxMilk.html).

Robens, E. and Aumann, D.C. (1988). Iodine-129 in the environment of a nuclear fuel reprocessing plant: I. ^{129}I and ^{127}I contents of soils, food crops and animal products. *J. Environ. Radioactivity.* 7. pg. 159-175.

- Between 1971-1985 this fuel reprocessing facility released about 7.3 x 10^9 Bq of ^{129}I to the environment.
- This article contains the first detailed analysis of ^{129}I in food crops, which generally range in 10^{-6} Bq/kg or lower. The peak value of ^{129}I concentration in leeks was noted as 1.1x10^{-6} Bq/g; tobacco was anomalous at 15x10^{-6} Bq/g.
- This volume begins an important series of reports on ^{129}I pathways.
- These are pre-Chernobyl peak concentrations.

United States Environmental Protection Agency. (1986). *Environmental Radiation Data: Report 44-45: October 1985-March 1986.* Report No. EPA520/5-86-018. US EPA, Washington D.C.

Date	Location	Media	Nuclide	Activity
12/18/85	Minot, ND	Milk	^{131}I	34 pCi/l pv
12/18/85	Memphis, TN	Milk	^{131}I	13 pCi/l

- Most pre-Chernobyl ^{131}I levels are recorded as below 10 pCi/l, which the EPA lists as their minimum detectable level for both radiocesium and radioiodine.

Aarkrog, A., et. al. (July, 1991). *Environmental radioactivity in Denmark in 1988 and 1989.* Riso-R-570. Riso National Laboratory, Roskilde, Denmark.

Date	Location	Media	Nuclide	Activity
1988-89	Danish waters	Fucus vesiculosus Fucus serratous	^{99}Tc	Range: 6.3 - 200 Becquerels/kg dry

- One of the first appearances of ^{99}Tc in radiological surveillance reports is the data in these Riso publications. Of 103 samples of Fucus vesiculosus and Fucus serratous taken in Danish waters in 1988 and 1989, 30 samples showed ^{99}Tc contamination with a range from 6.3 Becquerels/kg dry up to 200 Becquerels/kg dry (pg. 151-153).

- These low-levels of contamination may serve as a reference point for higher levels of Sellafield-derived technetium now being monitored in the Irish Sea by the Radiological Protection Institute of Ireland.

Aarkrog, A. et al. (February 1995). *Environmental radioactivity in Denmark in 1992 and 1993*. Riso-R-756(EN). Riso National Laboratory, Roskilde, Denmark.

Camplin, W.C., Tipple, J.R., Doddington, T.C., Thurston, L.M. and Hiller, R. (1990). A survey of tritium in sea water in Tees Bay, July 1986. *Fish. Res. Data Rep.* MAFF Direct. Fish. Res., Lowestoft. 23. pg. 1-23.

Collins, C. and Otlet, R. (1995). *Carbon-14 levels in UK foodstuffs*. MAFF Project B1408. Ministry of Agriculture, Fisheries and Food, London.

Cook, G.T., Begg, F.H., Naysmith, P., Scott, E.M. and McCartney, M. (October 1996). Anthropogenic ^{14}C marine geochemistry in the vicinity of a nuclear fuel reprocessing plant. *Oceanographic Literature Review*. 43(10). pg. 1060-1061.

Hisamatsu, S., Takzawa, Y. and Abe, T. (1987). Fallout ^{3}H ingestion in Akita, Japan. *Health Physics*. 53(3). pg. 287-293.

Date	Location	Media	Nuclide	Activity
1985	Akita, Japan	Daily dietary intake	[3]H	4.1 Bq/day

- Tissue bound tritium was 15% of the total, the same as reported for New York City, but lower than that reported for Italian diets.
- This is a typical pre-Chernobyl intake level for a relatively uncontaminated area.

Peirson, D.H. and Cambray, R.S. (1965). Fission product fallout from nuclear explosions of 1961 and 1962. *Nature*. 205. pg. 433-440.

Roland, R.F., Chmielewicz, Z.F., Weiner, B.A., Gross, A.M., Boening, O.P., Luck, J.V. and Bardos, T.J. (1960). Zinc-65 and Chromium-51 in foods and people. *Science*. 132(23). pg. 1895-1897.

Date	Location	Media	Nuclide	Activity
1959-60	Columbia River, WA	Irrigated soil	^{51}Cr	38.4 pCi/g (38,400 pCi/kg)
1959-60	Columbia River, WA	Irrigated alfalfa	^{65}Zn	89 pCi/g (8,900 pCi/kg)
1959-60	Columbia River, WA	Irrigated pasture grass	^{65}Zn	36 pCi/g (36,000 pCi/kg)
1959-60	Columbia River, WA	Milk	^{65}Zn	1.9 pCi/g (1,900 pCi/kg)

- Reflects extensive contamination of the Columbia River by Hanford Reservation effluents.

Smith, J.N., Ellis, K.M., Naes, K., Dahle, S. and Matishov, D. (May 1996). Sedimentation and mixing rates of radionuclides in Barents Sea sediments off Novaya Zemlya. *Oceanographic Literature Review*. 43(5). pg. 514.

Strand, P. (June 1995). Survey of artificial radionuclides in the Barents Sea and the Kara Sea. *Oceanographic Literature Review*. 42(6). pg. 500.

Valette-Silver, N.J. and Lauenstein, G.G. (May 1995). Radionuclide concentrations in bivalves collected along the coastal United States. *Marine Pollution Bulletin*. 30(5). pg. 320-331.

Naturally Occurring Radionuclides (NOR)

Baseline data pertaining to naturally occurring radionuclides (NOR) provide an important point of comparison for evaluating the significance of anthropogenic contamination from human activities such as weapons production and the generation of nuclear electricity. In many cases naturally occurring radionuclides are redistributed by human activities (e.g. the mining, milling and processing of uranium) and therefore are a distinct classification of anthropogenic radioactivity.

Camplin, W.C., Baxter, A.J. and Round, G.D. (1996). The radiological impact of disposals of natural radionuclides from a phosphate plant in the United Kingdom. *Environ. Int.* 22 Suppl. 1. pg. 5259-5270.

Hunt, G.J. and Allington, D.J. (1993). Absorption of environmental polonium-210 by the human gut. *J. Radiol. Prot.* 13(2). pg. 119-126.

Pentreath, R.J., Camplin, W.C. and Allington, D.J. (1989). Individual and collective dose rates from naturally-occurring radionuclides in seafood. In: *Radiation Protection*

Theory and Practice. Proc. 4th Int. Symp. Soc. Radiol. Prot. Malvern, 4-9 June 1989. Goldfinch, E.P. (Ed.), Institute of Physics, Bristol and New York. pg. 297-300.

Rollo, S.F.N., Camplin, W.C., Allington, D.J. and Young, A.K. (1992). Natural radionuclides in the UK marine environment. In: Proceedings of the Fifth International Symposium on Natural Radiation Environment, Saltzburg, September 22-28, 1991. *Radiat. Prot. Dosim.* 45(1/4). pg. 203-210.

US Radiation Data: Dietary Intake

Anthropogenic Radioactivity in Domestic Foods

As a result of the many nuclear weapons detonations, which began in 1950 and reached their peak in 1962, and which included large thermonuclear hydrogen bomb tests, worldwide contamination from stratospheric fallout became an object of widespread concern and resulted in a variety of studies of the dietary intake of key fallout nuclides such as strontium-90 and cesium-137. This section summarizes a number of dietary intake studies, including *Radiological Health and Data Reports*, which became the US *Radiation Data Reports* in the early 1970s. The second part of this section cites selected data pertaining to body burdens of radiocesium and other isotopes. The 1994 summary of radionuclides in domestic and imported foods, 1987-1992, annotated below (see Cunningham, 1994) contained the first announcement that 40% of targeted imported food samples were contaminated with Chernobyl-derived radiocesium.

The public health service initiated its institutional diet sampling program in 1961. The Atomic Energy Commission had already issued summaries of environmental radioactivity data for twenty-two AEC installations in *Radiological Health Data*, beginning in November, 1960. This publication was a response to wide-spread public concern about elevated levels of weapons-test-derived radioactive contamination of the food supply. The high levels of contamination documented between 1957 and 1964 in the following reports were not reached again until the advent of Chernobyl-derived contamination, the effect of which was felt primarily in foreign food supplies and had a minimal impact on domestic food production in the United States. Monitoring was discontinued in 1969 with the suspension of above ground weapons testing but was begun again in 1973 due to concerns about contamination from other sources. The publication of *Radiation Data Reports* was discontinued in 1976, with a more cursory and less detailed survey continued in the EPA publication *Environmental Radiation Data*.

United States Atomic Energy Commission. *Radiological Health Data*. (October 1963). Volume number unavailable. pg. 562.

Date	Location	Media	Nuclide	Activity
1957	Unknown	Milk	^{131}I	990 pCi/l yearly mean
1957	Unknown	Milk	^{140}Ba	530 pCi/l yearly mean
1959	St. Louis	Milk	^{137}Cs	75 pCi/l yearly mean
1959	Atlanta	Milk	^{137}Cs	68 pCi/l yearly mean
1961	Unknown	Milk	^{137}Cs	25 pCi/l yearly mean
Nov-Dec 1961	St. Louis	Milk	^{137}Cs	+80 pCi/l yearly mean
Nov-Dec 1961	St. Louis	Milk	^{90}Sr	33.3 pCi/l yearly mean
1962	Unknown	Raw milk	^{137}Cs	180 pCi/l yearly mean
1962	Unknown	Raw milk	^{90}Sr	45.6 pCi/l yearly mean
1962	Unknown	Raw milk	89Sr	335 pCi/l yearly mean
1962	Unknown	Raw milk	^{140}Ba	165 pCi/l yearly mean
1962	Unknown	Raw milk	^{131}I	535 pCi/l yearly mean

- Nuclear weapons testing was briefly discontinued in 1959-1960, with only three tests compared to 172 tests the previous four years. Tests were resumed with much larger weapons in 1961-1962, when the test ban treaty was implemented.
- Maximum fallout levels from the 1961-62 tests were reached throughout the hemisphere between 1962 and early 1965, as illustrated by both the Public Health Service diet studies and the Riso National Laboratory (Denmark) Cumulative Fallout Summary.

United States Department of Agriculture. (December 1963). Strontium-90 and cesium-137 content of beef and beef products 1960-62. *Radiological Health Data.* pg. 612.

Date	Location	Media	Nuclide	Activity
1960	Tacoma, WA	Beef rib meat	^{137}Cs	33 pCi/kg pv
1960	Atlanta, GA	Beef rib meat	^{137}Cs	32.4 pCi/kg pv

- For 76 samples, the average concentration was usually well below 10 pCi/kg. For 83 strontium-90 samples, the highest average was 16 pCi/kg, also in Tacoma, but most averages were below 10 pCi/kg.

Division of Radiological Health, Public Health Service. (September 1963). Radionuclides in institutional diet samples. *Radiological Health Data.* pg. 441-454.

Date	Location	Media	Nuclide	Activity
Jan-March 1963	US	Institutional average daily intake	^{137}Cs	peak range: 2,320 pCi/day
Jan-March 1963	US	Institutional average daily intake	^{137}Cs	mean range: 249 pCi/day

- This survey represents averages of the daily intake of radiocesium at a large number of institutions throughout the United States. The USFDA used these averages throughout the 1960s for assessing the impact of weapons test fallout. (p. 454).
- The peak average of 2,320 pCi/day (85.9 Bq/day) equals an annual dietary intake of 31,353 Bq; the average is almost an order of magnitude less.
- This peak value represents the maximum level of weapons testing contamination of the diet by this radionuclide (^{137}Cs).

United States Atomic Energy Commission. (Sept. 1963). Radioactivity in pasteurized milk. *Radiological Health Data.* pg. 441-454.

Date	Location	Media	Nuclide	Activity
Aug 1963	Little Rock, Ark.	Milk	^{90}Sr	peak range: 51 pCi/l
Aug 1963	Network average	Milk	^{90}Sr	mean range: 25.9 pCi/l
Aug 1963	Boston, MA	Milk	^{137}Cs	peak range: 380 pCi/l
Aug 1963	Network average	Milk	^{137}Cs	mean range: 150 pCi/l

- Most of the FDA surveys show the highest concentrations of fallout contamination in the eastern part of the United States; these elevated concentrations probably correlate with rainfall activity.

- The FDA reports also show frequently elevated levels of radiocesium in Southeast Florida for reasons which are not explained.

United States Atomic Energy Commission. (March 1964). Institutional daily dietary intake. *Radiological Health Data.*

- Resumption of USSR tests began in early September, 1961, followed several weeks later by the resumption of US testing. The institutional diet surveys show a sharp rise of ^{131}I intake to approximately 60 pCi/day in late September; this level of activity continued erratically (20-60 pCi/day) through the fall of 1963, when this reporting period terminates.
- Radiocesium levels had continued at moderately high levels during the weapons testing interruption and then rose gradually in 1962 to levels exceeding 150 pCi/day (mean range) and higher by late 1963.
- Strontium-89 activity levels followed approximately the same path as ^{131}I; strontium-90 levels followed the same intake pattern as cesium-137. See Figure Two, p.134.

United States Atomic Energy Commission. (1963). *Radiological Health Data.* pg. 591.

Date	Location	Media	Nuclide	Activity
April-June 1963	Boston, MA	Milk	^{137}Cs	380 pCi/l peak monthly average
April-June 1963	Little Rock, Ark.	Milk	^{90}Sr	51 pCi/l peak quarterly average

- During this period of extensive contamination of the food supply, average radiocesium concentrations in milk were running in the 100-250 pCi/l range for most US cities with the highest levels of contamination in the Northeast US Average concentrations of strontium-90 were frequently in the 25-40 pCi range. The highest strontium-89 average was also Little Rock, Ark. at 200 pCi/l.
- The survey of pasteurized milk for Oct. 1963 shows continuing high levels of contamination in milk with a peak quarterly average of 355 pCi/l in Manchester, NH. Average radioactivity concentrations continued at the same high levels as earlier in the year.

United States Atomic Energy Commission. (September 1963). *Radiological Health Data.* pg. 447.

ITEM	CESIUM-137 CONCENTRATIONS	
	TOP OF RANGE (pCi/kg)	AVERAGE (pCi/kg)
Vegetables	5, 540	570
Dairy products	1,300	133
Root vegetables	1,030	79
Fruits	553	79
Grain products	625	115
White potatoes	291	38
Coffee	652	431
Sea food	205	50
Spices	15,700	784
Egg substance	27	12
Tea	25,700	1,290

United States Atomic Energy Commission. (January 1965). *Radiological Health Data.* pg. 33-36.

- The institutional daily dietary intake (based on a composite 7 consecutive day sample in each month) shows radiocesium intake continuing at high levels with peak values up to 440 pCi/day in Columbia, MO in April 1964, continuing high levels in the Northeast and monthly nationwide averages of 165-200 pCi/day in April-June of 1964. Strontium-90 contamination continued with monthly averages near 40 pCi/day.

United States Atomic Energy Commission. (1965). *Radiological Health Data.*

- This report summarizes the average daily ^{137}Cs intake in twenty-three cities in February of 1963 to June of 1964 as follows.

142

February 1963	125 pCi/day
March 1963	158 pCi/day
April 1963	159 pCi/day
May 1963	209 pCi/day
June 1963	243 pCi/day
July 1963	324 pCi/day
August 1963	266 pCi/day
September 1963	220 pCi/day
February 1964	370 pCi/day
April 1964	434 pCi/day
May 1964	374 pCi/day
June 1964	339 pCi/day

- Nuclear detonations in Russia were terminated in 1962, but US test explosions continued during 1963, at least according to Figure Two in the *RHD Report* of March 1964 cited above (Aarkrog 1963, "Source Terms and Inventories of Anthropogenic Radionuclides" lists zero atmospheric nuclear explosions). The delay of over a year for ^{137}Cs to reach peak concentrations in the daily diet intake is significant in that similar delayed peak concentrations of ^{137}Cs will be evident in the post-Chernobyl FDA survey of imported food cited later in this section.
- Nuclear explosion yields (megatonnage) reached a peak in 1961-62. The fifty-six explosions which followed in 1964-74 only yielded about 20% of the energy of the more extensive testing done in the early 1960s. The decline in the megatonnage of the weapons test explosions is matched by a decline in the dietary intake of weapons-test-derived radioactivity throughout the late 1960s to the late 1970s.

US Department of Health, Education and Welfare. (1969). Estimated daily intake of radionuclides in California diets, November-December 1967 and January-September 1968. *Radiological Health, Data and Reports.* 10(5). pg. 208-211.

- Average daily intake of cesium-137 in California in an area upwind from many of the Nevada weapons tests dropped from an average slightly over 80 pCi per day to around 20 pCi per day by 1967 before beginning to arise again in 1968 to 30 pCi per day, probably under the influence of Chinese test explosions.

Magi, A., Snihs, J.O. and Swedjemark, G.A. (1970). Some measurements of radioactivity in Sweden caused by nuclear test explosions. *Radiological Health and Data Reports.* 11. pg. 487-509.

	Nationwide averages: Sweden (pCi/kg/yr)						
Date	Milk	Beef	Pork	Grain	Fish from salt waters	Fish from oligotrophic lakes	Reindeer meat
1962	120	380	190	305	100	2,000	17,000
1963	185	750	580	765	100	2,000	17,000
1964	180	760	845	310	100	5,000	39,000
1965	125	470	445	125	100	5,000	20,000
1966	70	230	235	65	100	5,000	14,000
1967	50	105	85	40	100	4,500	20,000
1968	40	95	85	40	100	4,000	17,000

- "No direct measurements on diet samples have been made." p. 494.

Author unknown. (Dec. 1971). *Radiological Health Data and Reports.* 12(12). pg. 614-30.

Date	Location	Media	Nuclide	Activity
Sept 1970-Aug 1971	SE Florida	12 month average in milk	^{137}Cs	68 pCi/l

- Other than elevated concentrations of ^{137}Cs in milk in all Florida reporting stations (seven), only six out of eighty-two other reporting stations had a twelve month average exceeding 20 pCi/l.

- The rather consistent elevated levels of radiocesium throughout Florida have yet to be explained to this editor, as there is no known connection between the phosphate industry, which causes elevated concentrations of some naturally occurring radionuclides, and cesium-137, a fission product.

Author unknown. (Dec. 1971). Estimated daily intake of radionuclides in Connecticut standard diet: January-December 1970. *Radiological Health Data and Reports.* 12(12) pg. 614-630.

- The daily radionuclide intakes of ^{137}Cs in the Connecticut standard diet between January and December of 1970 had a range of 20-70 pCi/day, with all months except three reporting an average of 40 pCi/day or less.

Author unknown. (April 1973). *Radiation Data and Reports.* 14(4).

- Continued elevated levels of ^{137}Cs are reported in Florida (peak concentrations to 53 pCi/l, 12 month average), with isolated elevated concentrations reported in Portland, ME (23 pCi/l), and in Minnesota (peak concentrations to 39 pCi/l); almost all other stations are reporting 10 pCi/l or less of radiocesium in milk. Strontium-90 concentrations also show the same pattern of decline following the test ban agreement.

Simpson, R.E., Baratta, E.J. and Jelinek, C.F. (1977). Radionuclides in Foods. *Journal of the Association of Official Analytical Chemists.* 60. pg. 1364-1368.

Radionuclides in total diet: summary of FY 74 data (domestic food supply)

Composite	Intake av., kg/day	Sp. act., pCi/kg	Cesium-137 Intake, pCi/day	Sp. act., pCi/kg	Strontium-90 Intake, pCi/day	Potassium ratio, g/kg sample
Dairy products	0.756	-	-	7.1	5.4	1.6
Meat, fish, poultry	0.290	3.1	0.9	1.7	0.5	2.4
Cereal	0.369	-	-	5.3	2.0	1.3
Potatoes	0.204	-	-	4.9	1.0	5.5
Leafy vegetables	0.059	4.7	0.3	9.9	0.6	1.8

Radionuclides in total diet: summary of FY 74 data (domestic food supply)						
			Cesium-137		Strontium-90	
Legumes	0.074	4.3	0.3	10.5	0.8	2.1
Root vegetables	0.034	4.2	0.1	5.4	0.2	1.8
Garden fruits	0.088	5.5	0.5	3.9	0.3	2.0
Fruits	0.217	-	-	2.7	0.6	1.5
Oils, fats	0.052	4.8	0.3	6.8	0.4	1.2
Sugars and adjuncts	0.082	-	-	4.5	0.4	0.6
Beverages, including drinking water	0.697	3.6	2.5	1.1	0.7	0.6
Total	2.922		4.9		12.9	

US Department of Health, Education and Welfare. (1979). *FY 75-FY 76 Radionuclides in Foods*. Food and Drug Administration, Washington, D.C.

- Domestic levels of cesium-137 are reported as non-detectable at all US reporting stations.
- This data was collected just prior to a 1978 Chinese test explosion which resulted in elevated levels of contamination in aquatic vegetation in the vicinity of Maine Yankee.
- The unreliability of the FDA data for this time frame is illustrated by the report of very elevated levels of radioiodine in milk in Massachusetts following a 1976 Chinese test explosion; this FDA report indicates radioiodine was not detectable in all dietary composites including dairy in 1976. (See Simpson, R.E., Shuman, F.G.D., Baratta, E.J. and Tanner, J.T. 1981. "Projected dose commitment from fallout contamination in milk resulting from the 1976 Chinese atmospheric nuclear weapons test.")

Cunningham, W.C., Stroube, W.B. and Baratta, E.J. (1989). Radionuclides in domestic and imported foods in the United States, 1983-1986. *J. Assoc. Off. Anal. Chem.* 72(1). pg. 15-18.

- The minimal data available in this brief report indicate that, between 1983-86, cesium-137 levels were below 2 Bq/kg (54 pCi/kg) in all samples except fish (up to 4 Bq/kg). The detection limit is noted as 2 Bq/kg; Figure One on p. 16 borders on the nonsensical and is nearly useless in providing information on the nuclide content of the reactor surveys in this report. (The samplings in this short report were done in the vicinities of domestic nuclear reactors with approximately 500 samples collected from 11 nuclear reactors.)
- This report notes that in regard to Chernobyl-derived contamination in imported foods, "surveillance efforts successfully targeted contaminated foods and that contamination levels were below levels of concern for all but one oregano and three cheese samples." This report also notes that the findings reported here include no results from the domestic total diet survey (TDS) samples collected since the Chernobyl nuclear accident.
- The inaccuracies and discrepancies along with the poor graphic quality of these FDA reports raise questions about the trustworthiness and the reliability of both the US total diet surveys and the surveys of imported foods. (See peak pulse analysis of Chernobyl-derived radiocesium in the secret FDA report summarized later in this section.)

Baratta, E.J. and Lumsden, E.M. (1977). *Isotopic analysis of Pu in food ash*. Laboratory Information Bulletin #2015. US Food & Drug Administration, Washington D.C.

Note: After 1980, the FDA combined their reports of anthropogenic radionuclides in domestic foods (Total Diet Study), reactor area surveys, and imported food surveys in a single, brief report.

Anthropogenic Radioactivity in Imported Foods

Simpson, R.E., Shuman, F.G.D., Baratta, E.J. and Tanner, J.T. (1981). Survey of radionuclides in foods, 1961-77. *Health Physics.* 40. pg. 529.

Food type	Intake kg/day	Cesium-137 Sp. act., pCi/kg FY 73	Cesium-137 Sp. act., pCi/kg FY 74	Cesium-137 Intake, pCi/day FY 74	Potassium content, g/kg sample FY 73	Potassium content, g/kg sample FY 74	Strontium-90 Sp. act., pCi/kg FY 73	Strontium-90 Sp. act., pCi/kg FY 74	Strontium-90 Intake, pCi/day FY 74
Tea	-	296.2	296.2 (231.6)	-	20.5	19.5	430.3	555.4	-
Tea Brew	0.002	196.7	172.7	0.35	10.2	15.1	40.3	44.7	0.09
Coffee	-	99.5 (49.8)	133.7 (73.6)	-	19.3	18.8	24.3	29.2	-
Coffee Brew	0.014	40.5 (23.1)	66.2 (40.4)	(0.57)	15.8	16.0	14.4	9.8	0.14
Canned fruit	0.073	2.1	4.0	0.29	0.95	1.3 (1.2)	4.6	2.8 (2.6)	(0.19)
Cashew nuts	-	78.3	71.0 (56.8)	-	6.0	6.1	8.6	5.4	-
Fish	0.023	70.4 (44.8)	51.2 (16.2)	(0.37)	2.9	2.7	1.6	3.3 (2.05)	(0.05)
Cocoa, all types	0.003	127.4 (116.8)	131.5 (124.8)	(0.37)	16.9	16.1	64.9	61.3	0.18
Cocoa, without chocolate mix	-	127.4 (121.4)	136.5 (128.9)	-	17.8	16.6	67.5	64.0	2.24
Cheese	0.032	-	47.2 (11.8)	(0.38)	0.63	0.87 (0.82)	47.7	70.1	2.24
Fresh Fruit	0.108	-	-	-	3.3	3.7	0.60	3.3 (1.7)	(0.18)

Radionuclides in imported foods: Summary of data

Stroube, W.B., Jelinek, C.F. and Baratta, E.J. (1985). Survey of radionuclides in foods, 1978-1982. *Health Physics.* 49(5). pg. 731-735.

Commodity	FY '79	FY '80	FY '81	FY '82
Cs-137: pCi/kg				
Cheese	18 +/- 16	12 +/- 24	21 +/- 35	92 +/- 384
Tea, dry	198 +/- 45	146 +/- 148	245 +/- 189	127 +/- 171
Tea, brew	118 +/- 46	114 +/- 84	171 +/- 300	140 +/- 121
Fish	26 +/- 9	40 +/- 89	179 +/- 580	10 +/- 18
Sr-90				
Cheese	31 +/- 15	31 +/- 51	34 +/- 50	33 +/- 76
Tea, dry	218 +/- 59	321 +/- 500	268 +/- 276	157 +/- 43
Tea, brew	21 +/- 6	27 +/- 47	21 +/- 15	16 +/- 12
Fish	1.3 +/- 0.6	1.0 +/- 1.8	2.4 +/- 6.0	2.2 +/- 2.4

Cunningham, W.C. and Anderson, D.L. (1994). Radionuclides in domestic and imported foods in the United States, 1987-1992. *Journal of AOAC International.* 77(6). pg. 1422-1427.

- The abstract to this article contains the following incorrect information which is contradicted by the contents of the report: "Approximately 2600 test portions of imported foods were analyzed for contamination associated with the Chernobyl nuclear accident. Concentrations of radionuclide were below limits of detection for the vast majority of the imported food test portions but were above the levels of concern for 23 portions. Since 1986, the fraction of imported food test portions having measurable amounts of contamination has steadily declined, as have the average concentrations of radionuclide activity." (pg. 1422).
- Over 7,000,000 imported food shipments were received during the time period covered by this report. "Food types most likely to have contamination were given preference during collection." (pg. 1426).

- "During FY86 and FY87, contamination was found in approximately 40% of the samples collected *and indicated that FDA inspectors were successfully targeting contaminated shipments*." (pg. 1426). [emphasis added.]
- The 1989 and 1990 Field Program summaries indicate that out of 307 samples tested in 1989, 24% were contaminated; in 1990, 25% of 293 samples were contaminated. No data is available for 1988. Chernobyl-derived contamination levels dropped to 8% for FY91 and 2% for FY92. (pg. 1426).
- Concentrations of radionuclide activity were thus not below the limit of detection for the vast majority of the imported food test samples as indicated in the abstract. Nor have concentrations steadily declined since 1986 since a perusal of the specific food samples tested by the FDA indicates that the peak pulse of contamination in imported foods occurred between Feb. 1 and Oct. 4, 1987. (See peak pulse analysis of Chernobyl-derived radiocesium in US imported foods.)
- "In spite of the general decline, contaminated foods were still occasionally found during FY91 and FY92; indeed, elk meat collected in FY91, contained the highest Cs contamination found since the Chernobyl accident occurred." (81,000 pCi/kg) (pg. 1426).
- Determination that the remaining 6,997,400 food shipments imported during this time frame contained no Chernobyl-derived radiocesium or other contamination was based, according to this report, on the mental acuity of knowledgeable FDA officials who, by a magical process not specified, were able to "successfully target contaminated shipments." (pg. 1426). The exceptional insights of FDA personnel during 1986-1987 are particularly remarkable since very little data about the distribution patterns of Chernobyl-derived fallout were available at this time.
- Another disturbing component about FDA reporting on Chernobyl-derived contamination in imported foods is the time frame of the reporting period: this article represents the first public disclosure by the federal government that, for a period of time (1986-1988), 40% of targeted imported food samples tested positive for Chernobyl-derived contamination. Not only is this information not included in the abstract of this article, but it was not available to the general public (those few who read the *Journal of AOAC International*) for over seven years after the peak concentrations of contamination occurred, which was, in itself, a year and a half after the accident date.
- This report is accompanied by summaries of the 1989-1990 field programs both of which contain the following note: "Past TDS [Total Diet Study] findings have shown it is not uncommon to find ^{137}Cs activity in the Range of 2 to 20 Bq/kg in a small number of samples." It should be noted that FDA total diet surveys of

domestic foods prior to the Chernobyl accident were showing almost no radiocesium contamination (See Cunningham 1989).

- This FDA summary may serve as a paradigm of what to expect in terms of timely information availability after a domestic nuclear accident. The Federal Emergency Management Agency (FEMA) expects contamination from a domestic nuclear accident not only to stay within a fifty mile "ingestion pathway" boundary, but to be of concern for only a few weeks after a domestic accident (Maine Radiological Emergency Response Plan (RERP)).
- Analysis of TDS [Total Diet Study] for domestic foods "showed that the radionuclide content ... was very low…The radionuclide content of the reactor-survey foods was low and no control measures were indicated." (pg. 1426).
- Another component of this summary as a paradigm is that the FDA continues to refer to "levels of concern" first described in the 1961 Federal Radiation Council protective action guidelines and reaffirmed after the Chernobyl accident: imported foods containing more than 10,000 pCi of a combination of ^{134}Cs (which was not prevalent during weapons testing fallout) and ^{137}Cs (ubiquitous in all nuclear industries) were (and will be) seized and destroyed. This summary makes no reference at all to the much higher FDA/FEMA protection action guidelines which would be in effect in the case of a domestic nuclear accident.

Body Burdens

Following is a sampling of the many thousands of articles on body burdens of weapons tests fallout nuclides.

Liden, K. (1961). Cesium-137 burdens in Swedish Laplanders and reindeer. *Acta Radiol.*, *56*, 365-390.

Cohn, S.H., Spencer, H., Samachson, J., Fedlstein, A., Gusmano, E. (1962). Influence of dietary stable strontium and calcium on the turnover of bone-fixed ^{85}Sr in man. *Proceedings of the Society for Experimental Biology and Medicine, 110,* 526-528.

- Strontium-90 exhibits long-term retention once incorporated into bone.
- "The turnover time for strontium in some of the 'long-term' compartments of bone may range up to values exceeding fifty years." (p. 528).

Hanson, W.C., Palmer, H.E., Griffin, B.I. (1964). Radioactivity in northern Alaskan Eskimos and their foods, summer 1962. *Health Phys.* 10. pg. 421-429.

Naversten, J. and Liden, K. (1964). *Half-life studies of radiocesium in humans. Assessment of Radioactivity in Man.* Report No. STI/PUB/84. Vienna: International Atomic Energy Agency.

Melandri, C. and Rimondi, O. (1964). *In vivo measurement of ^{137}Cs with a human body counter. Assessment of Radioactivity in Man*. Report NO. STI/PUB/84. Vienna: International Atomic Energy Agency.

Rundo, J. (1964). A survey of the metabolism of cesium in man. *Br. J. Radiol*. 37. pg. 108-14.

Ramsaev, P.V., Shamov, V.P., Troizkaj, M.N., Lebedev, O.V., Ibatulin, M.S. (1965). Indirect assessment of total body burden of ^{137}Cs in people. *Medizinskaj Radiol*. 6. pg. 22-28.

Magno, P.J., Kauffman, P.E. and Shleien, B. (1967). Plutonium in environmental and biological media. *Health Physics*. 13. pg. 1325-1330.

Date	Location	Media	Nuclide	Activity*
1965	Winchester, MA	human liver	239Pu	2.52 pCi/kg
1965	Winchester, MA	human lung	239Pu	1.13 pCi/kg

Boni, A.L. (1969). Variations in the retention and excretion of ^{137}Cs with age and sex. *Nature*. 222. pg. 1188-9.

Karches, G.J., Wheeler, J.K., Helgeson, G.L. and Kahn, B. (1969). Cesium-137 body burdens and biological half-life in children at Tampa, Florida and Lake Bluff, Illinois. *Health Physics*. 16. pg. 301-313.

Date	Location	Media	Nuclide	Activity*
October 1966	Tampa, Fl	Total Body Burdens - children	^{137}Cs	7.7 nCi TBB (240 Bq)
October 1966	Lake Bluff, Ill	TBB- children	^{137}Cs	3.4 nCi TBB (126 Bq)

Lloyd, D.R. Pendleton, R.C., Clark, D.O., Mays, C.W. and Goates, G.B. (1973). Cesium-137 in humans: A relationship to milk ^{137}Cs content. *Health Phys*. 24. pg. 23-36.

Lloyd, D.R. (1973). Cesium half-times in humans. *Health Phys*. 25. pg. 605-10.

Moghissi, A.A. and Mayes, M.G. (1973). Radiobioassay program of the institutional total diet sampling network III. Cesium -137 dose estimates and body burdens of children. *Radiation Data and Reports*. 14(4). Pg. 233-236.

- Cesium-137 body burdens in children showed average concentrations of 4.4 nCi (4,400 pCi) in 1966; 2.3 nCi in 1967; and 2.0 nCi in 1968.

Shukla, K.K., Dombrowski, C.S. and Cohn, S.H. (1973). Fallout ^{137}Cs levels in man over a 12 year period. *Health Phys.* 24(5). pg. 555-7.

Lewis, J.T., Gossman, L.W., Kereiakes, J.G. and Saenger, E.L. (1976). Cesium-137 body burdens and half-life in a group of adult males in Cincinnati, Ohio. *Health Phys.* 30. pg. 315-8.

Newton, D., Eagle, M.C. and Venn, J.B. (1977). The Cesium-137 content of man related to fallout in rain, 1957-76. *Int. J. Environ. Stud.* 11(2). pg. 83-90.

Glowiak, B.J., Pacyna, J. and Palczynski, R.J. (1977). Strontium-90 and caesium-137 contents in human teeth. *Environ. Pollut.* 14. pg. 101-111.

- "The cesium-137 activity in deciduous teeth was, on average, 0.09 pCi/g of ash and, in permanent teeth, 0.05 pCi/g of ash." (p. 101).

Thomas, R.G., Anderson, E.C. and Richmond, C.R. (1979). [137]Cs Whole-body content in a normal New Mexico population: 1956-1977. *Health Phys.* 37. pg. 588-91.

Fisenne, I., Cohen, N., Neton, J., Perry, J. (1980). Fallout plutonium in human tissues from New York City. *Radiation Research*. 83. pg. 162-168.

Szabo, A.S. (1980). Method for the theoretical determination of the biological half-lives of ^{137}Cs and ^{90}Sr in man, on the basis of K and Ca metabolism. *Radiochem. Radioanal. Letters*. 43(4). pg. 193-202.

Klusek, C.S. (1984). *Environmental Measurements Laboratory: Strontium-90 in human bone in the U.S., 1982*. EML-435. Department of Energy, New York, NY.

- One of many EML publications documenting weapons-testing-derived ^{90}Sr in the environment.
- Table 4 on page 18 summarizes the dietary intake of ^{90}Sr from 1954 to 1982; prior to the nuclear era, dietary intake of radiostrontium was zero. Peak intake of radiostrontium occurred in 1964 with ^{90}Sr intake levels reaching 29.8 pCi/gm of calcium in New York; ^{90}Sr intake in San Francisco generally runs 1/2 to 1/3 of that in the New York diet, New York being downwind from the Nevada Test Site. By 1982 ^{90}Sr intake had dropped to 5.6 pCi/gm of calcium in New York and 2.5 for San Francisco adults.

- Concentrations of ^{90}Sr in adult vertebrae show a similar but slightly delayed pattern with peak values noted again from New York at 2.2 pCi/gm of calcium for bone concentrations. The San Francisco maximum value was 1.2 pCi/gm of calcium; vertebrae concentrations show a much slower rate of decline than concentrations in milk: 1982 contamination levels in New York of 1.2 pCi/gm of calcium.

- Other EML publications contain extensive data on ^{90}Sr as well as on other nuclides. Also see the publications of the Riso National Laboratory, which contain the cumulative fallout record for ^{90}Sr.

Gjertsen, L., Lind, B. and Westerlund, E.A. (1984). *^{137}Cs in Norwegian Lapps*. Report No. SIS-1984:6. Oslo, Norway: Statens Inst. for Straalehygiene.

- Cesium-137 body burdens in Norwegian Lapps declined between 1965 and 1975 with an apparent half-life of four to five years. After 1975, the decline was slower due to Chinese nuclear weapons tests' fallout.

Schwarz, G. and Dunning, D.E., Jr. (1986). Indecision in estimates of dose from ingested ^{137}Cs due to variability in human biological characteristics. *Health Phys.* 43. pg. 631-45.

Westerlund, E.A., Berthelsen, T. and Berteig, L. (1987). Cesium-137 body burdens in Norwegian Lapps, 1965-1983. *Health Phys.* 52. pg. 171-7.

Gallelli, G., Orlando, P., Perdelli, F., De Flora, S., Malcontenti, R. and Bianchini, L. (1989). Assessment from autopsy sources of the internal dose due to ^{137}Cs and ^{134}Cs from the Chernobyl accident. *J. Environ. Radioactivity.* 9. pg. 131-143.

Clemente. G.F., Mariani, A.& Santaroni, G.P. (1990). Sex differences in Cs-metabolism in man. *Health Phys.* 23. pg. 394-5.

Date	Location	Media	Nuclide	Activity
June 1986 -June 1987	Genoa, Italy	Total Body Burden: Males	^{137}Cs	1245 Bq TBB
June 1986-June 1987	Genoa, Italy	Total Body Burden: Females	^{137}Cs	1017 Bq TBB

- These readings are less than some previously recorded pre-Chernobyl (fallout) body burdens. However, the Genoa area may have experienced much less of an impact from the Chernobyl accident than other areas of Italy. Due to the erratic Chernobyl fallout patterns, data from one location may not be interpreted as representative of any other location.

154

Gregory, J., Foster, K., Tyler, H. and Wiseman, M. (1990). *Dietary and nutritional survey of British adults in 1986-87*. HMSO, London.

Hoshi, M. et. al. (1994). ^{137}Cs concentration among children in areas contaminated with radioactive fallout from the Chernobyl accident: Mogilev and Gomel Oblasts, Belarus. *Health Physics*. 67(3). pp. 272-275.

V. Plume Source Points

Savannah River, South Carolina: Liquid High-Level Waste

The DOE Integrated Database (1994) lists the Savannah River facility as having the largest inventory of contained high-level wastes of any DOE weapons production facility, 534,500,000 curies of liquid high-level wastes. This does not include the 9,657 fuel assemblies listed in the US fuel assembly inventory (USDOE 1992). These wastes derive from the production of spent fuel at this location and its reprocessing for the purpose of extracting the plutonium for weapons fabrication. During the period of weapons production, huge additional uncontained quantities of liquid high-level and low-level wastes were released to the natural environment. The curic inventory of these uncontained releases probably exceeded 1 billion curies. The Savannah River plume constitutes what is one of the two largest releases of anthropogenic radioactivity occurring in the United States during the Cold War. Only the uncontained releases at the Hanford, Washington Reservation, an isolated and desiccated environment, have the potential to exceed the size of the Savannah River plume. The Savannah River release is also that plume likely to have the second largest health physics impact during the next few millennium. Of particular concern are the plutonium storage tanks in the F-canyon of Savannah River, as well as the extensive uncontained releases of spent-fuel-derived wastes which have historically occurred at the other fuel processing canyons and reactor basins at this facility. How will the humid climate, high rainfall, and numerous wetlands assist the transport of the long-lived radionuclides in this plume? How far afield from the lagoons and holding ponds of the Savannah River reservation will this contamination travel? What natural processes will make the plutonium in this plume, now predominantly in a biologically inert form, more available for uptake in pathways to humans? How will continuing activities at the Savannah River Plant (SRP) as well as the possibility of additional spent fuel reprocessing affect the size and duration of this plume?

Hanford Reservation, Washington: Liquid High-Level Waste

The Hanford Reservation in Washington State and the Savannah River Plant in South Carolina were the principal plutonium production facilities operated by the DOE for the purpose of fabricating nuclear weapons during the Cold War. The plutonium produced at Hanford Reservation was shipped to other weapons production facilities for refining and final fabrication into usable weapons. During this process, large quantities of spent fuel were created and then reprocessed to extract the plutonium, creating huge quantities of liquid high-level and low-level wastes. In 2002, the Hanford facility was reported to have over 110,000 fuel assemblies onsite. The DOE Integrated Database report (1994)

156

lists a current inventory of 348 million curies of contained high-level waste as well as additional quantities of stored or buried transuranic wastes at this site. As at the Savannah River facilities, huge additional quantities of uncontained liquid high-level and low-level wastes were produced and released to the natural environment. Not enough information is available to determine whether the uncontained releases at this location exceeded those at the Savannah River site which is closer to populated suburban and urban population centers.

The plutonium production facilities at Hanford lie alongside the Columbia River which was the recipient of significant, but unknown, quantities of reactor-derived liquid wastes. Aside from the large quantities of liquid high-level wastes now residing in leaking steel tanks, significant uncontained quantities of liquid wastes were released in shallow holding ponds which later dried out, in pits, and via shallow well and deepwell injection. The total uncontained release of weapons-production-derived wastes may exceed the total release at the Savannah River facility. In view of the total amount of missing military high-level wastes, the uncontained release of radioactivity to the natural environment at this location may exceed 1 billion curies of reprocessed spent-fuel-derived wastes and hundreds of millions of curies of low-level wastes. The location of these huge releases in a desiccated environment with low rainfall, little surface water other than the Columbia River, low water table, and lack of nearby population centers may serve to mitigate the health physics impact of what is likely the largest uncontained release of weapons-production-derived contamination in the United States. Unfortunately, unlike the Rocky Flats plutonium plume, there are very few environmental remediation solutions available to mitigate previous uncontained releases of radioactivity at this location. If the DOE fails to secure the contained tank wastes at this location or if criticality is reached in these vulnerable tanks, this release plume, which will continue to spread for hundreds of years, will be greatly enhanced.

Oak Ridge Reservation

The Oak Ridge Reservation includes a multiplicity of important plume source points, including the Oak Ridge National Laboratory, The K-25 Plant and the Y-12 Plant. The primary function of the latter two sites was the production of enriched uranium for the purpose of nuclear weapons production. The Oak Ridge Reservation may be the most complex plume source point among all the DOE weapons production laboratories. Plume source points include the X-10 Graphite Reactor which operated in the early years of the Cold War, as well as uranium contamination deriving from the K-25 Plant and the Y-12 Plant. Two buildings at the Y-12 Plant are currently considered among the top ten most dangerous DOE sites for processing or storing uranium due to inadequately designed vaults containing bomb grade uranium. The Oak Ridge Reservation is the location of the infamous hydrofracturing facility, which was designed specifically for the injection of highly radioactive reprocessed spent fuel wastes into underlying shale deposits in the form of a grout containing the unwanted wastes. At least 1.5 million curies of waste, and possibly much more, are contained in this component of the Oak Ridge plume. Other undocumented injections of liquid wastes occurred in the early years of operation of this facility. Maps contained in the *Baseline Environmental Management Report* (BEMR) (United States Department of Energy 1996) provide a graphic illustration of a series of interconnected "waste area groups" (WAG) which resulted in so much radioactive contamination being released to surface water supplies (White Oak Creek, White Oak Lake, etc.) that not only have special dams been constructed to slow the movement of surface contamination, but the Clinch River Basin has been declared a superfund site and is listed in the BEMR publication as a subject of DOE remediation efforts. The total curic content of uncontained releases of radioactive effluents to the environment at this location may never be known but could easily be in excess of 200 million curies of uranium processing-derived wastes as well as of "low-level" and mixed low-level wastes originating from reprocessed spent fuel and other weapons production facilities. Contamination of underground aquifers in the relatively highly populated areas of this section of Tennessee could result in a plume which equals or exceeds the size and the significance of the other major underground plumes at Hanford, SRP, and INEL.

Idaho National Engineering Laboratory

The Idaho National Engineering Laboratory (INEL) covers 890 sq. miles in southern Idaho along the edge of the Snake River Plain. The INEL is an important nuclear accident-in-progress with a variety of constituents including ten major operating areas in addition to the Argonne National Laboratory West and the Naval Reactor Facility, which are, in themselves, important plume source points. Of particular concern is the presence of large quantities of deteriorating irradiated reactor fuel and related corrosion products (sludge) in the 603 basin which is a component of the Idaho Chemical Processing Plant (ICPP). While removal of some of this deteriorating fuel has begun, this facility still remains the most dangerous of above-ground plume source points among all the 98 operable units which are subject to environmental remediation at this facility. Numerous other storage tanks, pits, trenches, evaporation ponds, "French drains," waste sumps, storage tanks, chemical washout areas, and other surface facilities and waste sites contribute to INEL as a plume source point with many constituents. The single largest component of INEL as a plume source point results, however, from shallow well and deep well injection of high-level as well as low-level mixed wastes which began with the establishment of this facility in the late 1940s.

158

Evaluation of the INEL site is complicated by continuing operations, a high level of secrecy and lack of documentation of disposal techniques in its early years of operation, and the multiplicity of source points of radiological contamination. These include an experimental breeder reactor, the power burst facility reactor, three or more test reactor areas, the ICPP and related tank farm, the waste calcine facility, the test area north including a manufacturing assembly and hot shop, and a radioactive waste management complex. Not enough information is available to determine which of these facilities contributed the largest quantities of radioactive wastes to the injection wells and "French drains" which were formerly utilized at this location. The *Baseline Environmental Management Report* (United States Department of Energy 1996) divides the ICPP into 14 units consisting of 93 potential release sites and then makes the following comment: "Most of the known contamination at the Idaho Chemical Processing Plant is below the surface of the soil." (pg. Idaho-29). The assessment of contamination in the Snake River Plain Aquifer, which has just begun, has detected volatile organic compounds in the aquifer 600 feet below the surface of the ground. The INEL site environmental report for 1995 limits discussion of the extensive shallow and deep well injections that occurred at INEL in the past to a few disposal wells. The complex ground water monitoring program activities, illustrated in Figure 5.1, page 5.4 of this report, which were implemented by the United States Geological Survey, reference the extensive undocumented disposal of liquid radioactive wastes of the past. While some components of the INEL facility may be the subject of successful remediation efforts with respect to the transfer of transuranic wastes to the Waste Isolation Pilot Project (WIPP) in New Mexico, the huge INEL-derived Snake River aquifer plume will likely rate among the top ten most significant source points of radioactive contamination in the US in the next millennium.

Tokai Uranium Processing Plant (Japan)

French, Howard W. (October 1, 1999). Japanese fuel plant spews radiation after accident. *The New York Times*.

- "In an indication that the accident was being brought under control, Japanese television said that at 6:30 AM today, the Science and Technology Agency reported that no radiation could be detected at 14 monitoring sites around the plant."
- One of many articles in the *Washington Post* and *The New York Times*, this relatively small accident has received a huge amount of publicity. The Washington Post indicates that approximately two dozen similar criticality accidents have taken place in the United States between 1945 and 1964.

Estimates of fuel reprocessing accidents range up to at least 60 on a world-wide basis since the beginning of the nuclear age.

- Tokai is "a re-conversion plant, where they process enriched uranium hexafluoride (UF6) to uranium dioxide (UO2) through various steps of chemical refinement. The criticality accident occurred in the process of converting uranyl nitrate solution (with 18.8% concentration of fissile U-235) to ammonium diranate (ADU) sediment." (personal communication, Dr. Hosokawa Komei, Dept of Resource Management & Society, Faculty of Agriculture, The University of Saga, 840-8502 Saga City, Japan).

- The fissile mass was 16 kg of highly enriched uranium destined for a fast breeder research reactor somewhere in Japan.

- The accident occurred when too much U-235 was poured into a settling basin designed for 2.4 kg, a typical human error accident.

- The accident lasted 17 hours before a water shield was drained which had served to keep the accident going by directing neutrons back into the fissile material. Workers were unable to drain the water by remote control, so they broke the pipes open and stopped the accident.

- An intense radiation field resulting from shine from the fissile material in the settling basin surrounded the facility. The fact that significant civilian population lives close to the facility complicated the accident response and ensured wide publicity.

- The accident occurred during a rainfall event which would serve to localize fallout from the plume which was produced by the accident. No information is yet available differentiating plume-derived shine from source point shine from the fissile material itself.

160

VI. Pathways

The first atomic explosion at Alamogordo, New Mexico at 5:29 A.M., July 16, 1945, ushered in an era of the systemic release of biologically significant radionuclides from anthropogenic sources. Those who created these devices of destruction never imagined the silent efficiency or the hemispheric thoroughness of the biogeochemical cycling which now expose all the inhabitants of the biosphere to these effluents.

A proliferation of anthropogenic sources of nuclear contamination, including the development of nuclear weapons, followed this first test explosion in 1945. The most obvious sources of contamination were the many nuclear weapons tests (1945-1980), but equally significant release sources were the weapons production facilities and fuel reprocessing sites which evolved with the development of military nuclear capabilities. The creation of atomic power stations was the inexorable result of the exploitation of the fission process for military purposes and constitutes an unfortunate footnote to the Cold War. The nuclear effluents released from these anthropogenic source points follow pathways, and create a baseline of nuclear contamination which can and must be documented to allow evaluation of the environmental impact of nuclear accidents such as Chernobyl and the future impact of releases from thousands of other potential source points of radioactive contamination.

Pathway Models: Nuclear weapons testing (1945-1978) resulted in local, tropospheric and stratospheric fallout patterns. Initially the low-yield "fat man" atomic weapons had only a modest input on stratospheric transport routes, but after the development of more powerful thermonuclear weapons in the mid-1950s (hydrogen bombs), **stratospheric fallout** became the principle mode of hemispheric transport of weapons tests fallout. Weapons testing stratospheric fallout occurred not only as a primary pulse in conjunction with a tropospheric component, but also as long-term fallout which continued in decreasing intensity over a period of decades, as documented by the Riso National Laboratories (Denmark) summary of cumulative fallout data reprinted in *Section IV. Baseline Information.*

In contrast to weapons testing pathways, Chernobyl contamination occurred primarily as a **tropospheric** injection of smoke and radionuclides which produced much higher than expected contamination in distant locations, as well as less than expected close-in fallout at the reactor accident site. The Chernobyl accident, which was **hemispheric** in its impact, serves as a model for the **tropospheric** dispersion of any major nuclear accident plume given the caveat that weather conditions and reactor design help dissipate the local impact of the fallout pattern. Weapons testing fallout, Sellafield fuel

161

reprocessing facility effluents, and later, the Chernobyl plume illustrate a fundamental reality about the biogeochemical pathways of effluents from a nuclear accident: radioactive contamination occurs not as one incident but as a series of pulses in time and space, impacting pathways to human consumption

Primary Pulse: Direct deposition of anthropogenic nuclear effluents in the form of rapidly moving air-borne pulses of radioiodine and vaporized radionuclides (e.g. radiocesium) resulting from major nuclear accidents such as Chernobyl, with total global tropospheric transport times of as little as two weeks. **Fallout from such events is associated with and maximized by rainfall (or snowfall) events which allow rapid transfer to human diet of radionuclides deposited directly in forage pathways (e.g. foliar deposition). Such transfer can occur within several days of the plume passage. Immersion, absorption, and inhalation are other exposure pathways.**

Secondary Pulse: The slower movement of radioactive contamination in the abiotic environment including delayed particulate fallout, the mobilization and uptake of existing fallout, and its bioaccumulation in pathways to human consumption. Passage and uptake of the secondary (indirect) pulse of contamination from abiotic media to biological media can vary in time from weeks to years.

Tertiary Pulse: The delayed redistribution of wind-blown deposition, the remobilization of existing fallout, the transport of surface contamination by human activities (vehicles, foot traffic, train, marine, and air transport, on clothing, and in manufacturing processes, etc.), and the incorporation of multiple modes of pathway contamination into processed foods and consumer products which may be transferred to areas unaffected by the primary and secondary pulses of an accident plume. Redistribution of wind-blown plutonium and other long-lived radionuclides from Chernobyl, Fukushima Daiichi, and military source points will continue for millennia (Pu-239 1/2 T = 24,131 y).

Liquid releases from facilities such as Sellafield follow plume pathways involving a slower dispersion of the primary pulse with less obvious secondary and tertiary pulses of delayed contamination of pathways to human consumption.

Post-Chernobyl World Health Organization (WHO) Pathways Summary:

Following the Chernobyl accident, WHO issued this outline of pathways exposure.

External:	Internal:
Ground shine	Ingestion
Cloud shine	Inhalation
Deposition on skin and clothing	Absorption from skin

Cloud Shine-Ground Shine:

Other important exposure pathways include cloud shine and ground shine from airborne and the terrestrially-deposited radioactivity derived from nuclear accidents. They result in **external** exposure via skin irradiation and absorption. The presence of cloud shine and ground shine assure the presence of **internal** exposure pathways by inhalation and ingestion. These rapidly moving pathway pulses, which have complex radionuclide composites, are a formidable challenge to accurate biological monitoring, the prerequisite of credible dose assessment.

Plume Pathway Models

Pathways of Nuclear Effluents to Man.[1]

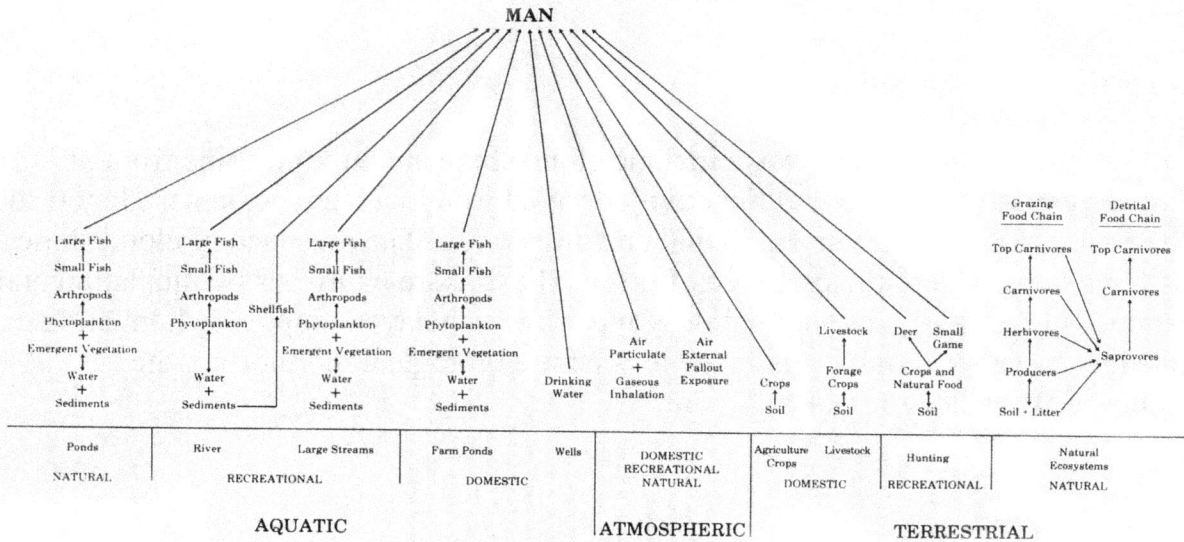

MAN

| | | | | | | | | | | | | Grazing Food Chain | Detrital Food Chain |

Top Carnivores   Top Carnivores

Large Fish   Large Fish   Large Fish   Large Fish   Carnivores   Carnivores

Small Fish   Small Fish   Small Fish   Small Fish

Arthropods   Arthropods   Arthropods   Arthropods   Livestock   Deer   Small Game   Herbivores

Shellfish

Phytoplankton   Phytoplankton   Phytoplankton   Phytoplankton   Air Particulate   Air External Fallout Exposure   Saprovores

\+   \+   \+

Emergent Vegetation   Emergent Vegetation   Emergent Vegetation   Producers

Water   Water   Water   Water   Drinking Water   Gaseous Inhalation   Crops   Forage Crops   Crops and Natural Food

\+   \+   \+   \+   Soil   Soil   Soil   Soil + Litter

Sediments   Sediments   Sediments   Sediments

| Ponds | River | Large Streams | Farm Ponds | Wells | DOMESTIC RECREATIONAL NATURAL | Agriculture Crops   Livestock | Hunting | Natural Ecosystems |
| NATURAL | RECREATIONAL | | DOMESTIC | | | DOMESTIC | RECREATIONAL | NATURAL |

AQUATIC   **ATMOSPHERIC**   **TERRESTRIAL**

1 Adapted from Platt, et al. "Empirical benefits derived from an Ecosystem Approach to Environmental Monitoring of a Nuclear Fuel Reprocessing Plant" IAEA-SM-172/31, B-268, p.678.

Figure 11. Pathways of nuclear effluents to man (Brack 1984, Appendix III pg. 62).

- This plume pathway model was adapted from Platt (1974) *Empirical benefits derived from an ecosystem approach to environmental monitoring of a nuclear fuel reprocessing plant.*

New Paradigms of Pathway Exposure

The accident at the Fukushima Daiichi complex has resulted in the emergence of a unique new variant of human exposure via the marine pathway. Appropriately titled the tsunami debris pathway, but also possibly called the North Pacific garbage patch pathway, this exposure route is a result of the prevailing winds, which, on March 15th, brought much of the source term release from the hydrogen explosion that destroyed the spent fuel pool at Reactor 4 into the western Pacific Ocean. All tsunami debris located off the coast of Japan has the potential to be significantly contaminated by the long-lived fission products that were emitted from this spent fuel pool. Due to the circulation of Pacific Ocean waters, some of this tsunami debris will become part of the famous North Pacific garbage patch. The shorelines of Hawaii and the mainland of the US have long been locations that receive tidal debris from this patch. Tidal debris contamination, along with contamination of all the marine species that are already dying from ingesting the plastics in the garbage patch, will be a long-term exposure pathway of interest for generations of fishermen and beach goers living in the Pacific Ocean coastal regions.

Accident Plume Pathway Timetable

Nuclear effluents move not only in space but also in time. The rapid tropospheric transfer of radionuclides as volatile gaseous and aerosol forms occurs much more quickly than the slower dispersion of stratospheric fallout. Re-suspension and remobilization of long-lived radionuclides occur long after the shorter lived radionuclides have decayed, and their movement through the biosphere can continue for thousands of years. In the first few days of a nuclear accident, the presence of ^{131}I and other short-lived nuclides overshadows the presence of all other radionuclides. As these nuclides decay, longer-lived isotopes such as ^{137}Cs emerge as the principle source of exposure. The surprising lesson of the Chernobyl accident is that in between the overwhelming domination of the radioiodine isotopes and in conjunction with the dispersion of radiocesium (Cs-137: $1/2T = 30.14$ y), numerous other biologically significant radionuclides such as ruthenium and tellurium also characterize an accident plume pathway as it silently moves across national boundaries. The list of indicator nuclides in the Plume Pathway Timetable, though incomplete, helps denote the complexity and duration of nuclear accidents which then can subject large population groups to low but biologically significant exposure to long-lived radionuclides for generations. The indicator nuclides listed in column one are present from the beginning of a release and provide exposure even while masked by the more intense activity levels of the shorter-lived nuclides. Long-term exposure is a function of radioactive and biological half-life as well as biological and mercantile availability. In the secondary and tertiary stages of a plume pulse, exposure is primarily from inhalation and ingestion of long-lived radionuclides. The total nuclide inventory of any source term release in a major nuclear accident will vary widely depending on the type of facility at which the accident occurs.

	Time	Principal exposure pathways	Exposure pathway distance
1 hour	short-lived	Inhalation, immersion	<50 miles
1 day	^{131}I, ^{132}Te, ^{99}Mo, ^{239}Nep	absorption	<1000 miles
1 week	^{103}Ru, ^{140}Ba, ^{95}Zr	Ingestion	2,000-5,000 miles
1 month	^{89}Sr, ^{134}Cs, ^{110m}Ag, ^{106}Ru	secondary pulse	hemispheric
1 year	^{154}Eu, ^{154}Ce, ^{90}Sr, ^{137}Cs, ^{241}Pu	tertiary pulse: remobilized long-lived radionuclides	
10 years	^{238,241}Pu		
100 years	^{241}Am		
1000 years	^{239}Pu		
10,000 years	^{99}Te, ^{237}Nep		
100,000 years	^{129}I		

Pathways: Bibliography

The following un-annotated citations are basic information sources for understanding the fundamentals of the *biogeochemical* cycling of radioactive contamination in the environment.

Aarkrog, A. (1971). Prediction models for strontium-90 and cesium-137 levels in the human food chain. *Health Physics.* 20. pg. 297-311.

Aarkrog, A. et al. (1987). Technetium-99 and cesium-134 as long distance tracers in Arctic waters. *Estuarine, Coastal Shelf Sci.* 24. pg. 637-647.

Agency for Toxic Substances and Disease Registry. (1990). *Toxicological profile for plutonium*. Public Health Service, US Department of Health and Human Services, ATSDR, Atlanta, GA.

Assimakopoulos, P.A., Ioannides, K.G., Pakou, A.A., Mantzios, A.S. and Pappas, C.P. (August 15, 1993). Transport of radiocaesium from a sheep's diet to its tissues. *Sci. Total Environ.* 136(1-2). pg. 1-11.

Barber, Ruth, Plumb, Mark A., Boulton, Emma, Roux, Isabelle and Dubrova, Yuri E. (May 14, 2002). Elevated mutation rates in the germ line of first- and second-generation offspring of irradiated male mice. *Proceedings of the National Academy of Sciences*. 10. pg. 1073. http://www.pnas.org/cgi/content/abstract/102015399v1.

- Review of article: Edelson, Ed. (May 6, 2002). Radiation causes mutations for several generations. *Health Scout News Reporter*. http://story.news.yahoo.com/news?tmpl=story&u=/hsn/20020507/hl_hsn/radiation_causes_mutations_for_several_generations.

Baxter, A.J. and Camplin, W.C. (1994). The use of caesium-137 to measure dispersion from discharge pipelines at nuclear sites in the UK. *Proc. Instn. Civ. Engrs. Wat., Marit. And Energy*. 106. pg. 281-288.

Belot, Y. (1986). Transfer of long-lived radionuclides through marine food chains: A review of transfer data. *J. Environ. Radioactivity*. 4. pg. 83-90.

Benoit, G., Rozan, T.F., Patton, P.C. and Arnold, C.L. (March 1999). Trace metals and radionuclides reveal sediment sources and accumulation rates in Jordan Cove, Connecticut. *Estuaries*. 22(1). pg. 65-80.

Boelskifte, S. and Dahlgaard, H. (1985). *Concentration factor of ^{60}Co for Fucus vesiculosus estimated by integrated water sampling*. Report No. DK-4000. Riso National Laboratory, Roskilde, Denmark.

Bunzl, K., Kracke, W. and Schimmack, W. (1992). Vertical Migration of plutonium-239 and -240, americium-241 and cesium-137 fallout in a forest soil under spruce. *Analyst*. 117. pg. 469-474.

Carvalho, F.P. and Fowler, S.W. (1985). Americium adoption on the surfaces of macrophytic algae. *J. Environ. Radioactivity*. 2. pg. 311-317.

Chamberlain, A.C. (1970). Interception and retention of radioactive aerosols by vegetation. *Atmos. Environ*. 4. pg. 57-78.

Churchill, J.H., Hess, C.T. and Smith, C.W. (1980). Measurement and computer modeling of radionuclide uptake by marine sediments near a nuclear power reactor. *Health Physics.* 38. pg. 327-340.

Comans, R.N.J., Middelburg, J.J., Zonderhuis, J., Woittiez, J.R.W., De Lange, G.J., Das, H.A. and Van Der Weijden, C.H. (1989). Mobilization of radiocesium in pore water of lake sediments. *Nature.* 339. pg. 367-369.

Cooke, A.I., Green, N., Rimmer, D.L., Weekes, T.E.C., Wilkins, B.T., Beresford, N.A. and Fenwick, J.D. (1996). Absorption of radiocaesium by sheep after ingestion of contaminated soils. *Science of the Total Environment.* 192(1). pg. 21-29.

Croom, J.M. and Ragsdale, H.L. (1980). A model of radiocesium cycling in a sand hills-Turkey Oak (*quercus laevis*) ecosystem. *Ecological Modeling.* 11. pg. 55-65.

Dahlgaard, H. and Boelskifte, S. (1985). "SENSI": A model describing the accumulation and time-integration of radioactive discharges in bioindicators (Fucus and Mytilus) including seasonal variation. Riso National Laboratory, Roskilde, Denmark.

Dahlgaard, H. and Boelskifte, S. (1992). SENSI: A model describing the accumulation and time-integration of radioactive discharges in the bioindicator *Fucus vesiculosus*. *Journal of Environmental Radioactivity.* 16(1). pg. 49-64.

Diabate, S. and Strack, S. (1993). Organically bound tritium. *Health Physics*. 65. pg. 698-712.

Ennis, E.M., Ward, G.M., Johnson, J.E. and Boamah, K.N. (1988). Transfer coefficients of selected radionuclides to animal products. II. Hen eggs and meat. *Health Physics.* 54. pg. 167-170.

Fisher, N.S., Olson, B.L. and Bowen, V.T. (1980). Plutonium uptake by marine phytoplankton in culture. *Limnol. Oceanogr.* 25(5). pg. 823-839.

Fisher, N.S., Cochran, J.K., Krishnaswami, S. and Livingston, H.D. (1988). Predicting the oceanic flux of radionuclides on sinking biogenic debris. *Nature.* 335(13). pg. 622-625.

Franke, B. (1986). Development of an adequate program of environmental radiation monitoring for the TMI nuclear power facility. Institute for Energy and Environmental Research, Takoma Park, MD.

Fulker, M.J., Jackson, D., Leonard, D.R., McKay, K. and John, C. (March 1998). Doses due to man-made radionuclides in terrestrial wild foods near Sellafield. *J. Radiological Protection.* 18(1), pg. 3-13.

Garten, C.T., Gardner, R.H. and Dahlman, R.C. (1978). A compartment model of plutonium dynamics in a deciduous forest ecosystem. *Health Physics.* 34. pg. 611-619.

Garten, C.T. (1995). Dispersal of radioactivity by wildlife from contaminated sites in a forested landscape. *Journal of Environmental Radioactivity.* 29(2). pg. 137-156.

Grytsyuk, N., Arapis, G., and Davydchuk, V. (2006). *Root uptake of ^{137}Cs by natural and semi-*natural *grasses as a function of texture and moisture of soils*. Journal of Environmental Radioactivity, Volume 85, Issue 1, Amsterdam, The Netherlands. pg. 48-58.

Gutknecht, J. (1961). Mechanism of radioactive zinc uptake by *ulva lactuca. Limnol. Oceanogr.* 6. pg. 426-431.

Hakanson, L., Andersson, T. and Nilsson, A. (1992). Radioactive cesium in fish in Swedish lakes 1986-1988 - General pattern related to fallout and lake characteristics. *Journal of Environmental Radioactivity.* 15(3). pg. 207-230.

Hakonson, T.E., Watters, P.L. and Hanson, W.C. (1981). The transport of plutonium in terrestrialecosystems. *Health Phys.* 40. pg. 63-69.

Handl, J. and Pfau, A. (1987). Feed-milk transfer of fission products following the Chernobyl accident. *Atomkernenergie.* 49(3). pg. 171-173.

Harvey, R.S. (1964). Uptake of radionuclides by fresh water algae and fish. *Health Physics.* 10. pg. 243-247.

Harvey, R.S. (1973). *Temperature effects on the sorption of radionuclides by aquatic organisms*. Report No. DP-MS-73-8. Prepared in connection with work under Contract No. AT(07-2)-1 with the US Atomic Energy Commission, Washington, D.C.

Hess, C.T., Smith, C.W. and Price, A.H. (1975). Model for the accumulation of radionuclides in oysters and sediments. *Nature.* 258(5532). pg. 225-226.

Hetherington, J.A. (1978). The uptake of plutonium nuclides by marine sediments. *Mar. Sci. Commun.* 4. pg. 239-274.

Hinton, T.G., McDonald, M., Ivanov, Y., Arkhipov, N. and Arkhipov, A. (1996). Foliar absorption of resuspended ^{137}Cs relative to other pathways of plant contamination. *Journal of Environmental Radioactivity.* 30(1). pg. 15-30.

Hinton, T.G., Stoll, J.M. and Tobler, L. (1995). Soil contamination of plant surfaces from grazing and rainfall interactions. *Journal of Environmental Radioactivity.* 29(1). pg. 11-26.

Honjo, S., Manganini, S.J. and Cole, J.J. (1982). Sedimentation of biogenic matter in the deep ocean. *Deep-Sea Research.* 29(5). pg. 609-625.

Hunt, G.J. (1998). Transfer across the human gut of environmental plutonium, americium, cobalt, caesium and technetium: Studies with cockles (*Cerostoderma edule*) from the Irish Sea. *J. Radiol. Prot.* 18(1). pg. 1-9.

Hunt, G.J., and Kershaw, P.J. (1990). Remobilisation of artificial radionuclides from the sediment of the Irish Sea. *J. Radiol. Prot.* 10(2). pg. 147-151.

Hunt, G.J., Hewitt, C.J. and Shepherd, J.G. (1982). The identification of critical groups and its application to fish and shellfish consumers in the coastal area of the north-east Irish Sea. *Hlth. Phys.* 43. pg. 875-889.

Hunt, G.J., Leonard, D.R.P. and Lovett, M.B. (1990). Transfer of environmental plutonium and americium across the human gut. *Sci. Total Environ.* 90. pg. 273-283.

Johanson, K.J., Bergstrom, R.V., Bothmer, S. and Karlen, G. (1990). Radiocesium in wildlife of a forest ecosystem in central Sweden. Transfer of radionuclides in natural and seminatural environments. Elsevier Applied Science, London. pg. 183-193.

Johnson, J.E., Ward, G.M., Ennis, M.E. and Boamah, K.N. (1988). Transfer coefficients of selected radionuclides to animal products. I. Comparison of milk and meat from dairy cows and goats. *Health Physics.* 54. pg. 161-166.

Kammerer, L., Hiersche, L. and Wirth, E. (1994). Uptake of radiocesium by different species of mushrooms. *Journal of Environmental Radioactivity.* 23(2). pg. 135-150.

Knowles, J.F., Smith, D.L. and Winpenny, K. (1998). A comparative study of the uptake, clearance and metabolism of technetium in lobster (*Homarus gammarus*) and edible crab (*Cancer pagurus*). *Radiation Protection Dosimetry.* 75. pg. 125-129.

Kudo, A., Ed. (2001). Radioactivity in the environment: Plutonium in the environment: Edited proceedings of the Second Invited International Symposium. Volume 1. Elsevier, NY, NY.

Larsen, R.J., Haagenson, P.L. and Reiss, N.M. (1989). Transport processes associated with the initial elevated concentrations of Chernobyl radioactivity in surface air in the United States. *Journal of Environmental Radioactivity* 10(1). 1-18.

Legget, R.W. (1986). Predicting the retention of Cs in individuals. *Health Physics.* 50. pg. 747-759.

McCarthy, J.F., Sanford, W.E. and Stafford, P.L. (November 11, 1998). Lanthanide field tracers demonstrate enhanced transport of transuranic radionuclides by natural organic matter. *Environmental Science and Technology.* Web edition <http://acsinfo.acs.org>. ASAP article.

Monte, L., Quaggia, S., Pompei, F. and Fratarcangeli, S. (1990). The Behavior of ^{137}Cs in some edible fruits. *Journal of Environmental Radioactivity.* 11(3). pg. 207-214.

Murphy, C.E., Jr. (1993). Tritium transport and cycling in the environment. *Health Phys.* 65(6). pg. 683-697.

Murphy, C.E. and Pendergast, M.M. (1979). Environmental transport and cycling of tritium in the vicinity of atmospheric releases.Behavior of tritium in the environment. IAEA-SM-232/80. International Atomic Energy Agency, Vienna.

Murphy, G.E. (1993). Tritium transport and cycling in the environment. *Health Physics.* 65. pg. 683-697.

Nelson, V.A. and Seymour, A.H. (1972). Oyster research with radionuclides: A review of selected literature. *Proceedings of the National Shellfisheries Association.* 62. pg. 89-93.

Odum, W.E. and Drifmeyer, J.E. (1978). Sorption of pollutants by plant detritus: A review. *Environmental Health Perspectives.* 27. pg. 133-137.

Patton, T.L. and Penrose, W.R. (1989). Fission product tin in sediments. *Journal of Environmental Radioactivity.* 10(3). pg. 201-212.

Pinder, J.E. III, Hinton, T.G., and Whicker, F.W. (2006). *Foliar uptake of cesium from the water column by aquatic macrophytes.* Journal of Environmental Radioactivity, Volume 85, Issue 1. Amsterdam, The Netherlands. pg. 23-47.

Platt, R. B., Palms, J. M., Ragsdale, H. L., Shure, D. J., Mayer, P. G. and Mohrbacher, J. A. (1974). Empirical benefits derived from an ecosystem approach to environmental monitoring of a nuclear fuel reprocessing plant. *Proceedings: Environmental behavior of radionuclides released in the nuclear industry.* International Atomic Energy Agency, Stationery Office Books, Vienna, Austria.

Potter, G.D., Mcintyre, D.R. and Pomeroy, D. (1969). Transport of fallout radionuclides in the grass-to-milk food chains studied with a germanium lithium-drifted detector. *Health Physics.* 16. pg. 197-300.

Prosser, S.L., Popplewell, D.S. and Lloyd, N.C. (1994). The radiological significance of ^{237}Np in the environment. *Journal of Environmental Radioactivity.* 23(2). pg. 123-134.

Price, K.R. (1973). A review of transuranic elements in soils, plants and animals. *J. Environ. Quality.* 2(1). pg. 62-66.

Raisbeck, G.M., Yiou, F., Zhou, Z.Q. and Kilius, L.R. (November 1995). ^{129}I from nuclear fuel reprocessing facilities at Sellafield (U.K.) and La Hague (France); potential as an oceanographic tracer. *Journal of Marine Systems*. 6(5-6). pg. 561-570.

Revsin, B.K. and Watson, J.E., Jr. (1993). Long-term environmental trends. Selection of sampling locations in a reactor-aquatic cooling system. *Health Physics*. 64. pg. 178-182.

Ritchie, J.C. and Ritchie, C.A. *Bibliography of publications of ^{137}Cesium studies related to erosion and sediment deposition*. USDA ARS Hydrology Laboratory, Beltsville, MD. http://hydrolab.arsusda.gov/cesium137bib.htm.

Robens, E., Hauschild, J. and Aumann, D.C. (1988). Iodine-129 in the environment of a nuclear fuel reprocessing plant: III. Soil to plant concentration factors for iodine-129 and iodine-127 and their transfer factors to milk, eggs and pork. *J. Environ. Radioactivity*. 8. pg. 37-52.

Schreckhise, R.G. and Cline, J.F. (1980). Comparative uptake and distribution of plutonium, americium, curium and neptunium in four plant species. *Health Physics*. 38. pg. 817-824.

Semioshkina, N., Voigt, G., Fesenko, S., Savinkov, A., and Mukusheva, M. (2006). *A pilot study on the transfer of 137Cs and 90Sr to horse milk and meat*. Journal of Environmental Radioactivity, Volume 85, Issue 1, Amsterdam, The Netherlands. pg. 84-93.

Shinn, Joseph H., Homan, Donald N. and Robison, William L. (July 1997). Resuspension studies in the Marshall Islands. *Health Physics*. 72(7).

Simkiss, K. (1993). Radiocesium in natural systems: A UK coordinated study. *Journal of Environmental Radioactivity*. 18(2). pg. 133-150.

Simmonds, J.R., Linsley, G.S. and Jones, J.A. (1979). *A general model for the transfer of radioactive materials* in *terrestrial food chains*. Report No. NRPB-R89. Harwell, Didcot, Oxon, National Radiological Protection Board.

Simmonds, J.R. and Linsley, G.S. (1981). A dynamic modeling system for the transfer of radioactivity in terrestrial food chains. *Nuclear Safety*. 22(6). pg. 766-777.

Skarlou, V., Nobeli, C., Anoussis, J., Haidouti, C. and Papanicolaou, E. (April 29, 1999). Transfer factors of ^{134}Cs for olive and orange trees grown on different soils. *Journal of Environmental Radioactivity*. 45(2). pg. 139-147.

Snoeijs, P. and Notter, M. (1993). Benthic diatoms as monitoring organisms for radionuclides in a brackish-water coastal environment. *Journal of Environmental Radioactivity*. 18(1). pg. 23-52.

172

Sternglass, E.J. and Gould, J.M. (1993). Breast cancer: Evidence for a relation to fission products in the diet. *Int. J. Health Services*. 23(4). pg. 783-804.

Swanson, S.M. (1985). Food chain transfer of U-series radionuclides in northern Saskatchewan aquatic system. *Health Physics*. 49. pg. 747-770.

Sweet, C.W., Murphy, C.E., Jr. and Lorenz, R. (1983). Environmental tritium transport from an atmospheric release of tritiated water. *Health Physics*. 44(1). pg. 13-18.

Swift, D.J. (1992). The accumulation of plutonium by the European Lobster (*Homarus gammarus* L.). *Journal of Environmental Radioactivity*. 16(1). pg. 1-24.

Swift, D. (1995). A laboratory study of ^{239,240}Pu, ^{241}Am and ^{243,244}Cm depuration by edible winkles (*Littorina litorea L.*) from the Cumbrian coast (NE Irish Sea) radiolabelled by the Sellafield discharges. *Journal of Environmental Radioactivity*. 27(1). pg. 13-33.

Takeda, H. and Iwakura T. (1993). Incorporation and distribution of tritium in rats exposed to tritiated rice or tritiated soybean. *J. Radiat. Res.* 33. pg. 309-318.

Tikhomirov, F.A. and Shcheglov, A.I. (December 11, 1994). Main investigation results on the forest radioecology in the Kyshtym and Chernobyl accident zones. *Sci. Total Environ.* 157(1-3). pg. 45-57.

United States Nuclear Regulatory Commission. (1975). Regulatory guide 4.1, Programs for monitoring radioactivity in the environs of nuclear power plants. US NRC, Washington, DC.

Voors, P.I. and Van Weers, A.W. (1989). Transfer of Chernobyl ^{134}Cs and ^{137}Cs in cows from silage to milk. *The Science of the Total Environment*. 85. pg. 179-188.

Walker, M.I., McKay, W.A. and Pattenden, N.J. (1986). Actinide enrichment in marine aerosols. *Nature*. 323(11). pg. 141-142.

Whicker, F.W., and Kircher, T.B. (1987). Pathway: A dynamic food-chain model to predict radionuclide ingestion after fallout deposition. *Health Physics*. 52(6). pg. 717-737.

Woodwell, G.M. (1963). The ecological effects of radiation. *Scientific American*. 208(6). pg. 40-49.

VII. Radiation Protection Guidelines

The literature on radiation protection guidelines is a quagmire of conflicting regulations. Particularly notable is the Federal Emergency Management Agency (FEMA) surface contamination guideline for authorized nuclear facility workers. This guideline of "300 counts per minute above background" contrasts sharply with the FDA-derived intervention levels summarized in *Section II. Basic Definitions and Concepts* of this *Handbook*, the second section of which is more fully described below. The FEMA guidelines contrast even more sharply with the pre-Chernobyl 1982 US Department of Human Services protection action guideline (PAG) cited below.

Committed Effective Dose Equivalent (CEDE)

The effective dose equivalent is the summation of the products of the dose equivalent received by specified tissues of the body and a tissue-specific weighting factor (H_{E50} = summation ($W_T H_{T50}$)). It is a risk-equivalent value, expressed in Sv or rem, that can be used to estimate the health-effects on an exposed individual. It is used in radiation safety because it implicitly includes the relative carcinogenic sensitivity of the various tissues (MARSSIM 1996, GL-3; United States Department of Health & Human Services 1997, 307).

Committed Dose Equivalent (H_{T50})

It is the dose equivalent to organs or tissues of reference (T) that will be received from an intake of radioactive material by an individual during the 50-year period following the intake (MARSSIM 1996, GL-3; United States Department of Health & Human Services 1997, 307).

Critical Group

The group of individuals reasonably expected to receive the greatest exposure to residual radioactivity for any applicable set of circumstances (MARSSIM, 1996 GL-4).

Derived Concentration Guides (DCGs)

The concentration that would result in a radiation dose equal to the DOE public dose limit of 100 millirems per year. "The DCGs consider only the inhalation of air, the ingestion of water, or submersion in air." The DOE DCGs raise the question that, if individuals receive a dose equal to the DCG for a particular nuclide, wouldn't they also be receiving substantial exposure for the other nuclides listed in the guide? The DCGs have nothing to do with ground deposition or dietary intake of radionuclides which provide an additional source of exposure following a nuclear accident.

Derived Intervention Level (DIL)

A protection action guideline issued by the Food and Drug Administration pertaining to contamination of human foodstuffs and based upon a committed effective dose equivalent of 5 mSv, or a committed dose equivalent to individual tissues and organs of 50 mSv, whichever is more limiting. The FDA DILs are contained in Radiation Protection Guidelines: *Accidental Radioactive Contamination of Human Food and Animal Feeds: Recommendations for State and Local Agencies*, the first section of which is reprinted in *Section II. Basic Definitions and Concepts* the second part is reprinted below. The new FDA guideline is especially noteworthy in extending the DILs to include a variety of radioisotopes not considered to be of much importance until after the Chernobyl accident (see table E6) e.g. ^{129}I: 56 Bq/kg (10 year old child).

1997 Revised FDA Radioactive Contamination Guideline: Part 2

FDA Recommended Derived Intervention Levels (DILs)

United States Food and Drug Administration (1997) Draft: *Accidental radioactive contamination of human food and animal feeds: Recommendations for state and local agencies.*

Recommended Derived Intervention Levels (DILs) or Criterion for Each Radionuclide Group All Components of the Diet						
Radionuclide Group	**(Bq/kg)**			**(pCi/kg)**		
Sr-90	160			4,300		
I-131	170			4,600		
Cs-134 + Cs-137	1,200			32,000		
Pu-238 + Pu-239 + Am-241	2			54		
Ru-103 + Ru-106	C_3 -------- + 6,800	C_6 -------- 450	< 1	C_3 ---------- + 180,000	C_6 --------- 12,000	< 1

(Table 2 pg. 16)

- "Typical precautionary actions include covering exposed products, moving animals to shelter, corralling livestock and providing protected feed and water."

(pg. 20). "The blending of contaminated food with uncontaminated food is not permitted because this is a violation of the Federal Food, Drug and Cosmetic Act (FDA 1991)." (pg. 22).

- "In 1986, FDA received a variety of foods collected locally by United States Embassy staff in Central and Eastern European countries. A total of 48 samples from Bulgaria, Czechoslovakia, Finland, Hungary, Poland, Romania, Russia, and Yugoslavia, were analyzed. Results for Ru-103, Ru-106, and Ba-140 are summarized in Table C-3." (pg. C-4). "In September 1986, 28 samples of spices from Turkey and Greece (not offered for import) were provided by the American Spice Trade Association (ASTA) for testing by FDA. ... a dilution factor of ten was applied to the concentrations for this category of foods [see table C-3, pg. C-10]." (pg. C-5).

- "The results support the expectation that concentrations of I-131 and Cs-134 + Cs-137 would serve as the main indicators of the need for protective actions for imported and local food. However, concentrations of Ru-106 were consistently in excess or at a significant fraction of the DIL, which suggests that Ru-106 should also serve as an indicator, i.e. be included as a principal radionuclide for nuclear reactor incidents... for local samples of fresh vegetables harvested during the first week of the incident, half of the samples had Ru-103 concentrations a significant fraction of the DIL and another quarter of the samples had Ru-103 concentrations in excess of the DIL. Consequently, it would be prudent to consider Ru-103 as a principal radionuclide for local deposition, particularly in the early phase of a nuclear reactor incident." (pg. C-6).

- "Also, the analytical method for determination of Sr-90 in food is lengthy compared to analysis for the gamma-ray emitting radionuclides, such that protective actions based on the concentration of Sr-90 could not be taken in a timely manner. Therefore, Sr-90 would not be an effective indicator of the need for protective actions in the early phase of a nuclear reactor incident." (pg. C-7).

- "During the first year after an accident, concentrations in local or imported food other than for I-131, Cs-134, Cs-137, Ru-103, and Ru-106 are expected to be significant only when one or more of these principal radionuclides has exceeded its DIL. Therefore, the food would already have been subject to protective action." (pg. C-7).

- "For food consumed by most members of the general public, ten percent of the dietary intakes was assumed to be contaminated. This assumption recognizes the ready availability of uncontaminated food from unaffected areas of the United States or through importation from other countries, and also that many factors could reduce or eliminate contamination of local food by the time it reaches the

176

market... FDA applied an additional factor of three to account for the fact that subpopulations might be more dependent on local food supplies. Therefore, during the immediate period after a nuclear accident, a value of 0.3 (i.e., thirty percent) is the fraction of food intake that FDA recommends should be presumed to be contaminated... For infants, (i.e., the 3-months and 1-year age groups) the diet consists of a high percentage of milk and the entire milk intake of some infants over a short period of time might come from supplies directly impacted by an accident. Therefore, f was set equal to 1.0 (100%) for the infant diet... DILs are presented in Table D-4 for Sr-90, I-131, Cs-134, Cs-137, Ru-103, Ru-106, Pu-238, Pu-239, and Am-241 for six population age groups and applicable PAGs." (pg. D-4, 5).

- "After a reactor accident, radionuclides other than the principal radionuclides may also be detected in the food supply... The DILs for fifteen other radionuclides were determined by the same procedure used in Appendix D." (pg. E-1).

- "Fractions of food intake assumed to be contaminated (f) are: 0.3 for all radionuclides except Te-132, I-133, and Np-239 in infant diets (i.e., the 3-month and 1-year age groups); 1.0 for Te-132, I-133 and Np-239 in infant diets." (pg. E-2).

- "During the immediate period after a nuclear reactor accident... Once food monitoring data is available, the recommended DILs or criterion for the principal radionuclides I-131, Cs-134 + Cs-137, and Ru-103 + Ru-106... should be used. The more complex radiochemical or gamma-ray spectrometric analyses for the fifteen other radionuclides listed in this Appendix would not be generally available." (pg. E-3,4).

	DERIVED INTERVENTION LEVELS (Bq/kg)					
Most limiting of Derived Intervention Levels for the 5 mSv H_E or 50 mSv H_T (individual radionuclides, by age group)						
Radionuclide	**3 months**	**1 year**	**5 years**	**10 years**	**15 years**	**Adult**
Sr-89	1400	2400	3600	4500	5800	87
Y-91	1200	1600	2300	3000	5300	59
Zr-95	4000	5000	7000	9700	14000	16000
Nb-95	12000	14000	19000	26000	35000	40000

Radionuclide	3 months	1 year	5 years	10 years	15 years	Adult
Te-132	4400	7300	35000	59000	89000	150000
I-129	110	76	72	56	68	84
I-133	7600	7000	30000	56000	79000	130000
Ba-140	6900	7900	11000	15000	27000	29000
Ce-141	7200	92	12000	18000	29000	34000
Ce-144	500	670	1100	1400	2300	2700
Np-227	4	37	27	22	16	15
Np-239	28000	36000	180000	260000	400000	460000
Pu-241	120	970	720	550	490	480
Cm-242	19	130	180	240	340	390
Cm-244	2	13	16	18	19	18

DERIVED INTERVENTION LEVELS (Bq/kg)
Most limiting of Derived Intervention Levels for the 5 mSv H_E or 50 mSv H_T
(individual radionuclides, by age group)

(Table E-6 pg. E-9)

- Neither the FDA nor any other U. S. Government agency has the capacity for the rapid and timely monitoring, analysis and reporting of radioactive contamination in any significant quantity of foodstuffs during a nuclear accident of any type. 1950s-era laboratory capacity remains unimproved in an era of funding shortages. In the event of a nuclear accident, neither the FDA nor any other agency would be able to determine whether intervention to prevent consumption of contaminated foodstuffs is justified or necessary. These revised DILs are a step in the right direction but have no real world credibility in the event that extensive foodstuffs monitoring becomes necessary.

- The peak pulse of Chernobyl-derived radiocesium in imported foods was observed by the FDA in 1987, ten to sixteen months after the accident began. At no time since the Chernobyl accident has the full body of raw data been available to the general public; the Center for Biological Monitoring obtained this information via a Freedom of Information Request. The FDA report on Chernobyl contamination took nine years to prepare and the result was only a few pages of clever disinformation.

- The Federal Emergency Management Agency (FEMA) has a surface contamination guideline of "300 counts per minute above background" for authorized persons entering an emergency operations center (EOC) during an accident at an NRC licensed nuclear facility. (Three hundred counts per minute (300 cpi) equals 660 pCi per person; it can be assumed that each authorized person has an absolute minimum of 1 square meter of surface area.)

U. S. Department of Health and Human Services. (1982). Accidental radioactive contamination of human food and animal feed; recommendations for state and local agencies. Docket No. 76N-0050. *Federal Register.* 47(205). pg. 47073-47084.

- This guideline applies only to contamination originating from a domestic nuclear accident and describes standards set forth by the Food and Drug Administration (FDA) as part of the development of guidelines for the Federal Emergency Management Agency (FEMA) to be used in the event of a domestic nuclear "incident."

- **This guideline does not apply to exposure from multiple pathways e.g. external radiation, inhalation or absorption; it only concerns exposure from the ingestion pathway.**

- This guideline consists of two components, "low impact protective actions (termed the Preventive PAG) at projected radiation doses of 0.5 rem whole body and 1.5 rem thyroid... The Emergency PAG action is recommended at projected radiation doses of 5 rem whole body and 15 rem thyroid." (pg. 47074)

- The most important component of the emergency protection action guidelines is that the projected radiation doses are assumed to take place in just a few days. It is assumed that exposure to the enormous amounts of radioactive contamination listed as the "response levels" in the emergency PAGs will not continue over a long period of time, despite the long radioactive half-life of ^{137}Cs, a primary constituent of accident deposition. The lesson of the Chernobyl accident in 1986 is that exposure to the longer-lived radionuclides in an accident plume pulse continue for far longer than the few days of exposure assumed in the FDA/FEMA guidelines.

- The following quotations are abstracted from the *Federal Register*:
  - "Assuming that initial contamination by I-131 and cesium or strontium... was at the preventive PAG level, radioactive decay and weathering would

reduce the levels so that protective actions could be ceased after one or two months." (pg. 47076).

- o "Although it may be desirable to consider total health effects, not just lethal effects, there is a lack of data for total health effects to use in such comparisons." (pg. 47077).
- o "FDA believes that, to establish a PAG, the primary concern is to provide adequate protection (or safe level of risk) for members of the public." (pg. 47077).
- o "Protective actions are appropriate when the health benefits associated with the reduction in exposure to be achieved are sufficient to offset the undesirable features of the protective actions." (pg. 47080).

- The FDA has established the following emergency protective action guidelines:

Emergency Protection Action Guideline (PAG)						
	I-131		Cs-134		Cs-137	
	Infant	Adult	Infant	Adult	Infant	Adult
Initial Deposition (microcurie/square meter)	1.3	18	20	40	30	50
Forage Concentration (microcurie/kilogram)	0.5	7	8	17	13	19
Peak Milk Intake (microcurie/liter)	0.15	2	1.5	3	2.4	4
Total Intake (microcurie/accident, 1-30 days)	0. 9	10	40	70	70	80

The above radiation protection guidelines expressed in microcuries have been reprinted below expressed in picocuries (1 microcurie = 1,000,000 picocuries) to provide contrast with the FEMA guidelines which are in counts per minute. (1 count/minute = 2.2 picocuries/minute), and the other post-Chernobyl era guidelines cited below.

Emergency Protection Action Guideline (PAG)						
	I-131		Cs-134		Cs-137	
	Infant	Adult	Infant	Adult	Infant	Adult
Initial Deposition (picocurie/square meter)	1,300,000	18,000,000	20,000,000	40,000,000	30,000,000	50,000,000
Forage Concentration (picocurie/kilogram)	500,000	7,000,000	8,000,000	17,000,000	13,000,000	19,000,000
Peak Milk Intake (picocurie/liter)	150,000	2,000,000	1,500,000	3,000,000	2,400,000	4,000,000
Total Intake (picocurie/accident, 1-30 days)	900,000	10,000,000	40,000,000	70,000,000	70,000,000	80,000,000

To convert picocuries to becquerels, divide by 27.

This older protection action guideline is included in our current radiation protection guidelines because, in the event of a serious nuclear accident at an NRC-licensed facility, the above 1982 guideline is the one likely to be used by the NRC and its licensees in informing the public about the relative risks of the resulting contamination. The NRC and its licensees have not acknowledged the publication of the 1997 FDA guidelines for contaminated food, nor is it likely that the FDA guidelines for contaminated food will be of any interest to the NRC and its licensees in an accident situation.

Shleien, B., Pharm, D., Schmidt, G.D. and Chiacchiernini, R.P. (1982). *Background for protective action recommendations: Accidental radioactive contamination of food and animal feeds.* HHS Publication FDA 82-8196. US Department of Health and Human Services, Public Health Service, Washington D.C.

This is a reiteration and elaboration of the 1982 DHHS Federal Register notice cited above.

ATSDR Toxicological Profile for Ionizing Radiation

United States Department of Health & Human Services. (September 1997). Draft for public comment: *Toxicological Profile for Ionizing Radiation.* Research Triangle

Institute for the US Dept. of Health & Human Services, Public Health Service, and Agency for Toxic Substances and Disease Registry, Atlanta, Georgia.

The Agency for Toxic Substances and Diseases Registry developed this toxicological profile as a component of the CERCLA, also known as the Superfund Act. This toxicological profile is probably the most important publication issued by the federal government pertaining to ionizing radiation.

- "The ATSDR toxicological profile succinctly characterizes the toxicologic and adverse health effects information for the hazardous substance described therein." [ionizing radiation] (pg. v).

- "The focus of the profiles is on health and toxicologic information; therefore, each toxicological profile begins with a public health statement that describes, in nontechnical language, a substance's relevant toxicological properties." (pg. v).

- The toxicological profile on ionizing radiation includes the following topics 1-8 plus the Glossary:

  1. Public health statement

  2. Principles of ionizing radiation

  3. Summary of health effects of ionizing radiation

  4. Radiation accidents

  5. Mechanisms of biological effects

  6. Sources of population exposure to ionizing radiation

  7. Regulations

  8. Observed health effects from radiation and radioactive material

- "No Minimal Risk Levels (MRLs) have been derived for any route of exposure in this profile at this time. However, ATSDR is currently in the process of examining and critically evaluating the large database of health effects caused by exposure to ionizing radiation. During this evaluation process, ATSDR is also examining many other factors, including (1) which specific studies would lend themselves to be most suitable for deriving an MRL, and (2) what health effect(s) an MRL should be based upon (cataract formation, reduction in IQ, etc.). Any MRLs that are derived will be integrated into the final version of this profile." (pg. 19).

- Table 2.2, Effective Half-Lives of Selected Radionuclides in Major Adult Body Organs, is significant in that it omits many of the biologically significant radionuclides associated with the nuclear fuel cycle. This is symptomatic of the

182

continuing aversion to a comprehensive documentation of all biologically significant radionuclides in all pathways to human consumption, not only in this profile, but in all federally sponsored publications on this subject.

- Chapter 2 includes a detailed description of equipment used to measure internally deposited as well as external ionizing radiation (pg. 53-69) and includes a listing of relevant Internet sites.

- Table 3.1 on page 72 is the ATSDR priority listing of radionuclides present at Department of Energy NPL sites. This listing is particularly important because it includes naturally occurring radionuclides (NOR) which have been remobilized by anthropogenic activities, especially in the nuclear fuel cycle. This priority listing also excludes some of the principle radionuclides of concern listed in the FDA's new proposed derived intervention level (DIL) guidelines. (See the 1997 *Revised FDA Radioactive Contamination Guideline* in *Section II.*) The failure to cite these guidelines in the references or to discuss or include these guidelines in the text is one of the principle shortcomings of this publication and should be remedied in the draft review process by their inclusion in the final report. Chapter 3 is otherwise an excellent general survey of the health effects of ionizing radiation.

- Chapter 4 is a grossly insufficient historical overview of nuclear accidents to date. The discussion of the first accident is symptomatic of the problem of accurate documentation of anthropogenic radioactivity in the environment: the loss of four nuclear weapons over Palomares, Spain in 1966. Maximum levels of surface contamination resulting from the spread of ^{239}Pu following the chemical explosion of two warheads is listed at just above 60,000 counts per minute per square meter (1,000 Bq/m^2). The same authors as those cited here have more recently reported soil contamination over a larger area with 2.2 ha exceeding 1,200,000 Bq/m^2 (72,000,000 cpm/m^2).

The excellent summary of the September 11, 1957, plutonium fire at the Rocky Flats Environmental Technology Site (RFETS) is a timely reminder that the contamination at there is not limited to the leakage from drums stored at pad 903 as so often implied by much of the current literature on this important source point.

Chapter 6 continues to perpetuate the obsolete model of average lifetime exposure to ionizing radiation. "Less than 1% of the total ionizing radiation to the U.S. population comes from occupational sources, nuclear fallout, the nuclear fuel cycle, or other miscellaneous exposures. The total average annual effective dose equivalent for the population of the United States, natural and anthropogenic, is approximately 360 mrem (3.6 mSv) and is described further in Chapter 1 of this profile (BEIR V 1990)." (pg.

211). *The obsolescence of this antiquated model of sources of exposure to* ionizing radiation has been dramatically illustrated by the National Cancer Institute Study: Estimating Thyroid Doses of I-131 Received by Americans from Nevada Atmospheric Nuclear Bomb Test. The following quotation from the Executive Summary of the NCI report graphically illustrates this point. "The overall average thyroid dose to the approximately 160 million people in the country during the 1950s was 2 rads. The uncertainty in this per capita dose is estimated to be a factor of 2, that is, the per capita dose may have been as small as 1 rad or as large as 4 rads, but 2 rads is the best estimate. The study also demonstrated that there were large variations in the thyroid dose received by subcategories of individuals. The primary factors contributing to this variation are county of residence, age at the time of exposure, and milk consumption patterns." (pg. 2).

- The exposure levels documented in the NCI report during the 1950s <u>and 1960s</u> as well as those experienced by many communities following the Chernobyl accident in 1986 demonstrate the obsolescence of the ancient paradigms which are perpetuated in the NCRP 1987 Report Number 93 cited as the basis of the pie chart on page 211 of the ATSDR profile. Prior to publishing a final copy of this report, the ATSDR needs to consider the irrelevance of ivory tower estimations of total average annual effective dose equivalents for large populations when in fact the real issue is the radically variable exposures of specific population groups and individuals to specific sources of ionizing radiation. A truthful 1960s era pie chart incorporating two rads of exposure to ^{131}I, not to mention the accompanying exposure to all the other longer-lived test explosion fission products and nuclear fuel cycle contaminants, would in no way have any resemblance to Figure 6-1 in this publication. This editor urges the authors of this profile to reconsider the wide range of exposures to the many sources of anthropogenic radioactivity which are generally *not* less than 1% of total exposure and which can only be documented by the tedious analysis of the TEDE resulting from all radionuclides in all pathways in the site-specific situation, rather than in a theoretical model.

- The discussion of exposure from the nuclear fuel cycle in Section 6.4.3, pages 231-235, constitutes only the briefest of introductions to this Pandora's box of sources of exposure to ionized radiation. In contrast, the discussion of exposure to radiopharmaceuticals used in medicine in the same chapter is excellent.

- Chapter 7 is a discussion of regulations and guidelines applicable to ionizing radiation, but as noted is grossly deficient for its failure to incorporate the FDA's 1997 revised radiation contamination guidelines.

- Chapter 8 discusses the health effects from radiation in four pathways: inhalation, "oral," dermal, and external exposure. The use of the term "oral" in place of the

184

more traditional ingestion pathway references the reluctance of the ATSDR as well as other government agencies including the NRC to execute comprehensive pathway analyses, of which the ingestion pathway is the most important.

- The glossary continues this ritual of evasion: no definitions are provided nor references made to terms such as bioaccumulation, bioindicators, forage pathways, long-lived radionuclides, concentration ratios, etc. In fact nowhere in this publication is there any adequate illustration of pathways of nuclear effluents to man, as illustrated in Platt, 1974. *Empirical Benefits Derived from an Ecosystem Approach to Environmental Monitoring of a Nuclear Fuel Reprocessing Plant.* pg. 678.

MARSSIM Draft Multi-Agency Radiation Survey and Site Investigation Manual (EPA, NRC, DOE)

United States Department of Energy, Environmental Protection Agency, Nuclear Regulatory Commission and Department of Defense. (December 6, 1996). Multi-Agency Radiation Survey and Site Investigation Manual (MARSSIM): Draft for public comment. NUREG-1575. EPA 402-R-96-018. NTIS-PB97-117659. Washington, D.C. http://www.epa.gov/radiation/cleanup.

"MARSSIM provides information on planning, conducting, evaluating, and documenting environmental radiological surveys for demonstrating compliance with dose-based regulations. The MARSSIM, when finalized, will be a multi-agency consensus document." (Federal Register, January 6, 1997, 62(3), pg. 736).

The MARSSIM is a notable landmark in the publication of US government radiological surveillance literature, summarizing as it does the policies, paradigms and paradoxes of federal radiological surveillance programs in the twilight of the nuclear era.

This manual summarizes the techniques and surveillance models to be used by the EPA, NRC, DOD and DOE in decommissioning or remediating a wide variety of contaminated weapons or nuclear electricity production sites or facilities. At the same time it expresses many of the conflicts and institutionalized rituals of evasion which allow these four federal agencies to avoid comprehensive radiological characterization of the environmental impact of many controversial facilities such as the Maine Yankee Atomic Power Station. See *Section IX. Nuclear Dada: The Maine Yankee Atomic Power Plant* for additional comments on the Maine Yankee facility.

Post-Chernobyl National Safety Guidelines for I-131 in Milk (WHO 1986)

The following WHO summary provides a selection of radiation protection guidelines in effect in Europe for I-131 at the time of the Chernobyl accident:

National Safety Guideline		
Country	becquerels/liter	picocuries/liter
Soviet Union	2,000	54,000
Poland	1,000	27,000
Sweden	2,000	54,000
Romania	185	4,975
Austria	370	10,000
Czechoslovakia	1,000	27,000
West Germany	500	13,500
Switzerland	3,700	99,900
Yugoslavia	-	-
Turkey	-	-
United Kingdom*	2,000	54,000
United States	555	15,000
Italy	500	13,500

- These diverse guidelines compare with the FDA/FEMA emergency PAG for domestic nuclear accidents which are 150,000 pCi/liter I-131 in milk for infants and 2,000,000 pCi/l I-131 in milk for adults.

- Following the Chernobyl accident, much controversy surrounded the temporary radiation standards of 600 Bq/kg set for radiocesium in the general food supply in the European community, which were finally set at 1,000 Bq/kg (= 60,000 cpm) for dairy products and 1,250 Bq/kg for other commodities, by the European Commission in May of 1987. The radiocesium limits for Britain were set at 1,000 Bq/kg for all imported foods. The 1987 guidelines are the ones currently in effect.

- For more information on I-131 and milk see NCI's 1997, *Estimated exposures and thyroid doses received by the American people from Iodine-131 in fallout following Nevada atmospheric nuclear bomb tests* and our special appendix: Contaminated milk: A paradigm (http://www.davistownmuseum.org/cbm/RadxMilk.html).

US Radiological Environmental Monitoring Reports (REMPs)

Berger, J.D. (1992). *Manual for conducting radiological surveys in support of license termination*. NUREG/CR-5849, Draft report for comment. US NRC, Washington, D.C. and Oak Ridge Associated Universities.

Berven, B.A., Cottrell, W.D., Leggett, R.W., Little, C.A., Myrick, T.E., Goldsmith, W.A. and Haywood, F.F. (1986). *Generic radiological characterization protocol for surveys conducted for DOE remedial action programs*. ORNL/TM-7850. Martin Marietta Energy Systems, Inc., Oak Ridge National Laboratory.

Boyns, P.K. and Sevart, M.D. (December, 1973). *Aerial radiological survey of the area surrounding the Dresden Nuclear Power Station, Morris, Illinois, September, 1968*. EGG-1183-1528. [NRC] EG and G, Inc., Las Vegas, Nevada.

- "Aerial radiation survey data consisting of exposure rates normalized to 3 feet above the ground plus gamma ray spectral charts, effluent characterization for operational sites (intensity rates and isotope constituents), and pertinent descriptive information of the installation." (abstract).
- "The report contains the data for the survey of the Dresden Nuclear Power Station and surrounding area, including an effluent plume tract." (abstract).

Daily, M.C., Huffert, A., Cardile, F. and Malaro, J.C. (August 1994). *Working draft regulatory guide on release criteria for decommissioning: NRC staff's draft for comment*. NUREG-1500. Division of Regulatory Applications, Office of Nuclear Regulatory Research, US NRC, Washington, D.C.

- This draft is a key component of the NRC regulatory puzzle pertaining to determining the release criteria for decommissioned facilities such as NRC "cleanup criteria is determined on a site specific basis, taking into account the radiological characteristics of both a residual radioactivity and background at the facility ... The process for establishing site specific data needs is currently performed by the licensee, but, in the future, data needs might have to be negotiated with local, state, or federal regulatory entities." (pg. 3).
- "Wherever possible the licensee should use actual measurements, rather than modeling, when determining the source term (i.e., residual radioactivity remaining at the site) upon which the calculated TEDE will be based." (pg. 6).

- "In each scenario, the critical group is an individual or relatively homogeneous group of individuals expected to receive the highest exposure within the assumptions of the particular scenario. The average member of the critical group is that individual who is assumed to represent the most likely exposure situation, based on prudently conservative exposure assumptions and parameter values within the model calculations." (pg. 6).
- "The NRC has developed models to provide generic dose conversion factors for residual radioactivity ... The low concentration levels, extended time periods for analysis, and multiple pathways of concern make model calculations the most defensible and cost effective approach." (pg. 9).

Fauver, D.N., Weber, M.F., Johnson, T.C. and Kinneman, J.D. (November 1995). *Site decommissioning management plan*. NUREG-1444. Supplement 1. Division of Waste Management, Office of Nuclear Material Safety and Safeguards, US Nuclear Regulatory Commission, Washington, D.C.

Gogolak, C.V., Huffert, A.M. and Powers, G.E. (August 1995). *A nonparametric statistical methodology for the design and analysis of final status decommissioning surveys: Draft report for comment*. NUREG-1505. Division of Regulatory Applications, Office of Nuclear Regulatory Research, US NRC, Washington, D.C.

Huffert, A.M., Meck, R.A. and Miller K.M. (August 1994). *Background as a residual radioactivity criterion for decommissioning: Appendix A to the generic environmental impact statement in support of rulemaking on radiological criteria for decommissioning of NRC-licensed nuclear facilities*. Draft report. NUREG-1501. Division of Regulatory Applications, Office of Nuclear Regulatory Research, US NRC, Washington, D.C.

- "Existing clean-up criteria for nuclear facilities are a patchwork of applicable regulation, guidance, and practices ... these regulations do not explicitly state which radiological criteria to apply to demonstrate that a site has been adequately remediated." (pg. 59).
- This flexibility allow licensees such as MYAPC huge leeway in characterizing the environmental impact of plant operations, as there is now no longer any need for an exact and scientific calculation of defacto background radiation levels, which now include radiation from all anthropogenic sources except the licensee.

Huffert, A.M., Abelquist, E.W. and Brown, W.S. (August 1995). *Minimum detectable concentrations with typical radiation survey instruments for various contaminants and field conditions*. NUREG-1507. Division of Regulatory Applications, Office of Nuclear Regulatory Research, US NRC, Washington, D.C.

Huffert, A.M. and Miller, K.M. (August 1995). *Measurement methods for radiological surveys in support of new decommissioning criteria: Draft report for comment.* NUREG-1506. Division of Regulatory Applications, Office of Nuclear Regulatory Research, US NRC, Washington, D.C.

Kennedy, W.E., Jr. and Strenge, D.L. (October 1992). *Residual radioactive contamination from decommissioning.* NUREG/CR-5512, Final report. Pacific Northwest Laboratory, US NRC.

- Another component in the labyrinth of NRC rules and regulations which allow licensees to decommission facilities without an accurate account of the actual environmental impact of facility operations.

Tichler, J., Doty, K. and Lucadamo, K. (1993). *Radioactive materials released from nuclear power plants: Annual report 1993.* Vol. 14. NUREG/CR-2907, BNL-NUREG-51581. Brookhaven National Laboratory, US NRC.

United States Nuclear Regulatory Commission. (September 1978). *Decommissioning of nuclear facilities - an annotated bibliography.* NUREG/CR-0131. Pacific Northwest Laboratory for US NRC, Washington, D.C.

United States Nuclear Regulatory Commission. (August 1979). *Decommissioning of nuclear facilities - a review and analysis of current regulations.* NUREG/CR-0671. Pacific Northwest Laboratory for US NRC, Washington, D.C.

United States Nuclear Regulatory Commission. (August, 1979). *Technology, safety and costs of decommissioning a reference pressurized water reactor power station.* NUREG/CR-0130 Addendum. Pacific Northwest Laboratory for US NRC, Washington, D.C.

United States Nuclear Regulatory Commission. (December, 1979). *Facilitation of decommissioning of light water reactors.* NUREG/CR-0569. Pacific Northwest Laboratory for US NRC, Washington, D.C.

United States Nuclear Regulatory Commission. (1988). *Final generic environmental impact statement on decommissioning of nuclear facilities.* NUREG-0586. US NRC, Washington, D.C.

- This NRC publication analyses the decommissioning process without addressing the environmental impact of the decommissioning process or levels of residual radioactivity which would result.

- Issues discussed include worker radiation exposure, amounts of waste generated, radiation exposure received by the public and other details pertaining to planning and executing the decommissioning process.
- This NRC publication raises a question which is not resolved in this or any following NRC publication: how can the TEDE of the general population be determined without knowledge of the environmental impact of plant operations based on (nonexistent) pathway analyses of all radionuclides in all ecosystems which are components of the source term release by a specific installation such as MYAPC?

United States Nuclear Regulatory Commission. (August, 1994). *Generic environmental impact statement in support of rulemaking on radiological criteria for decommissioning of NRC-licensed nuclear facilities: Main report: Draft report for comment.* NUREG-1496. Vol. 1. Division of Regulatory Applications, Office of Nuclear Regulatory Research, US NRC, Washington, D.C.

- A general overview of the context of what were at this point ongoing rulemaking changes for decommissioning including a discussion of regulatory alternatives and the various kinds of nuclear facilities affected by this proposed rulemaking. This volume includes lengthy cost analyses.
- This NRC publication was the first to introduce the concept of setting residual radioactivity levels resulting from plant operations and decommissioning (e.g. total TEDE at 25 mrem/yr).

United States Nuclear Regulatory Commission. (August, 1994). *Generic environmental impact statement in support of rulemaking on radiological criteria for decommissioning of NRC-licensed nuclear facilities. Appendices. Draft report for comment.* NUREG-1496. Vol. 2. Division of Regulatory Applications, Office of Nuclear Regulatory Research, US NRC, Washington, D.C.

United States Nuclear Regulatory Commission. (1995). *Proposed methodologies for measuring low levels of residual radioactivity for decommissioning.* NUREG-1506, draft report for comment. US NRC, Washington, D.C.

United States Nuclear Regulatory Commission. (1995). *Measurement methods for radiological surveys in support of new decommissioning criteria.* NUREG-1506. US NRC, Washington, D.C.

- For conducting radiological surveys for decommissioning, the DQO approach would, in general, entail the following:

1. Identify the critical radionuclides, their critical pathways, the contaminated media, and the types of measurements or samples that are needed.
2. Check default values of the concentrations for each identified radionuclide...
3. Determine whether the radionuclide is already present in the background and establish the needs of the statistical tests that will be used to demonstrate compliance with the dose limits and ALARA requirements.
4. Choose instrumentation/measurement methods based on detection limits as compared to the default concentrations for each radionuclide, as well as for estimating the site inventory, that is, the total amount of residual radioactivity present in the environmental media.
5. Establish numbers of personnel, types of expertise, and necessary training levels required to conduct measurements. Formulate a plan and then perform measurements. Assess measurements as the plan is executed. (pg. 2-2).

- "It is recommended that water bodies on site be included in a survey to support decommissioning, as both the water itself and, to a greater extent, the underlying sediment represent sinks for runoff of radionuclides from facility operations." (pg. 5-3).

United States Environmental Protection Agency. (January, 1996). Environmental *Radiation Data Report 76: October - December 1993*. EPA-402-R-96-004. National Air and Radiation Environmental Laboratory, US EPA, Montgomery, AL.

Boulding, J.R. (1993). *Description and sampling of contaminated soils: A field pocket guide*. EPA/625/12-91/00.

United States Environmental Protection Agency. (1988). *Guidance for conducting remedial investigations and feasibility studies under CERCLA, interim final*. EPA/540/G-89/004. OSWER Directive 9355.3-01. US EPA, Washington, D.C.

United States Environmental Protection Agency. (1988). *Superfund removal procedures*. OSWER Directive 9360.0-03B. Office of Emergency and Remedial Response. US EPA, Washington, D.C.

United States Environmental Protection Agency. (1991). *Site assessment information directory. Office of Emergency and Remedial Response*. US EPA, Washington, D.C.

United States Environmental Protection Agency. (May, 1992). *Manual of protective action guides and protective actions for nuclear incidents*. EPA 400-R-92-001. ANR-460. Office of Radiation Programs, US EPA, Washington, D.C.

- This manual includes the famous Federal Register Notice of October 22, 1982, which includes the emergency protective action guidelines (PAG) which were later incorporated into the FEMA guides. While the emergency protection action guide for a teenager or adult is 50 mCi/m2 (110 million counts per minute), the emergency PAG for authorized persons within an Emergency Operations Center (EOC) is 300 counts per minute above background total body contamination (shower, shave quickly, use a separate exit, keep out of the dining room, etc. etc.)
- Nonetheless, this publication is an excellent guide to what the EPA thinks should be the basic approach to any emergency radiological situation.
- "The estimation of radiation risks is not a fully mature science and the evaluation of radiation hazards will continue to change as additional information becomes available." (pg. B-19).

United States Environmental Protection Agency. (March 1993). *Environmental characteristics of EPA, NRC, and DOE sites contaminated with radioactive substances.* EPA/402-R-993-011. Radiation Protection Division, Office of Radiation & Indoor Air, US EPA, Washington, D.C.

United States Environmental Protection Agency. (January 1996). *Documenting ground water modeling at sites contaminated with radioactive substances.* EPA/540-R-96-003. Radiation Protection Division, Office of Radiation & Indoor Air, US EPA, Washington, D.C.

United States Environmental Protection Agency. (January 1996). *Three multimedia models used at hazardous and radioactive waste sites.* EPA/540-R-96-004. Radiation Protection Division, Office of Radiation & Indoor Air, US EPA, Washington, D.C.

United States Environmental Protection Agency. (June 1996). *Radiation exposure and risks assessment manual (RERAM).* EPA/402-R-96-016. Radiation Protection Division, Office of Radiation & Indoor Air, US EPA, Washington, D.C.

United States Environmental Protection Agency. (November 1996). *Technology screening guide for radioactively contaminated sites.* EPA/402-R-96-017. Radiation Protection Division, Office of Radiation & Indoor Air, US EPA, Washington, D.C.

Department of Energy (DOE) - National Laboratories Site Environmental Reports

Almost all these citations are either site-specific environmental monitoring reports, which are often done annually for the larger weapons production facilities now under remediation, or are survey guides. The DOE is one of four participants sponsoring the MARSSIM, a multi-agency radiation survey manual. This flawed document provides interesting insight on current thinking within the federal government on how radiological monitoring should be done.

Berven, B.A., Cottrell, W.D., Leggett, R.W., Little, C.A., Myrick, T.E., Goldsmith, W.A. and Haywood, F.F. (1987). *Procedures manual for the ORNL radiological survey activities (RASA) program.* ORNL/TM-8600. Martin Marietta Energy Systems, Inc., Oak Ridge National Laboratory.

Mork, H.M., Larson, K.H., Kowalewsky, B.W., Wood, R.A. and Paglia, D.E. (July 1966). *Project SEDAN. Part I. Characteristics of fallout from a deeply buried nuclear detonation from 7 to 70 miles from ground zero. Part II. Aerial radiometric survey (final rept.).* AEC-PNE-225F. Atomic Energy Commission, Washington, D.C. pp. 117.

- "Adequate samples of fallout from the detonation of a nuclear device buried in desert alluvium at 635 feet below ground surface were obtained to delineate the eastern part of the fallout pattern from 7 to 70 miles from ground zero." (abstract).
- "The Aerial Radiometric Surveys, CETO Project 62.80, determined the distribution of Sedan fallout to a distance of more than 200 miles from ground zero. The dose rate contours show the pattern to be asymmetric with a steep gradient west of the midline with a very gradual gradient on the east." (abstract).

Myrick, T.E. et. al. (1981). *State background radiation levels: Results of measurements taken during 1975-1979.* ORNL/TM 7343. Oak Ridge National Laboratory, Oak Ridge, TN.

United States Department of Energy. (1991). *Environmental regulatory guide of radiological effluent monitoring and environmental surveillance.* DOE/EH-0173T. US DOE, Washington, D.C.

United States Department of Energy. (1992). *Environmental implementation guide for radiological survey procedures manual, DOE report for comment.* Martin Marietta Energy Systems, Oak Ridge National Laboratory.

United States Department of Energy. (1994). *Decommissioning handbook.* DOE/EM-0142P. US DOE, Washington, D.C.

Radiological Monitoring Programs and Remediation Guides - Programs Outside the USA

Aarkrog, A., Botter-Jensen, L., Chen Qing Jang, Dahlgaard, H., Hansen, H., Holm E., Lauridsen, B., Nielsen, S.P. and Sogaard-Hansen, J. (1991). *Environmental radioactivity in Denmark in 1988 and 1989.* Riso National Laboratory, Roskilde, Denmark.

Aarkrog, A. (1992). *Source terms and inventories of anthropogenic radionuclides.* Report No. DK-4000. Riso National Laboratory, Roskilde, Denmark.

Commission of the European Communities. (1989). *Council regulation (Euratom) No 3954/87 laying down the maximum permitted levels of radioactive contamination of foodstuffs and feedingstuffs following a nuclear accident or any other case of radiological emergency.* Off. J. Eur. Commun., 11(L371), amended by Council Regulation 2218/89 Off. J. Eur. Commun., 1(L211).

International Atomic Energy Agency. (1996). *International basic safety standards for protection against ionizing radiation and for the safety of radiation sources.* Saf. Ser. No. 115. IAEA, Vienna.

International Commission on Radiological Protection. (1977). Recommendations of the International Commission on Radiological Protection. *Annal. ICRP.* 1(3). Pergamon Press, Oxford, ICRP Publ. 26.

International Commission on Radiological Protection. (1991). 1990 recommendations of the International Commission on Radiological Protection. *Annal. ICRP.* 21(1-3). Pergamon Press, Oxford, ICRP Publ. 60.

International Commission on Radiological Protection. (1993). Principles for intervention for protection of the public in a radiological emergency. *Annal. ICRP.* 22(4). Pergamon Press, Oxford, ICRP Publ. 63.

International Commission on Radiological Protection. (1994). Age-dependent doses to members of the public from intake of radionuclides: Part 2 ingestion dose coefficients. *Annal. ICRP.* 23(3/4). Pergamon Press, Oxford, ICRP Publ. 67.

International Commission on Radiological Protection. (1996). Age-dependent doses to members of the public from intake of radionuclides: Part 5 compilation of ingestion and inhalation dose coefficients. *Annal. ICRP.* 26(1). Elsevier Science, Oxford, ICRP Publ. 72.

International Commission on Radiological Protection. (1996). Conversion coefficients for use in radiological protection against external radiation. *Annal. ICRP*. 26(3/4). Elsevier Science, Oxford, ICRP Publ. 74.

Ministry of Agriculture, Fisheries and Food. (1996). *Radioactivity in food and the environment, 1995*. RIFE-1. MAFF, London.

Ministry of Agriculture, Fisheries and Food and Scottish Environment Protection Agency. (1997). *Radioactivity in food and the environment, 1996*. RIFE-2. MAFF and SEPA, London.

Ministry of Agriculture, Fisheries and Food and Scottish Environment Protection Agency. (September 1998). *Radioactivity in food and the environment, 1997*. RIFE-3. Centre for Environment, Fisheries and Aquaculture Science, MAFF and SEPA, London.

- This series of reports is among the most important and comprehensive examples of a surveillance program for anthropogenic radioactivity in the environment carried out by any governmental entity in the world. RIFE stands in startling contrast (despite the tradition of government 'secrecy' in the United Kingdom) to the witless and evasive monitoring efforts of the United States federal government including the NRC and its licensees and the EPA and DOE.

- Routinely included in the RIFE surveillance program are many radionuclides absent from US monitoring activities: ^{14}C, ^{65}Zn, ^{95}Zr, ^{95}Nb, ^{99}Tc, ^{103}Ru, ^{106}Ru, ^{110m}Ag, ^{125}Sb, ^{134}Cs, ^{144}Ce, ^{154}Eu, ^{155}Eu, ^{237}Np, ^{238}Pu, ^{239,240}Pu, ^{241}Pu, ^{241}Am, ^{242}Cm, and ^{244}Cm. Small plumes of many of these radioisotopes are associated with discharges from not only Sellafield but many other nuclear facilities in the United Kingdom.

- Also, desultory NRC monitoring efforts are in contrast to the wide variety of biological media surveyed in the RIFE report as well as the detailed nature of the overall surveillance program in the United Kingdom. No United States publication even comes close to the comprehensive presentation and assessment of sampling and measurements contained in these reports.

- The RIFE surveillance program constitutes, in essence, what is completely missing from United States monitoring efforts: the proverbial "10-61 analyses," so-called because the United States Code of Federal Regulations Section 10 Part 61 requires NRC licensees to document comprehensively the radiological content of all solid low-level waste destined for near surface landfills. At the same time, the NRC is completely negligent in requiring comprehensive documentation of the environmental impact of gaseous and liquid releases from US facilities, both routine and non-routine.

- This report is particularly important in documenting the huge impact of fuel reprocessing activities at the Sellafield facility. Of particular importance is the documentation of the ^{99}Tc (technetium) pulse in lobsters and other biota (see pg. 53, 55, 76, 93, 99) derived from the new Thorp installation in Sellafield, and ^{241}Am (americium) pulse and contamination from other isotopes derived from the Sellafield facility. For more information about the ^{99}Tc pulse, surf to Greenpeace.
- "The report demonstrates that foodstuffs and seafood produced in and around the United Kingdom in 1997 are radiologically safe to eat and that the exposure of consumers to artificially produced radioactivity via the food chain remains well below UK and EU limits." (executive summary).
- "Natural radionuclides are the most important source of exposure in the average diet of consumers. Man-made radionuclides contributed less than 5% of the dose." (executive summary).
- "Estimated doses to high-rate fish and shellfish consumers, in the vicinity of Sellafield, from artificial radionuclides in the diet have decreased from 14% (in 1996) to 10% (in 1997) of the EU dose limit of 1 millisievert. The decrease was largely due to changes in the consumption of shellfish by these people." (executive summary).
- "The highest dose to members of the public in the UK from both artificial and natural radioactivity was estimated to be 0.49 millisieverts to high-rate fish and shellfish consumers in the Whitehaven area." (executive summary).
- "The highest doses in Scotland were also attributable to liquid wastes from Sellafield and were received by a group of high-rate fish and shellfish consumers in Dumfries and Galloway. Their dose was 0.047 millisievert. Technetium-99 contributed the single largest dose to this group, 0.010 millisievert in 1997, a reduction from 0.019 millisievert in 1996 due to a decrease in the detected levels in *Nephrops*." (executive summary).
- "Levels of technetium-99 in lobsters from the vicinity of Sellafield were again above those specified in the EU post-accident intervention levels and were comparable to 1996 levels. The assessed dose to the most exposed group of seafood consumers from technetium-99 discharges was less than 5% of the EU dose limit.' (executive summary).

Simmonds, J.R., Lawson, G. and Mayall, A. (1995). *Radiation protection 72; Methodology for assessing the radiological consequences of routine releases of radionuclides to the environment*. Report EUR 15760 EN. Office for Official Publications of the European Community, Luxembourg.

Post-Chernobyl Radiation Protection Guidelines

The following radiological protection guidelines date from after the Chernobyl accident and were issued by various European international governmental organizations (EEC, NEA, IAEA, Euratom). Many of these are referenced in *Chernobyl: Ten years on: Radiological and Health Impact* (OECD 1995).

Boeri, G. and Viktorsson, C. (1988). *Emergency planning practices and criteria in the OECD countries after the Chernobyl accident: A critical review*. OECD/NEA, Paris.

Crick, Malcolm. (1996). Nuclear and radiation safety: Guidance for emergency response. *IAEA Bulletin.* 38(1). pg. 23.

- The intervention level table from this paper is reprinted below under the 1994 IAEA citation *Safety Series No. 109.*

European Commission. (1989). *Council Regulation (Euratom) No. 3954/87 of 22 December 1987*.

- Laying down maximum permitted levels of radioactive contamination of foodstuffs and of feedingstuffs following a nuclear accident or any other case of radiological emergency.

European Commission. (1989). *Council Regulation (Euratom) No. 944/89 of 12 April 1989*.

- Laying down maximum permitted levels of radioactive contamination in minor foodstuffs following a nuclear accident or any other case of radiological emergency.

European Commission. (1989). *Council Regulation (EEC) No. 2219/89 of 18 July 1989*.

- On the special conditions for exporting foodstuffs and feedingstuffs following a nuclear accident or any other case of radiological emergency.

Joint FAO/WHO Food Standards Programme. (1991). Levels for Radionuclides. *Codex Alimentarius*. Vol. 1 section 6.1.

IAEA. (1994). Guidelines for agricultural countermeasures following an accidental release of radionuclides. *IAEA/FAO Technical Report Series* No. 363. IAEA, Vienna.

IAEA. (1994). *Safety Series No. 109, Intervention criteria in a nuclear or radiation emergency*. STI/PUB/900. IAEA, Vienna. pp. 117.

- "This Safety Guide, published in 1994, represents the international consensus reached on principles for intervention and numerical values for generic intervention levels. These principles and values subsequently became the basis of intervention guidance in the *Basic Safety Standards for Protection against*

Ionizing Radiation and for the Safety of Radiation Sources, which have been issued jointly by the IAEA, FAO, ILO, NEA, PAHO, and WHO." (Crick 1996).

- The table below is excerpted from Crick, 1996.

Generic intervention levels in emergency response situations		
	Urgent protective actions	
Action	**Avertable dose (Generic intervention level)**	
Sheltering	10 mSv for a period of no more than 2 days	1,000 rem
Iodine prophylaxis	100 mGy (committed absorbed dose to the thyroid)	
Evacuation	50 mSv for a period of no more than 1 week	5,000 rem

Generic action levels for foodstuffs		
(From the CODEX Alimentarius Commission Guideline levels for radionuclides in food moving in international trade following accidental contamination)		
Radionuclides	Foods destined for general consumption (kBq/kg)	Milk infant foods and drinking water (kBq/kg)
^{134}Cs, ^{137}Cs, ^{103}Ru, ^{106}Ru, ^{89}Sr	1	1 (1,000 Bq/kg)
^{131}I		0.1 (100 Bq/kg)
^{90}Sr	0.1	
^{241}Am, ^{238}Pu, ^{239}Pu	0.01	0.01 (10 Bq/kg)

Long-term actions	
Action	**Avertable dose (generic intervention level)**
Initiating temporary relocation	30 mSv in a month
Terminating temporary relocation	10 mSv in a month
Considering permanent resettlement	1 Sv in a lifetime

The Swedish company Studsvik RadWaste has developed a process it claims will reduce activity levels to below the IAEA release limit of 1 Bq/g. This process is called strong ozone decontamination process (SODP) and is used to recycle metals from decommissioned steam generators.

IAEA. (1996). Safety Series No. 115, *International basic safety standards for protection against ionizing radiation and for the safety of radiation.* STI/PUB/996. Jointly sponsored by FAO, IAEA, ILO, OECD/NEA, PAHO, and WHO. IAEA, Vienna. pp. 353.

IAEA. (August, 1997). *Generic assessment procedures for determining protective actions during a reactor accident.* IAEA-TECDOC-955. IAEA, Vienna. pp. 259.

- A comprehensive guideline to nuclear accident management procedures. Unfortunately, this text doesn't include clear and comprehensible contamination guidelines.

International Commission on Radiological Protection. (1984). *Protection of the public in the event of major radiation accidents: Principles for planning.* ICRP Publication No. 40. Pergamon Press, Oxford.

International Commission on Radiological Protection. (1989). *Age dependent doses to members of the public from intake of radionuclides: Part I.* ICRP publication No. 56. Pergamon Press, Oxford.

International Commission on Radiological Protection. (1992). Principles for intervention for protection of the public in a radiological emergency. *Annals of the ICRP.* Publication 63. 22(4).

MacLachlan, Ann. (December 10, 1998). Belgium cancels reprocessing, orders review of nuclear future. *Nucleonics Week.* 39(50). pg. 1.

- "Measurements showed [the] surface contamination [of a spent fuel cask] exceeded the strict 4 becquerel/square centimeter limit in one place."

OECD. (1997). *Radiation in perspective: Applications, risks and protection.* ISBN 92-64-15483-3. Organization for *Economic* Cooperation and Development, Paris. pp. 94.

OECD/NEA. (1989). Emergency planning in case of nuclear accident: Technical aspects. *Proceedings of a Joint NEA/CEC Workshop, Brussels, June 1989.* OECD/NEA, Paris.

OECD/NEA. (1990). *Protection of the population in the event of a nuclear accident.* OECD/NEA, Paris.

OECD/NEA. (1995). Short-term countermeasures after a nuclear emergency. *Proceedings of an NEA Workshop, June 1994, Stockholm.* OECD/NEA, Paris.

Richards, J.I. et. al. (1996). Standards and criteria established by international organizations for agricultural aspects of radiological emergency situations. *Proc. of NEA Workshop on the Agricultural Issues Associated with Nuclear Emergencies, (June 1995).* OECD/NEA, Paris.

World Health Organization. (1988). *Derived intervention levels for radionuclides in food.* WHO, Geneva.

- "The *calculated* values are of necessity based on an effective dose of 5 mSv for each nuclide *in isolation in a single food category*, since it is not possible to generalize regarding which nuclides will be most important in each food category after an accident." (pg. 26).

Guideline values for derived intervention levels (Bq/kg)								
Class of Radionuclide	Cereals	Roots and tubers	Vegetables	Fruit	Meat	Milk	Fish	Drinking water
I: High dose per unit intake factor (10^{-6} Sv/Bq)	35	50	80	70	100	45	350	7
II. Low dose per unit intake factor (10^{-8} Sv/Bq)	3,500	5,000	8,000	7,000	10,000	4,500	35,000	700

(Table 5 pg. 26)

Guideline values for derived intervention levels for [milk in] infants (Bq/l)	
^{90}Sr	160
^{131}I	1600[a]
^{137}Cs	1800
^{239}Pu	7
[a] Based on a mean life of 11.5 days for ^{131}I and an organ dose of 50 mSv to the thyroid.	

(Table 6 pg. 27)

World Health Organization. (1989). *Nuclear accidents: Harmonization of the public health response*. WHO *Regional* Office for Europe, Copenhagen.

Time	Indicator nuclides	Total nuclide inventory	Exposure mode	Pathway distance
1 hour	short-lived	$\pm 1\times10^8$ Ci	Inhalation, immersion	<50 miles
1 day	^{131}I, ^{132}Te, ^{99}Mo, ^{239}Nep		absorption	<1000 miles
1 week	^{103}Ru, ^{140}Ba, ^{95}Zr		Ingestion	2,000-5,000 miles
1 month	^{89}Sr, ^{134}Cs, ^{110m}Ag, ^{106}Ru		secondary pulse	hemispheric
1 year	^{154}Eu, ^{154}Ce, ^{90}Sr, ^{137}Cs, ^{241}Pu		tertiary pulse: remobilized long-lived radionuclides	
10 years	^{238,241}Pu			
100 years	^{241}Am			
1000 years	^{239}Pu			
10,000 years	^{99}Te, ^{237}Nep			
100,000 years	^{129}I			

VIII. Chernobyl Fallout Data

The following information is excerpted from the Center for Biological Monitoring archives. Most of the following citations were printed in the hard copy edition of *Chernobyl Tenth Anniversary: Fallout Data 1986 – 1996.* The extensive documentation of Chernobyl fallout in many nations provides an important database for the evaluation of the ongoing nuclear disaster in Japan. Among the most important issues highlighted by this database are the complexity of the Chernobyl accident source term and its release inventory of many isotopes. Also of note are the high levels of short-lived isotopes, which are associated with the early stages of the accident, and the widespread terrestrial distribution of Chernobyl-derived tropospheric fallout. Also of particular relevance to Japan emissions is the hot particle component of the accident source term, an issue that will impact Fukushima Daiichi decommissioning activities and exposure risks for generations. While tropospheric fallout over terrestrial landscapes was much more extensive as a result of the Chernobyl accident in comparison with the Japan MIME, much larger discharges of both radioactive water and of relatively heavy hot particles are associated with the Japan source term releases.

No attempt is made in this *Handbook* to evaluate the health physics impact of either the Japan or the Chernobyl disasters. Despite the passage of a quarter of a century since the Chernobyl disaster, evaluation of the casualties from Chernobyl ranges from just a few deaths to 975,000 related deaths (Yablokov 2009). The two quotations below pertaining to the health physics impact of the Chernobyl accident illustrate an entrenched and fairly widespread perspective still held by numerous experts (see the comments of Frank N. von Hippel in the May 5, 2011 edition of *The New York Times*) who have and will continue to emphasize the insignificance of the health physics impact of the Fukushima Daiichi accident. That individual exposures and dose assessments are averaged over national and world populations is one of the more reprehensible of our numerous equivocations about exposure to anthropogenic radioactivity. There are now at least 500,000 spent fuel assemblies in American nuclear power and weapons production facilities spent fuel pools with at least another 1.5 million assemblies accumulated on a worldwide basis. Many of these are stored within "subprime" boiling water reactors such as those at Fukushima Daiichi; others are located at operational reactors with inadequate reactor vessel containment structures such as that at Chernobyl. This growing legacy of nuclear waste will certainly be the source of future nuclear accidents. The Chernobyl and Fukushima Daiichi databases will be useful tools for the evaluation of future radiation releases. The optimistic take on the Chernobyl accident is summarized by these two notable quotations:

1. "The dose and subsequent health risk are minimal." (EML 1986).

202

2. "The health physics impact of a source term release will not have a significant impact on the overall cancer rate." (Goldman 1987).

No new information has been posted in this archive since 2001.

Introduction

RADNET Editorial Comments

As a preview to the annotated citations pertaining to Chernobyl-derived fallout, the editor of RADNET offers the following comments and observations:

- Nuclear safety experts had not anticipated that a nuclear accident would release this large an inventory of radionuclides.
- These nuclides were dispersed further, more erratically, and in much greater quantities than had been anticipated prior to the accident.
- At the time the accident was occurring, and during the weeks and months that followed, there was a widespread lack of accurate information about the seriousness and the radiological impact (deposition levels) of the accident.
- During and after the accident, official information sources ranged from unreliable (Russian and French government sources) to inaccurate (IAEA, National Radiological Protection Board, etc.). Political considerations and partisan prejudice in favor of nuclear energy production combined with the lack of environmental monitoring information and skewed objective accident analysis with the result that the impact of the accident was and continues to be minimized.
- This underestimation of the extent of the Chernobyl accident continues today in most official versions in terms of where and in what quantity deposition from the accident occurred.
- Only a few locations were equipped with sufficient instrumentation to make accurate real-time nuclide-specific measurements of the passage of the fallout cloud and its erratic rainfall-associated deposition.
- Rainfall events were the fundamental mechanism responsible for the extremely high deposition levels in some locations, including areas located thousands of kilometers from the accident site. Dry deposition played a lesser role in the spread of Chernobyl fallout than in weapons testing fallout events.
- Only a minimum of information has been collected about the actual levels of the dietary intake of Chernobyl-derived radionuclides for persons living in areas with high fallout - greater than 1 Ci/km^2 (37,000 Bq/m^2).
- The failure to measure accurately the dietary intake of specific population groups in the most affected areas and the general tendency to average dose equivalents

over large population groups (including estimating projected deaths as a percentage of hemispheric death rates) is particularly reprehensible.

- A reconsideration of the accident ten years later can only conclude that accurate information is still unavailable about actual deposition levels over vast areas of the Northern Hemisphere where millions of residents do not have access to accurate radiological monitoring data (Turkey, Iran, Iraq, North Africa, etc.).

- Even in countries with modest to excellent radiological monitoring capabilities, accurate information about the impact of the accident was not available in a timely manner and, in some cases, has never been made available.

- The United States serves as an example of the problem of freedom of information. While most areas of the United States received only a minimum of Chernobyl-derived fallout, some locations (See Dibbs, Maryland) received fallout which exceeded weapons testing deposition. The radiological surveillance data collected by the EML (Environmental Measurements Laboratory) and the EPA were either limited to a very small number of locations or, in the case of the EPA, did not include ground deposition data (Bq/m^2) or accurate air concentrations expressed in $\mu Bq/m^3$ (microbecquerels).

- Extensive data collected by the National Reconnaissance Office pertaining to the Chernobyl accident is not yet available to the general public.

- We welcome your comments on our editorial opinions. We also solicit additional citations pertaining to Chernobyl fallout.

- Articles cited in this section but not annotated were not present at hand for review.

- We will add citations and data to RADNET as they become available.

- This section of RADNET combines some editorial content with the data citations.

Chernobyl General Bibliography

Fusco, Paul and Caris, Magdalena. (2001). *Chernobyl Legacy: Twenty four minutes and zero seconds anti meridian*. de.MO, Millbrook, NY.

- A chilling and moving photo tour of the legacy of the Chernobyl accident.
- The cesium contamination maps show fallout levels ranging up to 7,400,000 Bq/m^2 in a spotty pattern over thousands of square miles to the north and northeast of Chernobyl, with lesser quantities of deposition on the European component of the map found later in the text.
- The limited written text is poetic, informative, concise, and haunting.

NOTICE TO THE READER: Levels of contamination cited within the Chernobyl data base are peak concentrations unless otherwise noted. Ground deposition activities varied widely in most areas impacted by the Chernobyl accident: A location receiving, for example, 40,000 Bq/m^2 could be only a few kilometers from another location receiving an order of magnitude less deposition. Nurmijarvi, Finland, a location with real time data collection capabilities, recorded the highest air concentrations of any location cited in RADNET (over thirty Chernobyl-derived nuclides were observed); ground deposition activities at this location, while elevated, were typical of many locations receiving heavy rainfall associated fallout. The data cited for both ground deposition and contamination of abiotic and biotic media which follow are the highest readings in the survey being cited, unless otherwise indicated.

Estimated Release of Long-Lived Radionuclides from the Chernobyl Accident

Aarkrog, A. (1994). *Source terms and inventories of anthropogenic radionuclides*. Riso National Laboratory, Roskilde, Denmark.

Radionuclide	Total released radioactivity (Curies)
^{137}Cs	2,700,000
^{134}Cs	1,350,000
^{90}Sr	216,000

Radionuclide	Total released radioactivity (Curies)
^{106}Ru	948,000
^{144}Ce	2,430,000
^{110m}Ag	40,500
^{125}Sb	81,000
^{239,240}Pu	1,480
^{238}Pu	700
241Pu	135,000
^{241}Am	162
^{242}Cm	16,200
^{243,244}Cm	162

- This incomplete source term release will be updated with a more complete description of the total nuclide inventories released from the Chernobyl accident if and when the tenth anniversary report of the Chernobyl accident listing the revised release estimates is received from the OECD/NEA. The current estimates listed above derive from a world health organization report in 1989 which may underestimate the actual release activity during the accident. Many earlier reports contain even larger underestimations of the actual release during the accident, and, in fact, an exact source term estimate for all radionuclides released in the Chernobyl accident may never be possible. For a more detailed analysis of the release dynamics of the Chernobyl accident and the many mysteries surrounding exactly what transpired during the accident, see the publications of Alexander Sich, 1994 etc., which are reviewed in the following pages. It has taken almost a decade for an accurate analysis of the accident dynamics to emerge from the official evasions and misinformation which characterized the early reports on Chernobyl.

SIZES OF CONTAMINATED TERRITORIES IN THE FORMER USSR
(Measured in thousands of curies per square meter)

206

Aarkrog, A., Tsaturov, Y. and Polikarpov, G.G. (1993). *Sources to environmental radioactive contamination in the former USSR*. Riso National Laboratory, Roskilde, Denmark.

States	Sizes of contaminated territories, km^2			
	37-185 kBqm^2	185-555 kBqm^{-2}	0.55-1.5 MBqm^{-2}	>1.5 MBq^{-2}
Russia	48100	5450	2130	310
Byelorussia	29920	10170	4210	2150
Ukraine	37090	1990	820	640
Moldova	50	-	-	-
Total	115160	17160	7160	3100

- The Chernobyl accident, if contaminated areas outside the USSR are included, resulted in the deposition of long-lived radionuclides in excess of 37,000 Bq/m^2 (1 Ci/km^2) on ±200,000 km^2 of the world's surface. Areas impacted by iodine-131, ruthenium-103, tellurium-132, barium-140, and other short-lived isotopes (1/2 T = 1 week to 1 yr), along with the longer-lived isotopes, to levels exceeding 37,000 Bq/m^2, may have exceeded 1,000,000 km^2 in the weeks after the accident.
- The primitive maps reproduced in this publication show extensive contamination not only in Byelorussia, but also throughout central Russia. With each passing year, our knowledge of the extent of the deposition from the Chernobyl accident grows larger as more information is collected and collated and the parameters of Chernobyl-derived deposition in excess of one curie per square kilometer are expanded.
- An accurate radiometric survey of the hemispheric impact of the Chernobyl accident would probably reveal significant additional contamination in locations such as Turkey, Iran, Iraq, North Africa, and possibly even areas in the Far East and in North America.
- The National Reconnaissance Office has extensive additional radiological surveillance data pertaining to the Chernobyl accident which is not available to the general public because it is classified.
- The USSR contamination estimates were republished by Aarkrog, et al, 1993, in the above citation from a UNSCEAR publication which was citing a Russian source (Israel, Y.A., Tsaturov Y.S., et al., 1991).

207

Aarkrog, A., Angelopoulos, A., Calmet, D., Delfanti, R., Florou, H., Permattei, S., Risica, S. and Romero, L. (1993). *Radioactivity in Mediterranean waters: Report of working group II of CEC project MARINA-MED*. Riso National Laboratory, Roskilde, Denmark.

Date	Location	Media	Nuclide	Activity*
1984	Aegean Sea	Fish	^{137}Cs	0.53 Bq/kg mean value
1984	Tryrhenian Sea	Fish	^{137}Cs	0.10 Bq/kg mean value
1986	Aegean Sea	Fish	^{137}Cs	4.9 Bq/kg mean value
1986	Black Sea	Fish	^{137}Cs	2.0 Bq/kg mean value
1990	Black Sea	Fish	^{137}Cs	3.3 Bq/kg mean value
1985	Tyrrhenian Sea	Shellfish	^{137}Cs	0.36 Bq/kg mean value
1986	Tyrrhenian Sea	Shellfish	^{137}Cs	14.0 Bq/kg mean value
1990	Tyrrhenian Sea	Shellfish	^{137}Cs	3.2 Bq/kg mean value
1990	Black Sea	Surface sediments	^{137}Cs	164.0 Bq/kg mean value

- The Black Sea was more impacted by the Chernobyl accident than the other Mediterranean sea basins; it was still showing the cumulative effects of the accident in 1990.
- The data was collected by a number of countries adjacent to the Mediterranean Sea, and is an extensive summary of the mean values, with a sea-by-sea survey of the major Mediterranean basins.

Aarkrog, A. (1988). The radiological impact of the Chernobyl debris compared with that from nuclear weapons fallout. *J. Environ. Radioactivity.* 6. pg. 151-162.

- "Transfer factors are strongly influenced by seasonal and geographical distributions. For example, if 1,000 Bq of 137 per m^2 are deposited over a barley field three months before harvest, the concentration in the mature grain will be 1 Bq ^{137}Cs/kg. If on the other hand contamination, with the same deposition, occurs one month before harvest, the mature grain will contain approximately 100 Bq ^{137}Cs/kg." (pg.155).

- "The mean concentration in Danish grain in 1962-74 was 7.1 Bq ^{137}Cs/kg. In 1986 the mean level was 3.3 Bq." (pg. 157) This illustrates the efficiency and uniformity of stratospheric fallout contamination compared to the erratic distribution patterns of Chernobyl-derived radiocesium, which did not significantly affect Denmark during the growing season.

Andersson, K.G. and Roed, J. (1994). The behavior of Chernobyl ^{137}Cs, ^{134}Cs and ^{106}Ru in undisturbed soil: Implications for external radiation. *J. Environ. Radioactivity*. 22. pg. 183-196.

- "The URGENT computer model developed at Riso has shown that as much as 89% of the dose to urban populations came from contamination on the soil surface in open areas such as gardens and parks." (pg. 183).
- Cesium remained strongly bound in the topmost 2 cm of soil associated with a mineral fraction; ruthenium was associated with an organic fraction; external exposure is the primary exposure pathway four years after the initial deposition.

Andersson, K.G. and Roed, J. (2006). *Estimation of doses received in a dry-contaminated residential area in the Bryansk region, Russia, since the Chernobyl accident*. Journal of Environmental Radioactivity, Volume 85, Issues 2-3, Amsterdam, The Netherlands. pg. 228-240.

Anspaugh, L.R., Catlin, R.J. and Goldman, M. (1988). The global impact of the Chernobyl reactor accident. *Science*. 242. pg. 1513-1519.

- "By means of an integration of the environmental data, it is estimated that ~100 petabecquerels of cesium-137 (1PBq = 10^{15} Bq) were released during and subsequent to the accident." (pg. 1513).

Apsimon, H.M., Gudiksen, P., Khitrov, L., Rodhe, H. and Yoshikawa, T. (1988). Lessons from Chernobyl: Modeling the dispersal and deposition of radionuclides. *Environment*. 30(5) pg. 17-20.

- "Localized peaks of wet deposition (in excess of 100 kilobecquerels per square meter) occurred in parts of Central Scandinavia." (pg. 18).
- "Deposition of the most important long-lived nuclide, ^{137}Cs, did not decrease smoothly with travel distance but was enhanced when rain or snow interrupted the plume." (pg. 18).
- The estimate for plume transport atmospheric height ranged from 4 km to 10 km. (pg. 19).

Apsimon, H.M., MacDonald, H.F. and Wilson, J.J.N. (1986). An initial assessment of the Chernobyl-4 reactor accident release source. *J. Soc. Radiol. Prot.* 6(3) pg. 109-119.

- Long range atmospheric dispersion model, MESOS, was used to provide a preliminary estimate of the accident source term release; 15-20% of iodine, tellurium and cesium and 1% or less of ruthenium and other isotopes was the estimated release.
- Relatively low airborne concentrations of Chernobyl-derived radionuclides were observed in comparison to ground deposition levels noted by other researchers (See EML-460).
- This is another in a series of early underestimations of the severity of the Chernobyl accident and the extent of the erratic fallout patterns which characterized the plume pulse pathway.

Balter, Michael. (December 15, 1995). Radiation biology: Chernobyl's thyroid cancer toll. *Science.* 270(5243). pg. 1758.

- "Geneva--radiation scientists now accept that the large increase in childhood thyroid cancers, particularly in Belarus and Ukraine, is the result of radiation released by the Chernobyl nuclear accident. The new focus is on trying to explain why the cancer epidemic is so virulent." (abstract).

Bedyaev, S.T., et. al. (1991). The Chernobyl source term. *Proc. Seminar on Comparative Assessment of the Environmental Impact of Radionuclides Released During Three Major Nuclear Accidents: Kyshtym, Windscale, Chernobyl.* EVR-13574, CEC. pg. 71-91.

Beskorovajnyj, V.P., et. al. (1995). Radiation effects of collapse of structural elements of the sarcophagus. *Sarcophagus Safety '94: Proceedings of an International Conference, Zeleny Mys, Chernobyl, Ukraine, March 14-18, 1994.* OECD/NEA, Paris. pg. 196-202.

- This publication also includes the following titles:
  - "Hydrogeological Effects of the Principal Radioactive Waste Burial Sites Adjacent to the Chernobyl NPP."
  - "The Current State of the Regulations on the Safety of Unit 4 at the Chernobyl NPP."
  - "Hypothetical Accidents in the Sarcophagus."
  - "Design of a Shelter - Experience of Planning and Construction in 1986."
  - "Current State of the Sarcophagus and Safety Problems."

Beardsley, T. (1986). US analysis incomplete. *Nature.* 321. pg. 187.

- "One of the highest atmospheric air concentrations recorded outside the Eastern Bloc, was in Stockholm, where a level of 5,130 pCi (190,000,000 micro becquerels) of ^{131}I per cubic meter of air was found." (pg. 187).

Begichev, S.N., Borovoi, A.A., Burlakova, E.V., A. Y. Gagarinsky, Demin, V.F., Khodakovsky, I.L. and Khurlev, A.A. (1990). Radioactive releases due to the Chernobyl accident. In: *Fission product transport processes in reactor accidents.* J.T. Rogers (ed.). Hemisphere.

Beninson, D. and Lindell, B. (1986). *Chernobyl reactor accident: Report of a consultation, 6 May 1986.* Report No. ICP/CEH. World Health Organization, Copenhagen, Denmark.

- While this report contains little or no data, it does have a list of remedial actions and a preliminary review of some precautions taken by a number of countries affected by the Chernobyl accident. (Fig. 10, pg. 30).

Borovoi, A.A. and Sich, A.R. (1995). The Chernobyl accident revisited, part II: The state of the nuclear fuel located within the Chernobyl sarcophagus. *Nuclear Safety.* 36 (1).

- The second in a series of articles in Nuclear Safety by A.R. Sich and, in their totality, the best summary of the Chernobyl accident available in the literature.
- "Approximately 135 tonnes of the 190.3-tonne initial core fuel load (~71%) at Chernobyl Unit 4 melted and flowed into the lower regions of the reactor building to form various kinds of the now-solidified lava-like fuel-containing materials (LFCMs) or corium." (pg. 1).
- Excellent descriptions and photographs of the sarcophagus which was built over the ruined reactor after the accident, with a detailed analysis of the location of the melted and resolidified fuel in the lower regions of the reactor building.
- "Investigations conducted during 1986 to 1989 showed that previous notions concerning the extent of damage within Unit 4 as a result of the accident in most cases did not correspond to the actual state of the destroyed reactor." (pg. 8).
- Contents of the sarcophagus are listed as including the following: "1,270 and 1,350 tonnes of fuel-containing materials (FCMs) (material containing ~ 10.5% of partially 'burned' nuclear fuel), 64,000 m^3 of other radioactive material (concrete, building metal, etc.), approximately 10,000 tonnes of construction

metal, and 800 to 1,000 tonnes of contaminated water are located within the sarcophagus." (pg. 15).

- "A considerable amount of ^{137}Cs (35%) remains within the solidified remnants of the core....significantly higher than that retained at TMI-2 in the molten ceramic lower plenum debris (average of 3% retained) or in the upper plenum debris (average of 19% retained)." (pg. 29).

Burkart, W. et. al. (1991). Assessing Chernobyl's radiological consequences. *Nuclear Europe Worldscan.* 1(3-4). pg. 27-30.

Buzulukov, Y.P. and Dobrynin, Y.L. (1993). Release of radionuclides during the Chernobyl accident. In: *The Chernobyl Papers.* Merwin, S. E. and Balonov, M.I., (eds.) Research Enterprises, Richland, WA. 1. pg. 321.

Cambrai, R.S. et. al. (1987). Observations on radioactivity from the Chernobyl accident. *Nuclear Energy.* 26. pg 77.

Devell, L. et. al. The Chenobyl reactor accident source term: Development of a consensus view. *CSNI Report* in preparation. OECD/NEA, Paris.

Dickerson, M.H. and Sullivan, T.J. (1986). *ARAC response to the Chernobyl reactor accident.* (Under US Department of Energy Contract W-7405-Eng-48). Lawrence Livermore National Laboratory, Livermore, CA.

- This report illustrates the lack of centralized facility in the US for accurate "real-time" analysis of radioactive contamination and the failure of existing computer models to predict accurately the erratic ground deposition of Chernobyl fallout patterns.
- Deposition levels in Europe were grossly underestimated.
- "Detection of BA-140 and Zr-95 in Sweden implied a significant meltdown." (pg. 12).
- "An amount of 9000 pCi/l was estimated as the maximum expected I-131 concentration in milk for the U.S..." (pg. 13). (No contamination levels of this magnitude inside the US were noted in the citations reviewed to date for RADNET.)

Dickman, S. (1988). IAEA's verdict on Chernobyl. *Nature.* 333. pg. 285.

- "According to one IAEA official... on the basis of a study of 30,000 people living in the (Chernobyl) area, no adverse health effects on the general population had been attributed to the radiation." (pg. 285).

212

- "Although there are still a few hot spots, most of the area within 10-30 km from the reactor has returned to normal levels of activity." (pg. 285).
- Extraordinary misinformation from one of the most preeminent scientific journals; this IAEA editorial rhetoric is completely contradicted by other reports and data.

Editorial. Anxiety about reactor accident subsides. (May 8, 1986). *Nature.* 321. pg. 100.

- This news summary is the paradigm of misinformation and selective interpretation of inadequate data and is an example of the biased reporting that characterized much of the Chernobyl-related editorial content of Nature in the first few months after the Chernobyl accident. This biased editorial reporting contrasts with the many objective scientific reports and papers which Nature published after the Chernobyl accident.

Eremeev, V.N., Ivanov, L.M., Kirwan, A.D. Jr. and Margolina, T.M. (1995). Amount of ^{137}Cs and ^{134}Cs radionuclides in the Black Sea produced by the Chernobyl accident. *Journal of Environmental Radioactivity.* 27(1). pg. 49-63.

Gittus, J.H., Hicks, D., Bonell, P.G., Clough, P.N., Dunbar, I.H., Egan, M.J., Hall, A.N., Hayns, M.R., Nixon, W., Bulloch, R.S., Luckhurst, D.P., Maccabee, A.R., Edens, D.J. (1988). *The Chernobyl accident and its consequences.* Report No. NOR 4200. United Kingdom Atomic Energy Authority, London.

- Extensive discussion of how the accident happened.
- Little specific fallout data.
- Gross underestimation of radiological impact of the accident: inaccurate and overly generalized radiation dispersion maps.

Goldman, M. (1987). Recalculating the cost of Chernobyl. *Science.* 236 pg. 658-659.

- Global fatal cancers estimated at 39,000, most of them outside the Soviet Union.

Goldman, M. (1987). Chernobyl: A radiobiological perspective. *Science.* 238. pg. 622-623.

- Radiocesium release was calculated to be 2.4 million curies (US DOE).
- Global fatal cancer ratio assessment of up to 28,000 deaths.

- This is one of many fluctuating estimates of deaths resulting from Chernobyl, none of which will allegedly have a statistically significant impact on the overall cancer rate.
- *The New York Times* (1995, date unavailable) has reported a sharp drop in the life expectancy of the Russian population since the Chernobyl accident. What role Chernobyl played in the drop is unknown.

Gudiksen, P.H., Harvey, T.F. and Lange, R. (1989). Chernobyl source term, atmospheric dispersion and dose estimation. *Health Physics.* 57(5). pg. 697-706.

Gudiksen, P.H. and Lange, R. (1986). Atmospheric dispersion modeling of radioactivity releases from the Chernobyl event. Report No. UCRL- 95363, Preprint. Lawrence Livermore National Laboratory, Livermore, CA.

- This report illustrates the unreliability of computer models in estimating atmospheric dispersion from a nuclear accident, particularly in the early stages of an accident with limited data availability.
- Neither the calculated nor the measured deposition levels seem to match data collected by other researchers.

Hohenemser, C., Deicher, M., Ernst, A., Hofsass, H., Lindner, G. and Recknagel, E. (1986). Chernobyl: An early report. *Environment.* 28(5). pg. 6-43.

Date	Location	Media	Nuclide	Activity*
April 28, 1986	Forsmark, Sweden	Ground deposition	^{132}I	120,000 Bq/m^2
April 28, 1986	Forsmark, Sweden	Ground deposition	^{131}I	4,000 Bq/m^2
April 28, 1986	Forsmark, Sweden	Rainwater	^{132}I	839,000 Bq/l
April 30, 1986	Konstanz, Germany	Ground deposition	^{132}Te	87,000 Bq/m^2

- "In Konstanz the current ground activity of cesium-137 is estimated at 8,000-12,000 Bq/m^2, whereas the global weapons-testing fallout peak in West Germany was 800 Bq/m^2 in 1963." (pg. 36).
- "During passage of the cloud peak air radionuclide concentrations reached 100,000 times background levels in Poland and as high as 10,000 times background in Scotland." (One million times background equals 2,000 Bq/m^3.) (pg. 35).

214

Hotzl, H., Rosner, G. and Winkler, R. (1989). Long-term behavior of Chernobyl fallout in air and precipitation. *J. Environ. Radioactivity*. 10. pg. 157-171.

- "Ground level air concentrations... of ^{137}Cs in autumn 1986 were 100 times fallout values in 1985, and decreased by the end of 1987 to only 30 times the weapon fallout level. This very slow rate of decrease was not expected." (pg. 158).

Institut de Protection et de Surete Nucleaire. (1986). *The Tchernobyl accident*. Report No. IPSN 2/86, rev. 3. Institut de Protection et de Surete Nucleaire, Fontenay-aux-Roses.

Date	Location	Media	Nuclide	Activity*
April 26-May 6	Chernobyl	Total activity released per family	Noble gases	100%: 1x 10^8 Ci
April 26-May 6	Chernobyl	T.A.R.P.F.	Iodine	20%: 8.4 x 10^6
April 26-May 6	Chernobyl	T.A.R.P.F.	Cesium	15%: 1.2 x 10^6
April 26-May 6	Chernobyl	T.A.R.P.F.	Tellurium	15%: 1.0 x 10^7
April 26-May 6	Chernobyl	T.A.R.P.F.	Rutheniums and Rhodiums	4%: 1.6 x 10^7
April 26-May 6	Chernobyl	T.A.R.P.F.	Lanthanides	3%: 1.2 x 10^7
April 26-May 6	Chernobyl	T.A.R.P.F.	Zirconium	3%: 3.9 x 10^6
April 26-May 6	Chernobyl	T.A.R.P.F.	Actinides: alpha activity	3%: 2.3 x 10^4
			beta activity	3%: 2.3 x 10^6

- These preliminary source term estimates are for a core inventory with a cooling time of one hour. Total released activity is estimated at 1.58×10^8 Ci (158,000,000 Ci) including the noble gases. (pg. 73).
- "The Soviets distinguish between 4 phases in the main release which lasted 9 days." (pg.71).
  - Phase One: April 26: Mechanical dispersion of slightly enriched fuel (2.2×10^7 Ci).
  - Phase Two: April 27-May 1: Falling release level; diminishing graphite fire (2.2×10^7 Ci).
  - Phase Three: May 2-5: The core heats to a temperature exceeding 2000 degrees centigrade. Reactions occur between ^2O and graphite, fission product aerosols combine with graphite particles (2.7×10^7 Ci).
  - Phase Four: May 5-6: Rapid falloff in fission product emission due to halting of the fission process. (1×10^5 Ci).
- "Discharge of radioactive products into the atmosphere continued through the end of August at the rate of a few curies per day." (pg. 1).
- This revised early report still underestimates the source term release but is more accurate and comprehensive than the other reports presented at the IAEA conference at Vienna on August 25-29, 1986.

International Atomic Energy Agency. (1986). *The accident at Chernobyl nuclear power plant and its consequences.* Information compiled for the IAEA expert's meeting August 25-26, 1986, Vienna, Austria, USSR State Committee on the Utilization of Atomic Energy. (IAEA translation).

- This report is full of errors, incorrect descriptions of how the accident happened and incomplete or inaccurate information about the impact of the accident.
- A major blow to the credibility of the International Atomic Energy Agency and a graphic illustration of the unavailability of accurate information about the Chernobyl accident in the months after it occurred; much of the contents of this report can no longer be relied on to provide accurate information about the Chernobyl accident.

International Atomic Energy Agency. (1991). *The International Chernobyl Project - Assessment of radiological consequences and evaluation of protective measures.* Report by an International Advisory Committee. IAEA, Vienna.

International Atomic Energy Agency. (1991). *The International Chernobyl Project, surface contamination maps.* IAEA, Vienna.

International Atomic Energy Agency. (1991). *The International Chernobyl Project, technical report*. IAEA, Vienna.

Ilyin, L.A. and Pavlovskij, A.O. (1987). Radiological consequences of the Chernobyl accident in the Soviet Union and measures taken to mitigate their impact. *IAEA Bulletin* 4.

International Nuclear Safety Advisory Group. (1986). *INSAG summary report on the post-accident review meeting on the Chernobyl accident* (INSAG report to International Atomic Energy Agency general conference, Vienna, Austria, August 1986). Vienna. IAEA translation.

- This report includes the core inventory of radionuclides provided by Soviet authorities to the International Atomic Energy Agency at this conference and reprinted in RADNET under Warman, E.A. (1987) in this section.
- The core inventory of plutonium-239 at Chernobyl at the time of the accident is listed as 23,000 curies. For comparison with plutonium inventories at both US nuclear power facilities and at US DOE plutonium production facilities, RADNET readers are urged to refer to RADNET Section 11: Anthropogenic Radioactivity: Major Plume Source Points: US Military Source Points, Plutonium the first 50 years, and the annotations which follow this citation.

Jaworowski, Z. and Kownacka L. (1988). Tropospheric and stratospheric distributions of radioactive iodine and cesium after the Chernobyl accident. *J. Environ. Radioact.* 6. pg. 145-150.

Kirchner, G. and Noack, C.C. (1988). Core history and nuclide inventory of the Chernobyl core at the time of the accident. *Nuclear Safety*, 29(1). pg. 1-5.

- "Any calculation of the radionuclide inventory of the Chernobyl core at the time of the accident... requires the specification of burnup and detailed irradiation history of the reactor core prior to the accident - data not accessible as yet." (pg. 1).
- The reactor vessel inventory of nuclides in this report is listed in Table 2 and is the calculated concentrations of selected nuclides per ton of initial heavy metal at the time of the accident. A note at the bottom of the table lists fuel loading of the Chernobyl core at 192 tons.
- The calculated concentration of ^{137}Cs is listed as 1.6 x 10^{15} Bq/ton; ^{239}Pu is calculated at 4.7 x 10^{12} Bq/ton; 28 other nuclide concentrations are calculated in this table. (pg. 4).

Komarov, V.I. (1990). Radioactive contamination and decontamination in the 30 km zone surrounding the Chernobyl Nuclear Power Plant. Report No. IAEA-SM[-3]06/124. In: *Environmental contamination following a major nuclear accident, Vol. 2*. Report No. STI/PUB/825. International Atomic Energy Agency, Vienna.

Krey, P.W. (1986). International data exchange and cooperative research. In: *Environmental Measurements Laboratory: A compendium of the environmental measurements laboratory's research projects related to the Chernobyl nuclear accident: October 1, 1986*. Report No. EML-460. US Department of Energy, New York, NY. pg. 259-264.

- Chernobyl fallout conclusions (pg. 259-260):

  o "The dose and subsequent health risk to the population of Western Europe are minimal."
  o "Although fallout levels in Russia and Eastern Europe are not now known, circumstances would have been much worse had there been rain immediately following the accident."
  o "There was evidence of several pulses of Chernobyl fallout in Western Europe."
  o "The relative amount of gaseous ^{131}I was large and variable... deposited ^{131}I was distilled out of the soil back into the atmosphere during daylight hours."

Kryshev, I.I. (1995). Radioactive contamination of aquatic ecosystems following the Chernobyl accident. *J. Environ. Radioact.* 27(3). pg. 207-219.

Likhtarev, L.A. et. al. (1989). Radioactive contamination of water ecosystems and sources of drinking water. *Medical Aspects of the Chernobyl Accident*. TECDOC 516. IAEA, Vienna.

Morrey, M., Brown, J., Williams, J.A., Crick, M.J., Simmonds, J.R. and Hill, M.D. (1987). *A preliminary assessment of the radiological impact of the Chernobyl reactor accident on the population of the European community*. (Report from Health and Safety Directorate No. V/E/1 funded under CEC contract number 86 398). Commission of the European Communities, Luxembourg.

Date	Location	Media	Nuclide	Activity*
May 1986	S. Germany	Ground deposition	[131]I	240,000 Bq/m^2
May 1986	S. Germany	Milk	[131]I	17,000 Bq/l

- 17,000 Bq/l = 1,020,000 pCi/liter.
- This report contains detailed media specific Chernobyl-derived activity levels for many European countries as well as an interesting evaluation of the availability of environmental monitoring data at the time of this study (See Table B-2, pg. 40).
- This report comes in two sections. Section one contains dose assessments. Section two contains the appendices with all the environmental monitoring data, as well as additional dose estimates. Section two also contains information about countermeasures taken in each country.

Oak Ridge National Laboratory. *The use of Chernobyl fallout data to test model predictions of the transfer of* ^{131}I *and* ^{137}Cs *from the atmosphere through agricultural food chains*. Report CONF-910434-7. F. O. Hoffman Oak Ridge National Laboratory, TN.

OECD. (1987). *The radiological impact of the Chernobyl accident in OECD countries*. Organization for Economic Cooperation and Development, Paris.

- This is a lengthy and detailed review of the Chernobyl accident and its impact throughout the northern hemisphere. At first glance it would seem to be the definitive summary of the radiological impact of the Chernobyl accident, particularly in view of the polychrome radiometric maps which appear to document the fallout patterns in a number of countries (not all of the maps are in color, but the ones that are look very impressive). A close comparison of the maps with many of the papers and the data they contain cited in RADNET illustrate the continued underreporting of the actual radiological impact of the Chernobyl accident.
- The gross underestimation of fallout in the United Kingdom, much of which was initially estimated by unreliable surface fallout measurements, is a paradigm for how inaccurate even the most professional analysis of a nuclear accident can be.
- Data within many articles annotated in this website indicate that actual fallout levels are neither as low as generally indicated in this publication, nor as uniform as shown on many of the fallout maps.

OECD. (1989). The influence of seasonal conditions on the radiological consequences of a nuclear accident. *Proceedings of an NEA workshop, Paris, September 1988*. OECD/NEA, Paris.

OECD, NRC and IAEA. (May 1995). *Sarcophagus safety '9: The state of the Chernobyl Nuclear Power Plant Unit 4.* 66-95-10-1. ISBN 92-64-14437-4. Organization for Economic Cooperation and Development, Paris.

- "Nine years after the Chernobyl disaster, scientific data for remedial and recovery programmes still need to be assembled and evaluated. Many questions must be addressed before the site can be radiologically stabilized and environmental remediations can be found. Can the nuclear and radiation safety conditions of the site be assured? What is the state of integrity of the 'sarcophagus'? What is the nature and degree of the radioactive contamination?" (abstract).

OECD. (November 1995). *Chernobyl ten years on: Radiological and health impact: An assessment by the NEA Committee on Radiation Protection and Public Health.* Organization for Economic Cooperation and Development, Paris.

- This report is available on the Internet at URL: http://www.nea.fr/html/rp/chernobyl/chernobyl.html.
- This report is an update on the Chernobyl accident with a particular emphasis on the radiological and health impact; the bibliography of this report cites a large number of research projects pertaining to this subject and is probably the largest single compilation of health physics related data-derived from the Chernobyl accident available in one location.
- This report also includes the accident source term release as well as interesting maps denoting the "main spots" of ^{137}Cs contamination within the former Soviet Union. It is interesting to note that "main spots" are defined as those areas with a ground deposition greater than 555,000 becquerels/m^2 (555 kBq/m^2)(Fig. 5). The report notes that large areas of Ukraine and Belarus had ground deposition of ^{137}Cs over 40,000 becquerels/m^2 (40 kBq/m^2).
- "The most highly contaminated area was the 30-km zone surrounding the reactor where ^{137}Cs ground depositions generally exceeded 1,500 kBq/m^2 ... the ground depositions of ^{137}Cs in the most highly contaminated areas ... (The Bryansk-Belarus spot, centered 200 km to the North-northeast of the reactor) ... reached 5,000 kBq/m^2 " (5 million Bq/m^2).
- Minimal information is given about contamination outside the former Soviet Union.
- Figure 6 gives a graphic illustration of satellite-derived data of the areas covered by the main body of the radioactive cloud on various days during the release, as provided by the ARAC (Atmospheric Release Advisory Capability), Lawrence Livermore Laboratory, Livermore, CA. These satellite-derived photographs

220

provide an excellent overview of contamination dissemination but are not helpful in accurately describing actual ground deposition levels. The photographs in this report first appeared in a 1986 ARAC report: see Dickerson (1986) also in this section of RADNET.

- The remainder of this report is primarily concerned with:
    - (III) reactions of national authorities
    - (IV) dose estimates
    - (V) health impact
    - (VI) agricultural and environmental impacts (containing the above-mentioned maps)
    - (VII) potential residual risks
    - (VIII) lessons learned

Weapons Testing Fallout vs. Chernobyl Fallout vs. US Reactor Accident

Maximum annual weapons-testing-derived ^{137}Cs deposition: 1,000 Bq/m^2 (See Riso National Laboratory Cumulative Fallout Record)
OECD-NEA definition of "main" ^{137}Cs Chernobyl deposition: >555,000 Bq/m^2 (See above citation)
FDA-FEMA Emergency Action Guideline for radiocesium ground deposition following a nuclear reactor accident in the United States: 90 microcuries radiocesium/m^2 = 3,308,323 Bq/m^2 (begin destroying rather than storing contaminated food)

OECD. (1996). *The Chernobyl reactor accident source term*. Report No. OCDE/GD(96)12. Organization for Economic Cooperation and Development, Paris.

- The OECD Nuclear Energy Agency (NEA) is in the process of issuing an updated report on the Chernobyl accident and its radiological and health impact which will be issued on the occasion of the tenth anniversary of the accident. This OECD report on the reactor accident source term is one component of the larger report. It summarizes the research pertaining to the inventory of reactor nuclides and the percentage of these inventories released to the environment during the accident.

- This report contains an extensive bibliography which includes many publications pertaining to the reactor vessel inventories and source term releases, only a few of which are cited in RADNET.
- Reactor inventories for ^{137}Cs are estimated at between 2.2 x 10^{17} Bq and 2.9 x 10^{17} Bq; seven different reactor inventory estimates are included in this report.
- The percentage of the reactor inventory of cesium-137 estimated to have been released (source term release) is 33 + 10, indicating that, out of 6.95 x 10^6 to 7.84 x 10^6 curies of radiocesium, approximately 40% was released to the environment.

Parmentier, N. and Nenot, J-C. (1989). Radiation damage aspects of the Chernobyl accident. *Atmospheric Environment*. 23. pg. 771-775.

Powers, D.A., Kress, T.S. and Jankowski, M.W. (1987). The Chernobyl source term. *Nuclear Safety*. 28(1). pg. 10-28.

- "The prolonged second stage of the release is not... well understood. Physical and chemical processes not likely to develop during LWR accidents may be responsible for the release during this stage of the accident." (pg. 27).
- Another of the early misinterpretations of the extent of the Chernobyl source term release.

Rezzoug, S. Michel, H., Fernex, F., Barci-Funel, G., and Barci, V. (2006) Evaluation of ^{137}Cs fallout from the Chernobyl accident in a forest soil and its impact on Alpine Lake sediments, Mercantour Massif, S.E. France. *Journal of Environmental Radioactivity*, Volume 85, Issues 2-3, Amsterdam, The Netherlands. pg. 369-379.

Scheid, W., et. al. (1993). Chromosome aberrations in human lymphocytes apparently induced by Chernobyl fallout. ???? 64(5). pg. 531-534.

Scheid, W., Weber, J. and Traut, H. (1993). Chromosome aberrations induced in the lymphocytes of pilots and stewardesses. *Naturwissenschaften*. 80. pg. 528-530.

Shcherbak, Y. (April 1996). Ten years of the Chernobyl era. *Scientific American*.

Sich, A.R. (1994). Chernobyl accident management actions. *Nuclear Safety*. 35(1).

- The first in an important series of articles exploring what actually occurred during the release phase of the Chernobyl accident (April 26 through May 5).
- Startling information about the contradictions, misrepresentations and ineffectiveness of the accident management actions during and after the accident.
- The first clear analysis of the ineffectiveness of helicopter dropped materials and the flooding of the core with liquid (?) nitrogen in halting the accident.

- Excellent photographs and graphics give stark emphasis to the bizarre events which transpired during the accident.
- "71% of the initial 190.3 ton UO_2 fuel load was exposed to a high temperature oxidizing environment." (pg. 1).
- A frightening indictment of the inaccuracy of Soviet, IAEA and other early descriptions of the Chernobyl accident and an illustration of how long it can take to obtain accurate information about a serious nuclear accident and how it occurred.
- Mandatory reading for anyone trying to understand what really happened at Chernobyl, this is the first in a series of three articles by Sich in the Oak Ridge National Laboratory publication Nuclear Safety.

Sich, A.R. (1994). *The Chernobyl accident revisited: Source term analysis and reconstruction of events during the active phase*. (Ph.D. Thesis). Massachusetts Institute of Technology, Cambridge, MA.

Sich, A.R. (1995). The Chernobyl accident revisited, part II: The state of the nuclear fuel located within the Chernobyl sarcophagus phase. *Nuclear Safety*. 36(1). pg. 1-32.

- See Borovoi and Sich (1995) above, for a review of this citation.

Sich, A.R. (1996). The Chernobyl accident revisited, part III: Chernobyl source term release dynamics and reconstruction of events during the active phase. *Nuclear Safety*. 36(2). pg. 195-217.

- Appraisal of (inaccurate) Soviet release data is followed by an evaluation of new release data and a consideration of the active phase release dynamics.
- The source term release estimate (lower-bound activity releases for eight volatile isotopes) is followed by a reconstruction of events during the active phase and serves as a summary of Sich's accident release dynamic studies. Also see Sich (1996) Nuclear Engineering International for our RADNET citation summarizing Sich's accident analysis.
- Sich makes the following general observations at the beginning of this article:
  - "...iodine, cesium, and (to some extent) tellurium are considered to be the most important fission products in the early stages of a severe accident because they exhibit similar high volatility's and diffusion properties." (pg. 195).
  - "The less-volatile species may be divided broadly into three groups: the semivolatiles (tellurium and antimony), the low volatiles (strontium,

barium, and europium), and the refractories (molybdenum, ruthenium, zirconium, cerium, neptunium, etc.)." (pg. 195).

- o "What complicates time-dependent source term release analyses (especially for the case of Chernobyl's 10-day active phase) is that the longer lived fission products continue to decay until a stable product is formed. The physical and chemical states of the intermediate species in a given decay chain are important because their volatilities span the entire range noted previously." (pg. 196).

- Sich gives the release estimate for the eight most significant volatile isotopes as 92 MCi. "...substantially more than a total release of 50 MCi (excluding noble gases) claimed by the Soviets in Vienna in August 1986....if the contributions of all other longer lived radioisotopes are added, the total release may approach 150 MCi. In fact, if Np-239 (half-life 2.355 d) is considered and if it was released at the 3.2% fraction claimed by the Soviets, its contribution to the releases over the period of the active phase alone could reach 30 Mci." (pg. 208).

Sich, A.R. (1996). The Chernobyl active phase: Why the "official view" is wrong. *Nuclear Engineering International.* 40(501). pg. 22-25.

- Detailed analysis of the release dynamics of the accident; as the graphite component of the core (corium) burned, it allowed the remaining fuel to eat away the lower biological shield (LBS) and flow into the lower regions of the reactor building. (pg. 23).
- After nine days, the corium quickly solidified and the accident stopped without direct human intervention (helicopter dropped materials were ineffective). The decay heat dropped due to the uptake of surrounding materials (the stainless steel and serpentine components of the LBS) combined with rapid spreading of the melted fuel up to 40 m from the epicenter of the melted corium. (pg. 23).
- "A reconvergence of volatile and non-volatile behavior and a large release around 7.5-8.5 days may indicate when the LBS melted through." (pg. 25).
- The solidified, ceramic-like corium indicates this rapid cooling once the corium penetrated the lower biological shield and flowed into the lower regions of the reactor building.
- 65% of the radiocesium was released; the Soviet report of a 13% release was as unreliable as other early reports about the accident.

Sich, A.R. (1996). Through the looking glass. *Nuclear Engineering International.* 41(501). pg. 26-27.

224

- "He found research in the Zone to be poorly organized, encumbered with ideology, hampered by layer upon layer of bureaucracy and conducted in an atmosphere of conflict and mutual suspicion." (pg. 26).
- "The manner in which some international organizations have dealt with the accident over the past ten years has strengthened in me the conviction that, sadly, scientific inquiry and politics are inextricably linked…" (pg. 26).

Special issue: International Chernobyl Project. (1992). *J. Environ. Radioactivity.* 17(2-3).

- A number of articles from this special issue are cited in this section of RADNET, particularly under the subheading "Russia and former USSR."

United Nations. (August 29, 2003). *Optimizing the international effort to study, mitigate and minimize the consequences of the Chernobyl disaster: Report of the Secretary-General.* A/58/332. United Nations General Assembly. http://www.chernobyl.info/files/doc/UNRepOptimizingIntEff.pdf.

U. S. Department Of Energy. (1987). *Health and environmental consequences of the Chernobyl Nuclear Power Plant accident.* Report No. DOE/ER-0332. Committee on the Assessment of Health Consequences in Exposed Populations, U. S. Department of Energy, Washington, D.C.

- A compendium of the misinformation and underestimations within many of the early reports on the Chernobyl accident, this report includes the initial inaccurate source term release activities, a lack of media specific data, and summaries and conclusions based upon the inaccurate computer models of the time.
- The generalized conclusions about the health consequences of the Chernobyl accident in this and many other reports are simply speculation without a firm basis in an understanding of the radiological impact of the accident on specific population groups most affected by the erratic fallout patterns of the Chernobyl disaster.

US Nuclear Regulatory Commission. (1987). *Report on the accident at the Chernobyl nuclear power station.* Report No. NUREG-1250, Rev. 1. Government Printing Office, Washington, D.C.

- This report contains very little media specific data on Chernobyl fallout. Radionuclide deposition for Chester, NJ (5/6/86-6/2/86) is reported as (pCi/m^2): ^{131}I: 2,380; ^{137}Cs: 650; ^{134}Cs: 290; ^{103}Ru 720. (pg. 8-3).

- A detailed description of how the accident happened and of the design and construction of the reactor.

Volchok, H.L. and Chieco, N. (1986). *Environmental Measurements Laboratory: A compendium of the Environmental Measurements Laboratory's research projects related to the Chernobyl nuclear accident: Environmental report October 1, 1986.* Report No. EML-460. Department of Energy, New York, NY.

- This is a general summary of Chernobyl fallout data in the United States and in Sweden, with thirteen separate articles, the most important of which are cited in this website. See especially Hardy, et. al. (1986) in this Volume, Section 4, Sweden and Krey (1986) in this section.
- A bizarre documentation of the impact of the Chernobyl accident.

Warman, E.A. (1987). *Soviet and far-field radiation measurements and an inferred source term from Chernobyl.* Report No. TP87-13. Stone and Webster Engineering Corp, Boston, MA.

- One of the earliest reports to question the inaccurate source term reported by the Soviets.
- "Approximately 30-60% of the available radiocesium and at least 40-60% of the available radioiodine appear to have been released to the atmosphere from the accident." (pg. 1).
- "The radionuclide compositions observed outside the Soviet Union differ substantially from the Soviet source-term estimate, e.g., much more radioiodine and less nonvolatile radionuclides were observed in Europe than were estimated to have been released by the Soviets." (pg. 4).
- This is the first report to identify a second phase in the accident characterized by increased release of ^{132}Te, ^{103}Ru and ^{140}Ba.
- Warman's revision of the inaccurate Soviet source term release estimates were based upon a number of "far field" measurements taken after the accident in Finland (2), West Germany, Hungary and Greece, and summarized in chart form at the end of this report. Close inspection of isotopic ratios present in ground depositions and air samples led Warman to question, correctly, as it turned out, the inaccurate Soviet data.
- This is one of the few reports to include a core inventory of radionuclides at Chernobyl at the time of the accident (Taken by Warman from the International Safety Advisory Group (1986) report listed above):

Core Inventory of Radionuclides			
Radionuclide	**Half-Life**	**Inventory @ April 26**	
^{85}Kr	3,930	3.3×10^{16}	0.89
^{133}Xe	5.27	7.3×10^{18}	196
^{131}I	8.04	3.1×10^{18}	85
^{132}Te	3.25	3.3×10^{18}	90
^{134}Cs	750	1.9×10^{17}	5.0
^{137}Cs	1.1×10^{4}	2.9×10^{17}	7.8
^{99}Mo	2.8	7.3×10^{19}	1,980
^{95}Zr	65.6	4.9×10^{18}	135
^{103}Ru	39.5	5.0×10^{18}	133
^{106}Ru	368	2.0×10^{18}	54
^{140}Ba	12.8	5.3×10^{18}	142
^{141}Ce	32.5	5.6×10^{18}	152
^{144}Ce	284	3.2×10^{18}	86
^{89}Sr	53	2.3×10^{18}	62
^{90}Sr	1.02×10^{4}	2.0×10^{17}	5.4
^{239}Np	2.35	3.6×10^{18}	98
^{238}Pu	3.15×10^{4}	1.0×10^{15}	0.027
^{239}Pu	8.9×10^{6}	8.5×10^{14}	0.023
^{240}Pu	2.4×10^{6}	1.2×10^{15}	0.32
^{241}Pu	4,800	1.7×10^{17}	4.6
^{242}Cm	164	2.5×10^{16}	0.70

Webb, G.A.M., Simmonds, J.R. and Wilkins, B.T. (1986). Radiation levels in Eastern Europe. *Nature.* 321. pg. 821-822.

Date	Location	Media	Nuclide	Activity
29-30 April	Poland	Milk	^{131}I	2,000 Bq/l
1-4 May	Hungary	Milk	^{131}I	2,600 Bq/l

Williams, D. (1994). Chernobyl, eight years on. *Nature.* 371. pg. 556.

Wirth, E., van Egmond, N.D. and Suess, M.J. (1986). *Assessment of radiation dose commitment in Europe due to the Chernobyl accident: Report on a WHO meeting: Bilthoven, 25-27 June 1986.* Report No. ISH-HEFT 108. Institut fur Strahlenhygiene des Bundesgesundheitsamtes, Munchen.

- Cumulative deposition of iodine-131 in soil to May 8:
  - Byelorussia: 1,000,000 Bq/m^2 +
  - S. Germany: 130,000 Bq/m^2 pv
  - Austria: 150,000 Bq/m^2 pv
- This report uses two computer models (MESOS and GRID) for calculating deposition activity levels. These models appear to grossly underestimate Chernobyl fallout data in areas where comprehensive radiometric surveys are available.

World Health Organization. Health hazards from radiocesium following the Chernobyl nuclear accident: Report on a WHO meeting. *Environmental Health.* 24.

- "Six... pathways are possible by which exposure may occur following a nuclear accident..." (pg. 4).

External:	Internal:
Ground shine	Ingestion
Cloud shine	Inhalation
Deposition on skin and clothing	Absorption from skin

- "Root uptake of cesium will be substantially higher for acid soils with a low clay and a high organic matter content and may continue for many years in some soil conditions.... the external and internal doses will be roughly the same for the fifty year period after the accident." (pg. 8-9).
- "Direct exposure from deposited radionuclides together with the ingestion pathway was estimated to be three orders of magnitude greater than that from inhalation or exposure to airborne radionuclides (cloud shine)." (pg. 23).

World Health Organization. (September 8, 1986). *Working group on assessment of radiation dose commitment in Europe due to the Chernobyl accident: Bilthoven, 25-27 June 1986.* Report No. ICP/COR 129(s) Rev 1. 5134V. World Health Organization, Copenhagen, Denmark.

Date	Location	Media	Nuclide	Activity
May 1986	W. Europe	Ground deposition	^{131}I	+/- 1,000,000 Bq/m^2
May 1986	W. Europe	Ground deposition	^{137}Cs	+/- 140,000 Bq/m^2

- Large scale computerized dispersion models (MESOS and GRID) were used to reconstruct deposition patterns over Europe; these isolated areas of very high local deposition were located in the Ukraine, Central Scandinavia and Central Europe.
- "Exposure of the population occurs through three main pathways: inhalation of airborne material, external irradiation from material deposited on the ground and ingestion of contaminated foodstuff." (pg. 2).

WHO Regional Office for Europe. (1989). Health hazards from radiocesium following the Chernobyl nuclear accident: Report on a WHO working group. *J. Environ. Radioactivity.* 10(3). pg. 257-296.

- This publication contains no media specific data on the Chernobyl-derived radioactive fallout. It is a general survey of the pathways, radiological impact and risk assessment of radiocesium.

Hot Particles

Broda, R. (1987). Gamma spectroscopy analysis of hot particles from the Chernobyl fallout. *Acta Physica Polica.* B18. pg. 935-950.

- "Highly inhomogeneous distribution of Chernobyl fallout in Poland was one of the surprises contributing to serious difficulty in early estimates of the situation." (pg. 935).
- Ground level activity was observed to 360,000 Bq/m^2.
- Ground deposition hot spots were observed in the Masurian Lakes region as small as 30 cm in diameter with radioactivity 20 times higher than that in the larger areas surrounding them. This could have arisen as a result of erosion and washout of hot particles.
- Extensive deposition of hot particles was noted with some areas having one particle (+100 Bq) per ten square meters.
- Intense background radiation hindered the hot particle search in the Krakow area.
- One hot particle was nearly pure cerium; another had activity levels of 237,900 Bq of ^{103}Ru.
- "Extensive isotopic and activity ratio analysis of hot particles radionuclides in this report... gives extensive information about the age and fuel composition about the Chernobyl reactor and the processes taking place during the accident." (pg. 949).
- This is a landmark article in the literature of hot particle emissions. See the Poland listing for additional data from this article.

Jaracz, P., Mirowski, S., Trzcinska, A, Isajenko, K., Jagielak, J., Kempisty, T. and Jozefowicz, E.T. (1995). Calculations and measurements of ^{154}Eu and ^{155}Eu in "fuel-like" hot particles from Chernobyl fallout. *J. Environ. Radioactivity.* 26(1). pg. 83-98.

- ^{154}Eu and ^{155}Eu behave similarly to the nonvolatile radioisotopes ^{95}Zr and ^{95}Nb in hot particles (pg. 84), with small fractionation as compared to the large fractionation of ^{137}Cs and ^{106}Ru.

Raunemaa, T., Lehtinen, S., Saari, H. and Kulmala, M. (1987). 2-10 µm sized hot particles in Chernobyl fallout to Finland. *J. Aerosol. Sci.* 18(6). pg. 693-696.

- The first radioactivity noted in surface air was due to sedimenting of hot particles when the main plume stayed airborne at high elevations; most hot particle deposition was in SW parts of Finland. (pg. 693).
- Scots Pine Needles were used as bioindicators for hot particle distribution; typical peak concentration activity levels: ^{144}Ce: 63 Bq per particle; ^{141}Ce: 132 Bq; ^{103}Ru: 108 Bq; ^{106}Ru: 6.7 Bq; ^{95}Zr: 126 Bq; ^{95}Nb: 130 Bq.

Saari, H., Luokkanen, S., Kulmala, M., Lehtinen, S. and Raunemaa, T. (1989). Isolation and characterization of hot particles from Chernobyl fallout in Southwestern Finland. *Health Physics*. 57(6). pg. 975-984.

- Three types of activity composition were found in airborne hot particles from Chernobyl: The most common type contained [141]Ce, [144]Ce, [95]Zr and [95]Nb; the second type also included [103]Ru and [106]Ru; and the third type contained [103]Ru and [106]Ru only.

Sandalls, F.J., Segal, M.G. and Victorova, N. (1993). Hot particles from Chernobyl: A review. *J. Environ. Radioactivity*. 18(1). pg. 5-22.

- Two types of particles noted:
    1. "Mono or bielemental, since only one or two radionuclides were detected by gamma spectroscopy."
    2. "Particles or fragments of uranium oxide fuel containing a range of fission products found in fuel but often somewhat depleted in the non-volatile elements, such as sodium, cesium and ruthenium."
- Particle size ranges: 0.5 to 30 µm (Minsk).

Shubert, P. and Behrand, U. (1987). Investigations of radioactive particles from the Chernobyl fall-out. *Radiochimica Acta*. 41. pg. 149-155.

- Particles with high specific activity were investigated in Poland, Greece and Germany; many had a high proportion of [103]Ru and [106]Ru.
- Hot particles were noted from NE Poland with activity levels up to 139,000 Bq of [103,106]Ru. [106]Ru ratios were from 4.42 to 8.06.

Tcherkezian, V., Shkinev, V., Khitrov, L. and Kolesov, G. (1994). Experimental approach to Chernobyl hot particles. *J. Environ. Radioactivity*. 22(2). pg. 127-140.

- Hot particles contribution to the total activity in the 30 km zone was found to be not less than 65%.
- Linear size range: 0.5 to 100 µm.

van der Veen, J., van der Wijk, A., Mook, W.G. and de Meijer, R.J. (1986). Core fragments in Chernobyl fallout. *Nature*. 323. pg. 399-400.

- Core fragment hot spots measured in trousers and shoes by Dutch researchers of travelers returning from Kiev and Minsk... Gamma activity in hot spots to 850 Bq with smaller peaks of alpha emitting ^{42}Cm, ^{239,240}Pu, ^{243,244}Cm, ^{238}Pu and possibly ^{241}Am.

Chernobyl Plume: Country-By-Country Summary

Chernobyl fallout and plume maps online

- Overview - Geographical location and extent of radioactive contamination; http://www.chernobyl.info/en/Facts/MainOverview/Overview
- Evacuated zone; http://www.infoukes.com/history/chornobyl/gregorovich/figure03.gif
- Plume area map 1; http://www.taiwanwatch.org.tw/english/chernobyl-1.htm
- Plume area map 2; http://www.chernobyl.com/chern6.gif
- Northern hemisphere; http://www.grida.no/db/maps/prod/global/nc28_l.gif
- Northern Europe; http://www.grida.no/db/maps/prod/europe/nc29_l.gif
- Sweden; http://www.gr.is/nsfs/finck.htm
- Former Soviet Union area map 1; http://greenfield.fortunecity.com/flytrap/250/mapr1.gif
- Former Soviet Union area map 2; http://www.radiation.ru/eng/project/ChAES.htm#img4
- Belarus; http://www.belarusguide.com/chernobyl1/ctrace.html
- Briansk region; http://greenfield.fortunecity.com/flytrap/250/atlaseng.html
- Estonia; http://www.fi.tartu.ee/labs/tsl/Estonianenv.htm
- Europe; http://image.pathfinder.com/time/daily/chernobyl/images/europemap.gif
- United Kingdom map 1; http://www.infoukes.com/history/chornobyl/gregorovich/thumb06.gif
- United Kingdom map 2; http://www.antenna.nl/wise/349-50/eng.gif
- Scotland; http://www.scotland.gov.uk/library/stat-ses/sesm7-5.htm
- Greece; http://arcas.nuclear.ntua.gr/radmaps/page1.html
- Switzerland; http://www.hsk.psi.ch/pub_eng/gmap96.html
- A number of maps of radiation contamination in Europe can be found on the Humus website; http://www.progettohumus.it/RicercaProg/Mappe/Mappe.html. Some of the information on the links to the maps is in Italian but many of the maps have English place names.

Austria

Irlweck, K., Khademi, B., Henrich, E. and Kronraff, R. (1993). $^{239(240),238}$Pu, ^{90}Sr, ^{103}Ru and ^{137}Cs concentrations in surface air in Austria due to dispersion of Chernobyl releases over Europe. *J. Environ. Radioactivity.* 20(2). pg. 133-148.

Date	Location	Media	Nuclide	Activity
May 1986	Vienna	Air concentration	^{239,240}Pu	89 µBq/m^3
April 30, 1986	Vienna	Air concentration	^{137}Cs	9,700,000 µBq/m^3
April 30, 1986	Vienna	Air concentration	^{103}Ru	62,500,000 µBq/m^3

- Contamination by the volatile nuclides ^{137}Cs and ^{103}Ru preceded the peak pulse of plutonium by a few days.

Heinrich, G., Oswald, K. and Muller, J.J. (April 1, 1999). Lichens as monitors of radiocesium and radiostrontium in Austria. *J. Environ. Radioactivity.* 45(1). pg. 13-27.

Bangladesh

Mydans, S. (Friday, June 5, 1987). Specter of Chernobyl looms over Bangladesh. *The New York Times.* (page not available).

- "The Government announced that a 1,600-ton shipment of powdered milk from Poland, which was affected by the Chernobyl nuclear accident in the Ukraine in April 1986, showed unacceptably high levels of radioactivity... the shipment registered levels higher than the 300 Becquerels... it has been deemed to be unfit for consumption."

Black Sea

Buesseler, K.O. (1987). Chernobyl: Oceanographic studies in the Black Sea. *Oceanus.* pg. 23-30.

- "Chernobyl cesium-137 fallout input was roughly twice as large as the cesium-137 inventory deposited to this basin from weapons testing fallout." (pg. 26).
- Radionuclides identified included: ^{134}Cs, ^{137}Cs, ^{144}Ce, ^{141}Ce, ^{106}Ru, ^{103}Ru, ^{140}La, ^{140}Ba, ^{95}Nb and ^{129}Te.

- ^{137}Cs in surface water was elevated by a factor of over 20 over the previous pre-Chernobyl baseline (15 Bq/m^3 to 340 Bq/m^3).

Buesseler, K.O., Livingston, H.D., Honjo, S., Hay, B.J. Manganini, S.J., Degens, E., Ittekkot, V., Izdar, E. and Konuk, T. (1987). Chernobyl radionuclides in a Black Sea sediment trap. *Nature.* 329(29). pg. 825-828.

Date	Location	Media	Nuclide	Activity
Summer of 1986	Black Sea	Sediment traps	^{144}Ce	12,000 Bq/kg
Summer of 1986	Black Sea	Sediment traps	^{106}Ru	12,600 Bq/kg
Summer of 1986	Black Sea	Sediment traps	^{137}Cs	1,900 Bq/kg

- The sediment trap was located at a depth of 1,071 meters in the Southern Black Sea.

Livingston, H.D., Clarke, W.R., Honjo, S., Izdar, E. Konuk, T., Degens, E. and Ittekkot, V. (1986). Chernobyl fallout studies in the Black Sea and other ocean areas. In: *Environmental Measurements Laboratory: A compendium of the Environmental Measurements Laboratory's research projects related to the Chernobyl nuclear accident: October 1, 1986.* Report No. EML-460. US Department of Energy, New York, NY. pg. 214-223.

- Preliminary cesium data in ocean water was 852 d.p.m. per 100 liters = 142 Bq/m^2. Cesium-134 contamination is reported as about 50% of cesium-137 levels.

Bulgaria

Pourchet, M., Veltchev, K. and Candaudap, F. (October 1997). Spatial distribution of Chernobyl contamination over Bulgaria. *International Symposium OM2: Observation of the Mountain Environment in Europe, Borovets (Bulgaria), October 15-17, 1997.*

- "The main part of the radioactive contamination reached Bulgaria in the period from May 1st to 10th 1986. According to available measurements the average surface air radioactivity in this time interval was between 30 - 160 Bq/m^{-3}. The

234

maximum value was measured on May 1th, 1986. A secondary maximum (peak) of radioactivity was registered on May 9th, 1986." (pg. 2).

- This is the first summary of Chernobyl fallout deposition in Bulgaria received or located by RADNET.
- Excellent maps of Chernobyl ^{134}Cs and ^{137}Cs deposition. Peak values of ^{137}Cs deposition to 81,800 Bq/m^{-2} with a mean deposition of the most contaminated areas of 30,400 Bq/m^{-2} (Map 4). This contrasts with peak weapons testing cumulative ^{137}Cs deposition of 10,053 Bq/m^{-2}.

Canada

Department of National Health and Welfare. (1986). *Environmental radioactivity in Canada*. (Radiological monitoring annual report). Department of National Health and Welfare, Ottawa.

- An extensive survey of imported foods showed little impact from the Chernobyl accident. However, the surveys in this report were terminated prior to the peak pulse of Chernobyl-derived radiocesium reported by the USFDA in foods imported into the US
- Unusually high readings of radioiodine were reported in British Columbia in May of 1986. At Revelstoke, B.C. (pg. 63) an anomalous reading of 251 Bq/m^3 (251,000,000 μBq/m^3) was noted on May 13. This is the only RADNET citation reporting ^{131}I higher than the 223,000,000 μBq/m^3 recorded at Nurmijarvi, Finland on April 28, 1986.
- Vancouver also had high readings of 107 Bq/m^3 on May 16-17, 90 Bq/m^3 on May 17-18, and 115 Bq/m^3 on May 19-20. 176 Bq/m^3 were reported in Quebec on May 5-6. Numerous other elevated readings of radioiodine were reported.
- Unlike Chernobyl fallout data in the European environment, high levels of radioiodine in Canada were not associated with excessive deposition of cesium or ruthenium nuclides; however, ground deposition data measured in contamination per square meter is not available in this report.

Joshi, S.R. (1987). Early Canadian results on the long-range transport of Chernobyl radioactivity. *The Science of the Total Environment*. 63. pg. 125-137.

- This report notes high pre-Chernobyl ^{137}Cs and ^{65}Zn deposits from a 4-10-86 underground nuclear weapons test leak in Nevada.
- Table one notes major differences between Chernobyl fallout and weapons testing fallout: Chernobyl was characterized by the relatively intermittent release

of a full range of radionuclides at relatively low temperatures with very heavy local fallout from tropospheric transport. Weapons testing fallout was at a high temperature with more uniform stratospheric transport and longer residence time (1-10 years), with much less pronounced local fallout.

- "1963 fallout maximum (for Canada): 1.3 kBq m^{-2} (40-50° N)" (pg. 126).

Joshi, S.R. (1988). The fallout of Chernobyl radioactivity in Central Ontario, Canada. *J. Environ. Radioactivity.* 6. pg. 203-211.

Date	Location	Media	Nuclide	Activity
May 1986	Central Ontario	Rainfall	^{137}Cs	325 mBq/l

- "Chernobyl-derived radionuclides (^{103}Ru, ^{106}Ru, ^{134}Cs and ^{137}Cs) were consistently measurable until about mid-June... with a mean tropospheric residence time of about fourteen days for the four radionuclides." (pg. 203).

Roy, J.C., Cote, J.E., Mahfoud, A., Villeneuve, S. and Turcotte, J. (1988). On the transport of Chernobyl radioactivity to Eastern Canada. *J. Environ. Radioactivity.* 6. pg. 121-130.

- "Three waves of airborne radioactivity entered Eastern Canada (Quebec) on 6 May and around 14 May respectively via the Arctic, and 25 and 26 May via Pacific route." (pg. 121).
- "The removal in floc is high (70-90%) for ^{7}Be, ^{59}Fe, ^{95}Nb, ^{95}Zr, ^{103}Ru, ^{106}Ru and ^{131}I, intermediate (40-60%) for ^{140}La, ^{141}Ce and ^{144}Ce and low (25% and less) for ^{54}Mn, ^{60}Co, ^{65}Zn and ^{140}Ba, and variable for ^{137}Cs (50-90%)." (pg. 123).
- "Activity ratios of Chernobyl products are nearly the same in air and river water samples." (pg. 124).

Taylor, H.W., Svoboda, J., Henry, G.H.R. and Wein, R.W. (1988). Post-Chernobyl ^{134}Cs and ^{137}Cs levels at some localities in Northern Canada. *Arctic.* 41. pg. 293-296.

Czechoslovakia

Kliment, V. (1991). Contamination of pork by cesium radioisotopes. *J. Environ. Radioactivity.* 13(2). pg. 117-124.

- Pork was contaminated immediately after Chernobyl when whey, a milk by-product, was put in pig feed instead of uncontaminated 1985 cereal.
- Typical (mean) values of ^{137}Cs in pigs May 1986-July 1987: 15-25 Bq/kg.
- Average 1986 values for ^{137}Cs in wheat 16 Bq/kg; in barley 7.2 Bq/kg (pg. 120).

Kliment, V. and Bucina, I. (1990). Contamination of food in Czechoslovakia by cesium radioisotopes from the Chernobyl accident. *J. Environ. Radioactivity.* 12(2). pg. 167-178.

Date	Location	Media	Nuclide	Activity
June 1986	Czechoslovakia	Baby milk	^{137}Cs	110 Bq/l mean value
June 1986	Czechoslovakia	Baby milk	^{134}Cs	55 Bq/l mean value
June 1986	Czechoslovakia	Pork	^{137}Cs	45 Bq/kg mean value

- Mean values in all food samples gradually dropped over the next two years, e.g. baby milk to +/- 3.0 Bq/l.

Denmark

Online radiation fallout maps of Northern Europe:
http://www.grida.no/db/maps/prod/europe/nc29_l.gif and
http://www.mv.slu.se/ma/radio/radio/chern/cs-dep2.gif.

Numerous research papers published by the Riso National Laboratory in Roskilde, Denmark, pertaining to Chernobyl are cited at the beginning of this section in the General Bibliography. Please also refer to the Riso publications cited in RADNET Section 11 that also contain information on Chernobyl and the other source points discussed in that section. Other citations are in the Riso fallout summary in RADNET, Section 8, Baseline Data.

Aarkrog, A. (1988). Studies of Chernobyl debris in Denmark. *Environment International.* 14. pg. 149-155.

Aarkrog, A. (1989). Radioecological lessons learned from Chernobyl. *Proceeding of the XVeth Regional Congress of IPRA, Visby, Gotland, Sweden, 10-14 Sept., 1989.* pg. 129-134.

Aarkrog, A., et. al. (July, 1991). *Environmental radioactivity in Denmark in 1988 and 1989.* Riso-R-570. Riso National Laboratory, Roskilde, Denmark.

- See annotations in RADNET Section 8: Anthropogenic Radioactivity: Baseline Data: Section 6: Other Nuclides.

Aarkrog, A. et. al. (February 1995). *Environmental radioactivity in Denmark in 1992 and 1993*. Riso-R-756(EN). Riso National Laboratory, Roskilde, Denmark.

- Riso reports continue to be one of the best basic references documenting lingering anthropogenic radioactive contamination of the environment.
- The pervasive impact of the Chernobyl accident is clearly documented by the low levels of ^{137}Cs in the human diet, the human body, and in the environment in 1992 and 1993 compared with 1986.
- In the event of another Chernobyl-type nuclear accident at any location, the database of the Riso National Laboratory will provide an important index of environmental contamination by anthropogenic radioactivity. Will the present golden age of low-levels of hemispheric wide contamination continue in the next millennium, and if not, what will be the source points of the radioactive plumes of the future?

Table 5.9.4.B. Estimate of the mean content of ^{137}Cs in the human diet in 1993

Type of food	Annual quantity in kg	Bq ^{137}Cs per kg	Total Bq [137]Cs	Percentage of total Bq ^{137}Cs in food
Milk and cream	164.0	0.067	10.99	9.5
Cheese	9.1	0.048	0.44	0.4
Grain products	80.3	0.100	8.01	6.9
Potatoes	73.0	0.046	3.36	2.9
Vegetables	43.8	0.020	0.88	0.8
Fruit	51.1	0.019	0.97	0.8
Meat	54.7	0.23	12.58	10.8
Eggs	10.9	0.030	0.33	0.3

Table 5.9.4.B. Estimate of the mean content of ^{137}Cs in the human diet in 1993

Type of food	Annual quantity in kg	Bq ^{137}Cs per kg	Total Bq ^{137}Cs	Percentage of total Bq ^{137}Cs in food
Fish	10.9	6.74	73.5	63.1
Coffee and tea	5.5	0.95	5.23	4.5
Drinking water	548	0	0	0
Total			116.29	

- This table's total of 116.29 Bq for ^{137}Cs contrasts with 124.15 Bq in 1992. (Table 5.9.4.A, pg. 76).
- "Strontium-90 and Cesium-137 in Humans. The ^{90}Sr mean content in adult human bone (vertebrae) collected in 1992 was 18 Bq (kg Ca)$^{-1}$. Whole body measurements of ^{137}Cs were resumed after the Chernobyl accident. The measured mean level in 1990 was 359 Bq ^{137}Cs (kg K)$^{-1}$." (pg. 108).

Appendix D.3 Fallout rates and accumulated fallout (Bq ^{137}Cs m^{-2}) in Denmark

	Denmark		Jutland		Islands	
Year	di	Ai	di	Ai	di	Ai
1986	1210.000	3725.984	1340.000	4137.847	1080.000	3314.232
1987	29.000	3669.280	32.000	4047.674	26.000	3263.994
1988	11.900	3597.161	13.400	3994.768	10.300	3199.562
1989	3.500	3518.480	4.510	3907.998	2.530	3129.007
1990	2.63	3440.744	3.85	3822.564	1.41	3058.968
1991	1.63	3363.805	1.92	3737.194	1.36	2990.480
1992	0.98	3287.987	1.17	3653.041	0.79	2922.994
1993	0.96	3213.881	1.39	3571.026	0.53	2856.796

- In the above table, di indicates annual deposition and Ai indicates cumulative deposition. (pg. 122).

Andersson, K.G. and Roed, J. (1994). The behavior of Chernobyl ^{137}Cs, ^{134}Cs and ^{106}Ru in undisturbed soil: Implications for external radiation. *J. Environ. Radioactivity.* 22. pg. 183-196.

- "Four years after the Chernobyl accident it was found that most of the cesium remained firmly fixed in the topmost 2 cm. In the deeper layers of the soil it was less strongly bound." (pg. 183).
- "The ruthenium was found to be less strongly bound than the cesium and had penetrated a little deeper." (pg. 183).

Estonia

Online radiation fallout maps of Northern Europe;
http://www.grida.no/db/maps/prod/europe/nc29_l.gif and
http://www.mv.slu.se/ma/radio/radio/chern/cs-dep2.gif.

Maps of Estonia; http://www.fi.tartu.ee/labs/tsl/Estonianenv.htm.

Realo, E., Jogi, J., Koch, R. and Realo, K. (1995). Studies on radiocesium in Estonian soils. *J. Environ. Radioactivity.* 29. pg. 111-120.

Date	Location	Media	Nuclide	Activity
1986	Estonia	Ground deposition	^{137}Cs	40,000 Bq/m^2

- "The distribution of the Chernobyl deposition is extremely uneven." (pg. 111).

Finland

Online radiation fallout maps of Northern Europe;
http://www.grida.no/db/maps/prod/europe/nc29_l.gif and
http://www.mv.slu.se/ma/radio/radio/chern/cs-dep2.gif.

Arvela, H., Blomqvist, L., Lemmela, H., Savolainen, A.L. and Sarkkula, S. (1987). *Environmental gamma radiation measurements in Finland and the influence of meteorological conditions after the Chernobyl accident in 1986: Supplement 10 to annual report STUK-A55*. Report No. STUK-A65. Finnish Centre for Radiation and Nuclear Safety, Helsinki.

- The external dose rate peaked briefly at Uusikaupunki at 4.0 micro sieverts per hour on April 19, 1986; the background rate is 0.11.
- ^{137}Cs surface activity to 100,000 Bq/m^2.

Finnish Centre for Radiation and Nuclear Safety. (1986). *Interim report on fallout situation in Finland from April 26 to May 4 1986*. Report No. STUK-B-VALO 44. Finnish Centre for Radiation and Nuclear Safety, Helsinki.

- External gamma dose rate to 384.7 µRh^{-1} on April 29, 1986 at Uusikaupunki.
- ^{131}I to 105,000 Bq/m^2 on 29 April at Lieto, and 122,700 Bq/m^2 at Jyvaskyla.
- ^{132}Te to 113,000 Bq/m^2; ^{132}I to 98,300 Bq/m^2; ^{137}Cs to 8,800 Bq/m^2: Lieto, 29 April.

Finnish Centre for Radiation and Nuclear Safety. (1986). *Second interim report radiation situation in Finland from 5 to 16 May 1986*. Report No. STUK-B-VALO 45. Finnish Centre for Radiation and Nuclear Safety, Helsinki.

Date	Location	Media	Nuclide	Activity
May 6-7, 1986	S. Finland	Ground deposition	^{137}Cs	40,000 Bq/m^2
May 6-7, 1986	S. Finland	Ground deposition	^{134}Cs	24,000 Bq/m^2
May 6-7, 1986	S. Finland	Ground deposition	^{140}La	16,000 Bq/m^2

- This report contains a lengthy list of suggested restrictions (pg. 9-11) including use of respirators in soil cultivation work.

Finnish Centre for Radiation and Nuclear Safety. (1987). *Chernobyl and Finland*. Ministry of Trade and Industry, Helsinki.

- Preliminary estimates of the annual dietary intake of cesium-137 to 26,000 Bq (May 1, 1986 - April 30, 1987); maximum body burdens estimated up to 10,000 Bq by December 1986.
- Hot particles were an unexpected feature of Chernobyl fallout in Finland.
- Fresh water fish had peak concentrations up to 3,000 Bq/kg in 1986.

Hellmuth, K.H. (1987). *Rapid determination of strontium-89 and strontium-90 - Experiences and results with various methods after the Chernobyl accident in 1986.* Report No. STUK-A70. Finnish Centre for Radiation and Nuclear Safety, Helsinki.

- No additional Chernobyl-derived strontium-90 was found in milk in Finland; the pre-Chernobyl mean was 0.099 Bq/l.
- Maximum deposition of strontium-90 in Chernobyl fallout was 250 Bq/m^2; many areas were below the detection limit.

Ikaheimonen, T.K., Ilus, E.I. and Saxen, R. (1988). *Finnish studies on radioactivity in the Baltic Sea in 1987*: *Supplement 8 to Annual Report 1987 No. STUK-A74*. Report No. STUK-A82. Finnish Centre for Radiation and Nuclear Safety, Helsinki.

- Surface water concentration range after Chernobyl was 100-400 Bq/m^3.
- Sedimentation of Chernobyl fallout into bottom sediments of open sea basins only began to increase in 1987; peak values went to 18,000 Bq/m^2.
- Pike and cod ^{137}Cs range 15-30 Bq/kg, about five times the pre-Chernobyl level.

Ilus, E., Sjoblom, K.L., Saxen, R., Aaltonen, H. and Taipale, T.K. (1987). *Finnish studies on radioactivity in the Baltic Sea after the Chernobyl accident in 1986. Supplement 11 to Annual Report STUK-A55*. Report No. STUK-A66. Finnish Centre for Radiation and Nuclear Safety, Helsinki, Finland.

Date	Location	Media	Nuclide	Activity
1986	Baltic Sea	Fish	^{134}Cs	96 Bq/kg
1986	Baltic Sea	Fish	^{137}Cs	190 Bq/kg
1986	Baltic Sea	Plankton	Gross beta	2,600 Bq/kg
1986	Baltic Sea	Plankton	^{239}Np	3,900 Bq/kg
1986	Baltic Sea	Fucus vesiculosus	^{131}I	2,900 Bq/kg
1986	Baltic Sea	Fucus vesiculosus	^{137}Cs	4,900 Bq/kg
1986	Baltic Sea	Fucus vesiculosus	^{103}Ru	5,900 Bq/kg
1986	Baltic Sea	Sediment	^{239,240}Pu	4.3 Bq/kg

Ilus, E., Klemola, S., Sjoblom, K.L. and Ikaheimonen, T.K. (1988). *Radioactivity of Fucus vesiculosus along the Finnish coast in 1987: Supplement 9 to Annual Report 1987 (STUK-A74)*. Report No. STUK-A83. Finnish Centre for Radiation and Nuclear Safety, Helsinki.

- ^{137}Cs activity concentrations were highest in May 1986, to 4,900 Bq/kg dry weight.
- 1987 peak values were 16-17% of the 1986 values; some local power station effluents were noted.
- Rather uniform distribution pre-Chernobyl ^{137}Cs levels were noted in Fucus, averaging around 10-15 Bq/kg; including samples taken in the vicinity of nuclear power stations.
- Concentration factors from water to Fucus range from a low of 400 to a high of 2,000.

Ilus, E., Sjoblom, K.L., Hannele, A., Klemola, S. and Arvela, H. (1987). *Monitoring of radioactivity in the environs of Finnish nuclear power stations in 1986: Supplement 12 to annual report STUK-A55)*. Report No. STUK-A67. Finnish Centre for Radiation and Nuclear Safety, Helsinki.

Date	Location	Media	Nuclide	Activity
1986	Finland: Loviisa and Olkiluoto	Ground deposition, both nuclear stations	^{131}I	100,000 Bq/m^2
1986	Loviisa	Hair moss	^{137}Cs	28,000 Bq
1986	Loviisa	Hair moss	^{103}Ru	18,000 Bq
1986	Loviisa	Hair moss	^{89}Sr	3,500 Bq
1986	Olkiluoto	Annual deposition	^{137}Cs	23,000 Bq/m^2

- "Concentration of locally discharged nuclides was low in comparison with Chernobyl fallout nuclides." (pg. 35).
- "At the beginning of December an increase in cesium concentration was detected at both stations... the most probable reason for observations is a resuspension caused by snowflakes in the surroundings." (pg. 60).

Lang, S., Raunemaa, T. Kulmala, M. and Rauhamaa, M. (1988). Latitudinal and longitudinal distribution of the Chernobyl fallout in Finland and deposition characteristics. *J. Aerosol. Sci.* 19(7). pg. 1191-1194.

Date	Location	Media	Nuclide	Activity
May-Dec 1986	Central Finland	Pine needles	^{137}Cs	30,000 Bq/kg
May-Dec 1986	Central Finland	Pine needles	^{141}Ce	40,000 Bq/kg
May-Dec 1986	Central Finland	Pine needles	^{103}Ru	35,000 Bq/kg

- "Radionuclides can be categorized to soluble or insoluble species which behave differently in wet or dry deposition." (pg. 1191).

Puhakainen, M., Rahola, T. and Suomela, M. (1987). *Radioactivity of sludge after the Chernobyl accident in 1986: Supplement 13 to Annual Report STUK-A55.* Report No. STUK-A68. Finnish Centre for Radiation and Nuclear Safety, Helsinki.

Date	Location	Media	Nuclide	Activity
May 1986	Finland	Sewage sludge	^{131}I	8,200 Bq/kg dry weight
May 1986	Finland	Sewage sludge	^{103}Ru	12,000 Bq/kg d.w.
July-Aug. 1986	Finland	Sewage sludge	^{137}Cs	12,000 Bq/kg d.w.

- "Sludge dating from May 1986, containing the greatest amount of radioactive material should not be used on fields as a soil improvement agent." (pg.13-14).
- Areas of rain and runoff had increased levels of sludge contamination.

Rahola, T., Suomela, M., Illukka, E., Puhakainen, M. and Pusa, S. (1987). *Radioactivity of people in Finland after the Chernobyl accident in 1986: Supplement 9 to annual report STUK-A55.* Report No. STUK-A64. Finnish Centre for Radiation and Nuclear Safety, Helsinki.

- Whole body counting measurement of 624 persons in six different groups in Finland in April-Dec. 1986:
    - At the end of 1986 the mean ^{134}Cs body burden was 730 Bq.

- The ^{137}Cs mean body burden increased from 150 Bq to 1,500 Bq in December.
  - Peak body burdens were: ^{134}Cs: 6,300 Bq. ^{137}Cs: 13,000 Bq.
- The two main routes for internal contamination were inhalation and, especially, ingestion.
- The minimum detectable activity (MDA) for ^{134}Cs or ^{137}Cs is 30 Bq when the nuclides are measured separately; the MDA for ^{131}I is 20 Bq. (pg. 9).

Rantavaara, A., Nygren, T., Nygren, K. and Hyvonen, T. (1987). *Radioactivity of game meat in Finland after the Chernobyl accident in 1986: Supplement 7 to Annual Report STUK-A55*. Report No. STUK-A62. Finnish Centre for Radiation and Nuclear Safety, Helsinki.

Date	Location	Media	Nuclide	Activity
June 1986	Finland	Moose	^{137}Cs	1,610 Bq/kg
Sept. 1986	Finland	White tailed deer	^{137}Cs	1,954 Bq/kg
Sept. 1986	Finland	Arctic hare	^{137}Cs	1,888 Bq/kg
Aug. 1986	Finland	Goldeneye (waterfowl)	^{134,137}Cs	10,469 Bq/kg
Aug. 1986	Finland	Teal (waterfowl)	^{134,137}Cs	6,666 Bq/kg

Rantavaara, A. (1987). *Radioactivity of vegetables and mushrooms in Finland after the Chernobyl accident in 1986: Supplement 4 to Annual Report STUK-A55*. Report No. STUK-A59. Finnish Centre for Radiation and Nuclear Safety, Helsinki.

Date	Location	Media	Nuclide	Activity
May-Oct. 1986	Finland	Leafy vegetables	^{103}Ru	400 Bq/kg
May-Oct. 1986	Finland	Herbs	^{132}Te	730 Bq/kg
May-Oct. 1986	Finland	Cranberry	^{137}Cs	530 Bq/kg
May-Oct. 1986	Finland	Mushrooms	^{137}Cs	6,680 Bq/kg

- The wide variations in Chernobyl fallout are reflected in the wide variations of contamination in vegetables and mushrooms which range from very low to moderate contamination in most species except mushrooms.

Rantavaara, A. and Haukka, S. (1987). *Radioactivity of milk, meat, cereals and other agricultural products in Finland after the Chernobyl accident in 1986: Supplement 3 to Annual Report STUK-A55*. Report No. STUK-A58. Finnish Centre for Radiation and Nuclear Safety, Helsinki.

- Low to moderate mean concentrations of contamination in milk were reported: ^{131}I: 37 Bq/l; ^{137}Cs: 65 Bq/l.
- The national mean content of daily intake of ^{137}Cs from May-Dec 1986: 20 Bq/day (7,300 Bq/year).
- The 1985 average daily intake was 0.4 Bq/d (146 Bq/yr). The 1963 intake was 25 Bq/d (9,125 Bq/yr); this was the highest level of dietary intake of radiocesium resulting from nuclear weapons testing.
- Individuals in areas most affected by Chernobyl fallout had dietary intake levels of radiocesium far in excess of the national mean daily intake of 20 Bq/d.

Reponen, A., Jantunen, M., Paatero, J. and Jaakkola, T. (1993). Plutonium fallout in Southern Finland after the Chernobyl accident. *J. Environ. Radioactivity.* 21(2). pg. 119-130.

Date	Location	Media	Nuclide	Activity
1986	Lieksa	Ground deposition	^{239,240}Pu	17.93 Bq/m^2
1986	Koylio	Peat	^{239,240}Pu	1.779 Bq/kg
1986	Kankaanpaa	Ground deposition	^{238}Pu	6.369 Bq/m^2

- "The plutonium of Chernobyl origin correlates rather well with the non-volatile group and not at all with the volatile group. The volatile group comprises the nuclides ^{137}Cs, ^{134}Cs, ^{131}I, ^{132}Te, and the non-volatile group ^{95}Zr, ^{141}Ce, and Chernobyl-Pu." (pg. 119).

Rissanen, K., Rahola, T., Illukka, E. and Alfthan, A. (1987). *Radioactivity of reindeer, game and fish in Finnish Lapland after the Chernobyl accident in 1986: Supplement 8 to annual report STUK-A55*. Report No. STUK-A63. Finnish Centre for Radiation and Nuclear Safety, Helsinki.

- "Fallout in Finnish Lapland was considerably less than deposition in S. Finland... the lichen-reindeer-man food chain, located in a nutritionally deficient area, efficiently enhanced the uptake of radiocesium." (pg. 3).
- 9,300 reindeer were sampled with a mean concentration of: ^{137}Cs: 720 Bq/kg fresh weight; ^{134}Cs: 230 Bq/kg fresh weight.
- Mean ^{137}Cs levels in fish from lakes in N. Finland ranged from 18-280 Bq/kg; ^{134}Cs median range was 3-100 Bq/kg.

Saxen, R. and Rantavaara, A. (1987). *Radioactivity of fresh water fish in Finland after the Chernobyl accident in 1986: Supplement 6 to Annual Report STUK-A55.* Report No. STUK-A61. Finnish Centre for Radiation and Nuclear Safety, Helsinki.

Date	Location	Media	Nuclide	Activity
1986	Finland	Perch	^{137}Cs	16,000 Bq/kg
1986	Finland	Pike	^{137}Cs	10,000 Bq/kg
1986	Finland	Whitefish	^{137}Cs	7,100 Bq/kg
1986	Finland	Bream	^{137}Cs	4,500 Bq/kg
1986	Finland	Vendace	^{137}Cs	2,000 Bq/kg

- Highest concentration of radiocesium was found in the areas with the highest Chernobyl fallout.
- The smaller the lake the higher the concentration of radioactivity.

Saxen, R. and Aaltonen, H. (1987). *Radioactivity of surface water in Finland after the Chernobyl accident in 1986: Supplement 5 to Annual Report STUK-A55.* Report No. STUK-A60. Finnish Centre for Radiation and Nuclear Safety, Helsinki.

- The highest concentration of ^{137}Cs (5,300 Bq/m^3) found in 1986 was about 1,000 times higher than the average concentration of ^{137}Cs in surface water in 1985, and 10-80 times higher than the highest values detected after the weapons test period in the 1960's.
- A hot spot of 11,000 Bq/m^3 of ^{89}Sr was found in 1986.

Saxen, R., Taipale, T.K. and Aaltonen, H. (1987). *Radioactivity of wet and dry deposition and soil in Finland after the Chernobyl accident in 1986: Supplement 2 to*

Annual Report STUK-A55. Report No. STUK-A57. Finnish Centre for Radiation and Nuclear Safety, Helsinki.

Date	Location	Media	Nuclide	Activity
1986	Finland	Ground deposition	^{239,240}Pu	32 Bq/m^2
1986	Finland	Ground deposition	^{89}Sr	7,200 Bq/m^2
1986	Finland	Ground deposition	^{90}Sr	450 Bq/m^2

- The dominant alpha-emitting nuclide was ^{242}Cm, but levels were small, 0.02% of ^{137}Cs; for ^{239,240}Pu, 0.01% of ^{137}Cs.
- Chernobyl fallout was unevenly distributed in Finland: Variations in ^{137}Cs fallout ranged from 140 Bq/m^2 to 32,000 Bq/m^2.

Sinkko, K., Aaltonen, H., Mustonen, R., Taipale, T.K. and Juutilainen, J. (1987). *Airborne radioactivity in Finland after the Chernobyl accident in 1986: Supplement 1 to Annual Report STUK-A55*. Report No. STUK-A56. Finnish Centre for Radiation and Nuclear Safety, Helsinki.

- This report contains comprehensive real-time nuclide specific fallout data about the Chernobyl accident. The radionuclide concentrations noted in ground level air at Nurmijarvi on April 28 reached peak concentrations between 15.10 and 22.10; the record compiled by the Finnish Centre for Radiation and Nuclear Safety of the passage of the fallout cloud from Chernobyl at this location is one of the most important documents in the literature of radiological surveillance.
- The radionuclide concentrations listed below are measured in µBq/m^3; their normal activity levels are either zero or measured as just a few millionths of a becquerel of activity in the vicinity of nuclear power plants. The use of the reporting level µBq/m^3 in conjunction with real-time nuclide specific data collection, rather than as an averaged composite, graphically illustrates the extraordinary rise and fall of environmental radioactivity levels of the most important Chernobyl-derived nuclides as the plume pulse impacted this location.
- STUK documented the following activity levels in ground level air at Nurmijarvi on April 28:

^{95}Zr: 380,000 µBq/m^3	^{99}Mo: 2,440,000 µBq/m^3	^{103}Ru: 2,880,000 µBq/m^3
^{106}Ru: 630,000 µBq/m^3	^{110m}Ag: 130,000 µBq/m^3	^{115}Cd: 400,000 µBq/m^3
^{125}Sb: 253,000 µBq/m^3	^{127}Sb: 1,650,000 µBq/m^3	^{129m}Te: 4,000,000 µBq/m^3
^{131m}Te: 1,700,000 µBq/m^3	^{132}Te: 33,000,000 µBq/m^3	^{131}I: 223,000,000 µBq/m^3
^{133}I: 48,000,000 µBq/m^3	^{134}Cs: 7,200,000 µBq/m^3	^{136}Cs: 2,740,000 µBq/m^3
^{137}Cs: 11,900,000 µBq/m^3	^{140}Ba: 7,000,000 µBq/m^3	^{141}Ce: 570,000 µBq/m^3
^{143}Ce: 240,000 µBq/m^3	^{147}Nd: 150,000 µBq/m^3	^{239}Np: 1,900,000 µBq/m^3

- In the hours after 22.10 on the 28th of April, air concentrations of these nuclides began a rapid decline, in some cases reaching zero within 24 hours; in other cases not reaching zero for several weeks; or in the case of some of the longer lived radionuclides such as cesium, continuing throughout the summer at levels between 100 to 300 µBq/m^3.
- The two other reporting stations in Finland at Helsinki and Rovaniemi with real-time monitoring capabilities reported significantly lower peak concentrations of Chernobyl-derived radionuclides during the passage of the first plume.
- For a point of comparison air concentrations of cesium-137 measured by the Riso National Laboratory in Denmark in 1988 had a peak arithmetic mean of 2.04 µBq/m^3 out of two locations (see Riso R570, pg. 37). The highest recorded level of strontium-90 from weapons test fallout recorded in Denmark in the early 1960's was just over 1,000 µBq/m^3 (micro becquerels per cubic meter).
- The extraordinary pulse of Chernobyl-derived fallout recorded at Nurmijarvi reminds us that fallout events consist not of one nuclide, but of many nuclides; the Chernobyl accident has introduced a whole series of unfamiliar (to the layperson) radionuclides which now must be considered along with the more familiar radioiodines and radiocesiums in the evaluation of nuclear accidents of the order of magnitude of the Chernobyl accident.
- The report of high levels of Chernobyl-derived fallout at Nurmijarvi are also anomalous in that ground deposition maps give no hint that such high air concentrations occurred at this location.

Sjoblom, Kirsti-Liisa, Klemola, S., Ilus, E., Arvela, H. and Blomqvist, L. (June 1989). *Monitoring of radioactivity in the environs of Finnish nuclear power stations in 1987:*

Supplement 5 to Annual Report STUK-A74. Finnish Centre for Radiation and Nuclear Safety, Helsinki, Finland.

France

Barci, G., Dalmasso, J. and Ardisson, G. (1987). Letters:Chernobyl fallout measurements in some Mediterranean biotas. *J. Radioanal. Nucl. Chem.* 117(6). pg. 337-346.

- The Nice, France area experienced low levels of ground deposition; lichens had peak radiocesium concentrations of 4,983 Bq/kg.
- "The seaweed Spaerococcus exhibit(ed) a strong specific activity for iodine and ruthenium and poor concentration for cesium nuclides." (pg. 337).

Calmet, D., Charmasson, S., Gontier, G., Meinesz, A. and Boudouresque, C.F. (1991). Chernobyl radionuclides in the Mediterranean seagrass *Posdonia oceanica*, 1986-1987. *J. Environ. Radioactivity.* 13(2). pg. 157-174.

Date	Location	Nuclide	Media	Activity
May 28, 1986	Villefranche	^{103}Ru	Adult leaves	626 Bq/kg dry weight
May 28, 1986	Villefranche	^{137}Cs	Adult leaves	32 Bq/kg dry weight

- Preferential contamination of adult leaves.
- "Because of their rapid accumulation of radionuclides, they [adult leaves] may be a particularly interesting sentinel accumulator in the event of a nuclear accident."
- These data were collected in an area of relatively low Chernobyl contamination.

Coles, P. (1987). French suspect information on radiation levels. *Nature.* 329(96). pg. 475.

- "Following the Chernobyl accident the SCPRI (Service Central de Protection Contre les Radiations Ionisantes) initially denied that the radioactive cloud had passed over France." (pg. 475).
- "Doubts over the accuracy of SCPRI figures led to the creation of an independent body CRIIRAD which periodically monitors levels of radiation in foodstuffs." (pg. 475).
- Cesium-134,137 in mushrooms was recorded at levels up to 24,000 Bq/kg, but no baseline data was available for interpreting this level of contamination.

Commission de Recherche et d'Information Independantes sur la Radioactivite (CRII-RAD). (May 1998). *Contamination radioactive de l'Arc Alpin.* CRII-RAD, Valence, France.

- Printed in French only. A most important survey of Chernobyl-derived contamination in the higher altitudes of the Alps in France and Italy (500 to 2260 meters) in the "massif du mercantour."
- Soil contamination (^{137}Cs) levels noted up to 545,000 Bq/kg.
- More information about this study will be posted as soon as we can locate an English translation.

Martin, J. and Thomas, A. (1990). Origins, concentrations and distributions of artificial radionuclides discharged by the Rhone River to the Mediterranean Sea. *Journal of Environmental Radioactivity.* 2. pg. 105-139.

Melieres, Marie A., Pourchet, Michel, Pinglot, Jean F., Bouchea, Robert and Piboule, Michel. (June 20, 1988). Chernobyl, ^{134}Cs, ^{137}Cs, and ^{210}Pb in high mountain lake sediment: Measurements and modeling of mixing process. *Journal of Geophysical Research.* 93(D6). pg. 7055-7061.

- "A value of 0.042 Bq cm^{-2} has been estimated for the atmospheric ^{137}Cs fallout originating in Chernobyl release on the Alpine area: it represents 15% of the residual ^{137}Cs activity due to nuclear tests." (pg. 7055).

Thomas, A.J. and Martin, J.M. (1986). First assessment of Chernobyl radioactive plume over Paris. *Nature.* 321(26). pg. 817-819.

Date	Location	Media	Nuclide	Activity
April 29-30, 1986	Paris	Ground deposition	^{137}Cs	1,537 Bq/m^2
April 29-30, 1986	Paris	Rainfall	^{132}Te	7,400 Bq/l
April 29-30, 1986	Paris	Rainfall	^{137}Cs	700 Bq/l
April 29-30, 1986	Paris	Air concentration	^{239,240}Pu	0.004 mBq/m^3

- "During 1984, total ^{239,240}Pu activity was much smaller (10-40 nBq/m^3)." (pg. 818).

Germany

Bilo, M., Steffens, W. and Fuhr, F. (1993). Uptake of $^{134/137}$Cs in soil by cereals as a function of several soil parameters of three soil types in Upper Swabia and North Rhine-Westphalia (FRG). *J. Environ. Radioactivity.* 19(1). pg. 25-40.

Date	Location	Media	Nuclide	Activity
1986	Upper Swabia	Ground deposition	^{137}Cs	43,000 Bq/m^2
1986	Upper Swabia	Winter barley straw	^{137}Cs	7.74 Bq/kg
1986	Upper Swabia	Winter barley grain	^{137}Cs	3.26 Bq/kg

- Mean values in straw and grain, 1.64 Bq/kg + 0.52 Bq/kg respectively.

Brooke, J. (January 10, 1988). After Chernobyl, Africans ask if food is hot. *The New York Times*.

- Powdered milk shipments from West Germany to Angola showed levels to 6000 Bq/k (160,000 pCi/kg).
- Powdered milk shipped to Ghana, 5,460 Bq/kg.
- Powdered milk shipments to Egypt, 2,400 Bq/kg.
- Shipments returned to West Germany (Bavaria).
- Nuclide not specified; probably ^{137}Cs.

Bunzl, K. and Kracke, W. (1988). Transfer of Chernobyl-derived ^{134}Cs, ^{137}Cs, ^{131}I and ^{103}Ru from flowers to honey and pollen. *J. Environ. Radioactivity.* 6. pg. 261-269.

Date	Location	Media	Nuclide	Activity
May 1986	Germany	Pollen	^{137}Cs	+1,000 Bq/kg
May 1986	Germany	Honey	^{131}I	+14,000 Bq/kg
May 1986	Germany	Honey	^{103}Ru	+750 Bq/kg

- "Main activity deposition occurred in the afternoon of April 30, ^{137}Cs: 17,400 Bq/m^2; ^{131}I: 85,000 Bq/m^2; and ^{103}Ru: 24,000 Bq/m^2." (pg. 262).

252

- ^{134,137}Cs were more concentrated in pollen than in honey with peak concentrations in honey to 500 Bq/kg in an area with relatively low cesium deposition.

Bunzl, K. and Kracke, W. (1990). Simultaneous determination of Pu-238, Pu-239 + Pu-240, Pu-241, Am-241, Cm-242, Sr-89 and Sr-90 in vegetation samples, and application to Chernobyl Fallout contaminated grass. *J. Radioanalytical and Nuclear Chemistry, Articles.* 138. pg. 83-91.

Clooth, G. and Aumann, D.C. (1990). Environmental transfer parameters and radiological impact of the Chernobyl fallout in and around Bonn. *J. Environ. Radioactivity.* 12(2). pg. 97-120.

- Bonn escaped significant Chernobyl fallout. ^{137}Cs to 1,383 Bq/m^2 (highest of six locations).
- Geometric mean for soil-to-plant concentration factor for ^{137}Cs into pasture = 4.2 x 10^{-2} (concentration of radionuclides in plant, wet weight, divided by concentration of radionuclides in soil, dry weight).

Elstner, E.F., Fink, R., Holl, W., Lengfelder, E. and Ziegler, H. (1987). Natural and Chernobyl-caused radioactivity in mushrooms, mosses, and soil-samples of defined biotops in SW Bavaria. *Oecologia* 73. pg. 553-558.

Date	Location	Media	Nuclide	Activity
Sept. 1986	SW Bavaria	Mushrooms	^{137}Cs	8,300 Bq/kg
Sept. 1986	SW Bavaria	Moss layers	^{137}Cs	12,370 Bq/kg
Sept. 1986	SW Bavaria	Needlelayers	^{137}Cs	2,591 Bq/kg

- The 1985 peak concentration was 89 Bq/kg in mushrooms.
- Contamination with ^{134}Cs averaged about 40% of ^{137}Cs levels.

Gogolak, C.V., Winkelmann, I., Weimer, S., Wolff, S. and Klopfer, P. (1986). Observations of Chernobyl fallout in Germany by *in situ* gamma-ray spectrometry. In: *Environmental Measurements Laboratory: A compendium of the Environmental Measurements Laboratory's research projects related to the Chernobyl nuclear accident: October 1, 1986.* Report No. EML-460. US Department of Energy, New York, NY.pg. 244-258.

Date	Location	Media	Nuclide	Activity
June 3, 1986	Munich	Cumulative wet and dry deposition	^{132}Te	120,000 Bq/m^2
June 3, 1986	Munich	Cumulative wet and dry deposition	^{131}I	92,000 Bq/m^2
June 3, 1986	Munich	Cumulative wet and dry deposition	^{137}Cs	19,000 Bq/m^2

Heinzl, J., Korschinek, G. and Nolte, E. (1988). Some measurements on Chernobyl. *Physica Scripta*. 37. pg. 314-316.

Date	Location	Media	Nuclide	Activity
May 1986	Munich	Dry moss	^{134,137}Cs	30,000 Bq/kg
May 1986	Munich	Fawn meat	^{134,137}Cs	3,200 Bq/kg

Hennies, H.H. (1986). Radiation measurements in Germany resulting from the Chernobyl accident. *Nuclear Europe*. 7-8. pg. 22-25.

- "Strontium-90, which used to dominate after earlier bomb tests was present in only approximately 1% of ^{137}Cs." (pg. 22). ^{131}I dominated activity measurements at 36.6% with ^{132}Te at 15.6% and ^{132}I at 15.6%.

Hohenemser, C., Deicher, M., Hofsass, H., Lindner, G., Recknagel, E. and Budnick, J.I. (1986). Agricultural impact of Chernobyl: A warning. *Nature*. 321. pg. 817.

- "From a health perspective the expected accumulations will contaminate barns, lead to significant exposure of farm workers, and may pose special threats to children." (pg. 817).
- Hot particles were noted (1-2 μm) with single-isotope source strengths of 1,000-10,000 Bq, which would pose a particular hazard in stored hay.

Hotzl, H., Rosner, G. and Winkler, R. (1989). Long-term behavior of Chernobyl fallout in air and precipitation. *J. Environ. Radioactivity*. 10(2). pg. 157-172.

- "^{134}Cs and ^{137}Cs air concentration levels decreased exponentially with a half-time of about 250 days; .... airborne activity of ^{106}Ru decreased with a half-time of about 150 days." (pg. 157).

Hotzl, H., Rosner, G. and Winkler, R. (1987). Ground depositions and air concentrations of Chernobyl fallout radionuclides at Munich-Neuherberg. *Radiochimica Acta*. 41. pg. 181-190.

Date	Location	Media	Nuclide	Activity
April 30, 1986	Munich	Air concentration	Gross beta	100 Bq/m^3
April 30, 1986	Munich	Ground deposition	^{131}I	85,000 Bq/m^2
April 30, 1986	Munich	Ground deposition	^{132}Te	102,000 Bq/m^2

- Gamma dose rate rose from a normal of 7-8 μRh^{-1} to 108 μRh^{-1}.

Kammerer, L., Hiersche, L. and Wirth, E. (1994). Uptake of radiocesium by different species of mushrooms. *J. Environ. Radioactivity*. 23. pg. 135-150.

- "In South Bavaria... A total of 364 samples from 83 different fungal species...were taken and analyzed." (pg. 135).
- "The ^{137}Cs activity concentration in the fruiting bodies ranges between 2 and 15,000 Bq/kg fresh weight, depending on the living habits and the species of mushroom." (pg. 135).
- "During the last five years, no significant decrease of the ^{137}Cs activity in the fruiting bodies studied has been observed." (pg. 135).

Kempe, S. and Nies, H. (1987). Chernobyl nuclide record from a North Sea sediment trap. *Nature*. 329. pg. 828-831.

- The highest total specific activity of nuclides of depositing sediments reached 670,000 Bq/kg, with ^{103}Ru being the nuclide of greatest prevalence.
- Levels of deposited activity varied by a factor of 30 or more over distances of less than 100 km because of rainfall.

Kutschera, W. (1988). Measurement of ^{129}I/^{131}I ratio in Chernobyl fallout. *Physica Scripta*. 37. pg. 310-313.

- Isotopic ratio of ^{129}I/^{131}I was equal to 19 (+/- 5).
- ^{131}I activity in human thyroids ranged from 91-227 Bq in four persons on May 6, 1986, dropping to a range of 15-36 Bq for seven people on May 27.

Tagliabue, J. (January 31, 1987). A nuclear taint in milk sets off German dispute. *The New York Times.*

- 50 train cars of powdered milk destined for Egypt showed contamination of nearly 6000 Bq/kg.
- Other powdered milk in Cologne to 2400 Bq/kg.
- Contamination originated from Chernobyl-derived radiocesium in alpine meadows.
- Nuclides not specified, probably ^{137}Cs.

Tschiersch, J. and Georgi, B. (1987). Chernobyl fallout size distribution in urban areas. *J. Aerosol. Sci..* 18(6). pg. 689-692.

Greece

Online radiation fallout maps of Greece; http://arcas.nuclear.ntua.gr/radmaps/page1.html.

Assikmakopoulos, P.A., Ioannides, K.G., Pakou and Pparadopoulou, C.V. (1987). Transport of radioisotopes iodine-131, cesium-134, and cesium-137 from the fallout following the accident at the Chernobyl nuclear reactor into cheese and other cheese making products. *J Dairy Sci.* 70. pg. 1338-1343.

Date	Location	Media	Nuclide	Activity
May 1986	Spirus, Greece	Sheep milk	^{131}I	18,000 Bq/l (486,000 pCi/l)
May 1986	Spirus, Greece	Yogurt	^{131}I	6,000 Bq/kg (162,000 pCi/kg)

Kritidis, P. and Florou, H. (1991). Environmental study of radioactive cesium in Greek lake fish after the Chernobyl accident. *J. Environ. Radioactivity* 28(3). pg. 285-294.

- "The local deposition of cesium varies significantly (to 45,000 Bq/m^2); the bioaccumulation of cesium by the examined species seems to depend rather on the fish species than on the local environmental parameters." (pg. 285).

Liritzis, Y. (1987). The Chernobyl fallout in Greece and its effects on the dating of archaeological materials. *Nuclear Instruments and Methods in Physics Research.* pg. 534-537.

256

Date	Location	Media	Nuclide	Activity
May 2-6, 1986	Ptolemaida	Ground deposition	^{137}Cs	19,000 Bq/m^2
May 2-6, 1986	Ptolemaida	Ground deposition	^{134}Cs	9,000 Bq/m^2
May 2-6, 1986	Ptolemaida	Ground deposition	^{106}Ru	12,000 Bq/m^2
May 2-6, 1986	Megalopoli	Ground deposition	104Ce	4,000 Bq/m^2

- Normal background radiation, noted as 4 μRh^{-1}, reached 40 μRh^{-1} in some areas of Greece.

Papastefanou, C., Manolopoulou, M. and Charalambous, S. (1988). Cesium-137 in soils from Chernobyl fallout. *Health Physics.* 55(6). pg. 985-987.

- Ground deposition is noted as 24,000 Bq/m^2; this report doesn't indicate whether this is an average or a peak concentration.
- Peak concentrations of cesium-137 in soil is reported at 7,671 Bq/kg with all 28 soil samples showing significantly elevated radiocesium levels.

Papastefanou, C., Manolopoulou, M. and Charalambous, S. (1988). Silver-110m and ^{125}Sb in Chernobyl fallout. *The Science of the Total Environment.* 72. pg. 81-85.

- Two unusual nuclides were reported in fallout in Greece: "the ^{110m}Ag concentrations ranged from 4.5-46.1 Bq/kg in soils... and ^{125}Sb concentrations ranged from 15.6-284.6 Bq/kg in soils." Contamination levels in grass were one order of magnitude less than that in soils. (pg. 81).
- Peak concentrations of these nuclides were noted in December, "eight months after the accident, when the short-lived nuclides had decayed, and the specific activity of cesium nuclides had dropped to very low levels predominantly because of natural removal rather than decay." (pg. 82).

Papastefanou, C., Manolopoulou, M. and Charamlambous, S. (1988). Radiation measurements and radioecological aspects of fallout from the Chernobyl reactor accident. *J. Environ. Radioactivity.* 7. pg. 49-64.

Date	Location	Media	Nuclide	Activity
May 5-6, 1986	Thessaloniki	Total wet deposition	^{103}Ru	48,256 Bq/m^2

Date	Location	Media	Nuclide	Activity
May 5-6, 1986	Thessaloniki	Total wet deposition	^{131}I	117,278 Bq/m^2
May 5-6, 1986	Thessaloniki	Total wet deposition	^{132}Te	70,700 Bq/m^2
May 5-6, 1986	Thessaloniki	Total wet deposition	^{132}I	64,686 Bq/m^2
May 5-6, 1986	Thessaloniki	Total wet deposition	^{134}Cs	12,276 Bq/m^2
May 5-6, 1986	Thessaloniki	Total wet deposition	^{137}Cs	23,900 Bq/m^2
May 5-6, 1986	Thessaloniki	Total wet deposition	^{140}Ba	35,580 Bq/m^2
May 5-6, 1986	Thessaloniki	Total wet deposition	^{140}La	15,470 Bq/m^2

- The average consumption rates for the total yearly dietary intake of ^{137}Cs in Greece from May of 1986 to May of 1987 is 82,072 Bq/yr. The food samples used in this survey originated from all parts of Greece. (pg. 61).

Papastefanou, C., Manolopoulou, M. and Sawdis, T. (1989). Lichens and mosses: biological monitors of radioactive fallout from the Chernobyl reactor accident. *J. Environ. Radioactivity.* 9. pg. 199-207.

Date	Location	Media	Nuclide	Activity
1986	Northeastern Greece	Lichen	^{137}Cs	14,560 Bq/kg, dry weight

- "Locally high wet depositions included ^{131}I (117,000 Bq/m^2), ^{103}Ru (48,000 Bq/m^2), and ^{137}Cs (24,000 Bq/m^2)." (pg. 199).

Simopoulos, S.E. (1989). Soil sampling and ^{137}Cs analysis of the Chernobyl fallout in Greece. *Appl. Radiat. Isot.* 40(7). pg. 607-613.

- "The results show that ^{137}Cs fallout from Chernobyl presents a remarkable geographical variability. The evaluated ground activity due to ^{137}Cs deposition ranges between 0.01 and 137 kBq/m^2." (pg. 607).

258

Greenland

Davidson, C.I., Harrington, J.R., Stephenson, M.J., Monaghan, M.C., Pudykiewicz, J. and Schell, W.R. (1987). Radioactive cesium from the Chernobyl accident in the Greenland ice sheet. *Science.* 237. pg. 633-634.

- Wet and dry deposition of ^{134}Cs and ^{137}Cs reported as: 0.072 mCi/km^2 and 0.22 mCi/km^2.
- "The measured airborne concentrations... indicate that the radioactive cloud from Chernobyl spread rather uniformly across North America in the weeks after the accident." (pg. 634).

Pourchet, M., Pinglot, J.F. and Reynaud, L. (1988). Identification of Chernobyl fall-out as a new reference level in northern hemisphere glaciers. *Journal of Glaciology.* 34(117). pg. 183-187.

Date	Location	Media	Nuclide	Activity
10-11 May 1986	Austfonna, Greenland	Snowfall	Gross beta activity	8.84 Bq/kg
21 May 1986	Mt. Mont Blanc	Snowfall	Gross beta activity	8.78 Bq/kg

- "Very high total beta activity reported in snow." (pg. 183).
- In January 1987, additional snowfall reports are 10.05 Bq/kg and 23.46 Bq/kg, both near Mont Blanc.

Italy

Astori, E. et. al. (April 1, 1999). Surface contamination of radiocesium measured and calculated in South Piemonte (Italy). *J. Environ. Radioactivity.* 45(1). pg. 29-38.

Baldini, E., Bettoli, M.B. and Tubertini, O. (1987). Measurements on Chernobyl fallout in forest vegetation. *Inorganica Chimica Acta.* 140. pg. 331-333.

Date	Location	Media	Nuclide	Activity
July 10, 1986	N. Italy	Lichens	^{137}Cs	12,100 Bq/kg
July 10, 1986	N. Italy	Spruce twigs	^{137}Cs	3,750 Bq/kg

Date	Location	Media	Nuclide	Activity
July 10, 1986	N. Italy	Scotch pine twigs	[137]Cs	2,520 Bq/kg
July 10, 1986	N. Italy	Larch twigs	[137]Cs	1,600 Bq/kg

- Lesser amounts of [141]Ce, [144]Ce and [140]Ba-[140]La were noted.

Battiston, G.A., Degetto, S., Gerbasi, R., Sbrignadello, G. and Tositti, L. (1987). The deposition of Chernobyl fallout in north-east Italy. *Inorganica Chimica Acta.* 140. pg. 327-329.

Date	Location	Media	Nuclide	Activity
May 1, 1986	Padua	Air concentration, average values	[131]I	15,577 mBq/m^3
May 1, 1986	Padua	Air concentration, av	[132]I ([132]Te)	10,550 mBq/m^3
May 1, 1986	Padua	Air concentration, av	[103]Ru	4,070 mBq/m^3
May 1, 1986	Padua	Air concentration, av	[134,137]Cs	2,712 mBq/m^3

- "[137]Cs deposition in N. Italy ranged from 60,000 Bq/m^2 down to 2,000-5,000 Bq/m^2 on the coastal plain." (pg. 329).
- 15,577 millibecquerels (mBq) equals 15,577,000 microbecquerels (µBq).

Battiston, G.A., Degetto, S., Gerbasi, R., Sbrignadello, G. and Tositti, L. (1988). Fallout distribution in Padua and northeast Italy after the Chernobyl nuclear reactor accident. *J. Environ. Radioactivity.* 8. pg. 183-191.

Date	Location	Media	Nuclide	Activity
30 April 1986	NE Italy	Air	[131]I	28.6 Bq/m^3
30 April 1986	NE Italy	Air	[132]Te	19.2 Bq/m^3
May 1986	Padua	Soil	[137]Cs	45,000 Bq/m^2

- 28.6 becquerels (Bq) equals 28,600,000 microbecquerels (µBq).
- Average cumulative deposit from weapons test fallout from [239]Pu noted as 62 Bq/m^2.

Battiston, G.A., Degetto, S., Gerbasi, R. and Sbrignadello, G. (1989). Radioactivity in mushrooms in northeast Italy following the Chernobyl accident. *J. Environ. Radioactivity.* 9. pg. 53-60.

- Peak concentration of ^{137}Cs in edible mushrooms noted as 27,626 Bq/kg of which 21,978 Bq were Chernobyl-derived.
- Total ^{134,137}Cs deposition to 60,000 Bq/m^2 and higher (1:2 initial isotope ratio).

Boccolini, A., Gentili, A., Guidi, P., Sabbatini, V. and Toso, A. (1988). Short communication: Observation of silver 110m in the marine mollusc *Pinna nobilis*. J. *Environ. Radioactivity.* 6. pg. 191-193.

- High activity ratios of ^{110m}Ag to ^{137}Cs in the marine mollusk Pinna nobilis were recorded in the range of 15-36 Bq/kg, compared to that in dust in air filters, 0.4-0.5, illustrating the tendency of this very large bivalve mollusk to concentrate ^{110m}Ag present in seawater.

Bonnazzola, G.C., et. al. (1993). Profiles and downward migration of ^{137}Cs and ^{106}Ru deposited on Italian soils after the Chernobyl accident. *Health Physics.* 64(5). pg. 479-484.

Capra, E., Drigo, A. and Menin, A. (1989). Cesium-137 urinary excretion by northeastern (Pordenone) Italian people following the Chernobyl nuclear accident. *Health Physics.* 57(1). pg. 99-106.

Date	Location	Media	Nuclide	Activity
May 20, 1986	Pordenone	Milk	^{137}Cs	254 Bq/dm^3
Sept 30, 1987	Pordenone	Meat	^{137}Cs	395 Bq/kg

- Peak urinary excretion of ^{137}Cs occurred 300 to 425 days after main fallout cloud passage on May 5, 1986: pv 15-20 Bq/day.

Gentili, A., Gremigni, G. and Sabbatini, V. (1991). Ag-110m in fungi in central Italy after the Chernobyl accident. *J. Environ. Radioactivity.* 13(1). pg. 75-78.

- ^{110}Ag activity in fungi collected in autumn 1986 was not detectable.
- "Samples collected in 1988 and 1989 showed a remarkable accumulation of ^{110m}Ag (to 28 Bq/kg on Dec. 16, 1988)." (pg. 75).

- "[110m]Ag was indeed present in soils in 1986 but that it was not available to fungi because of its chemical form and because of lack of rainfall." (pg. 75).

Lofti, M., Notaro, M., Piermattei, S., Tommasino, L. and Azimi-Garakani, D. (1990). Radiocesium contents of meat in Italy after the Chernobyl accident and their changes during the cooking process. *J. Environ. Radioactivity.* 12(2). pg. 179-184.

- Radiocesium content of lamb is higher than any other meats considered. (pg. 179).
- "When meat is cooked in salt water (14%), the activity decreases by as much as 80%." (pg. 179).

Orlando, P., Gallelli, G., Perdelli, F. DeFlora, S. and Malcontenti, R. (1986). Alimentary restrictions and ^{131}I in human thyroids. *Nature.* 324. pg. 23.

Date	Location	Media	Nuclide	Activity
May 3- June 16, 1986	Genoa, Italy	51 adults, thyroid	[131]I	6.5 Bq/g, average
May 11, 1986	Genoa, Italy	Milk	[131]I	135,000 Bq/kg

- The average weight of a thyroid is 20 grams.

Roca, V., Napolitano, Speranza, P.R. and Gialanella, G. (1989). Analysis of radioactivity levels in soils and crops from the Campania region (south Italy) after the Chernobyl accident. *J. Environ. Radioactivity.* 9. pg. 117-129.

Date	Location	Media	Nuclide	Activity
20 May - 5 June 1986	S. Italy	Green wheat	[137]Cs	800 Bq/kg
June-Oct. 1986	S. Italy	Ground deposition	[137]Cs	8,100 Bq/m^2
June-Oct. 1986	S. Italy	Ground deposition	[103]Ru	11,900 Bq/m^2

Spezzano, P., Bortoluzzi, S., Giacomelli, R. and Massironi, L. (1994). Seasonal variations of ^{137}Cs activities in the Dora Baltea River (Northwest Italy) after the Chernobyl accident. *J. Environ. Radioactivity.* 22(1). pg. 77-88.

- "Increased water concentrations of ^{137}Cs in summer were attributed to cesium deposition and accumulation on snow covered surfaces in winter and a delayed release in summer during ice and snow melting." (pg. 77).

Spezzano, P. and Giacomelli, R. (1991). Transport of ^{131}I and ^{137}Cs from air to cows' milk produced in north-western Italian farms following the Chernobyl accident. *J. Environ. Radioactivity.* 13(3). pg. 235-250.

- "Predicted values were 2.6, 2.1, and 5.6 times higher than observed results for ^{131}I, and 4.3, 3.7 and 16 times higher for ^{137}Cs, for vegetation to air, milk to vegetation, and milk to air ratios, respectively." (pg. 235).

Spezzano, P. and Giacomelli, R. (1990). Radionuclide concentrations in air and their deposition at Saluggia (northwest Italy) following the Chernobyl nuclear accident. *J. Environ. Radioactivity.* 12(1). pg. 79-92.

Date	Location	Media	Nuclide	Activity
30 Apr-May 8	Saluggia	Ground deposition	^{137}Cs	11,000 Bq/m^2
May 2	Saluggia	Air concentration	^{131}I	8,510 Bq/m^3

- "97% of the total deposition occurred between April 30 and May 7." (pg. 89).
- "Wet deposition... was the dominant process in controlling deposition to the ground for the particulate radionuclides." (pg. 89).

Japan

Aoyama, M., Hirose, K. and Sugimura, Y. (1991). The temporal variation of stratospheric fallout derived from the Chernobyl accident. *J. Environ. Radioactivity.* 13(2). pg. 103-116.

- Chernobyl-derived stratospheric fallout continued until the end of 1988.
- Annual ^{137}Cs deposits were 135, 0.95, and 0.57 Bq/m^2, with 80% of this from Chernobyl.
- Peak weapons fallout of ^{137}Cs was in June 1963 at Tokyo: 548 Bq/m^2.
- Resuspension of ^{137}Cs contributed 2% and 34% to the annual deposition in 1987 and 1988 respectively. (pg. 113).

Aoyama, M., Hirose, K. and Sugimura, Y. (1987). Deposition of gamma-emitting nuclides in Japan after the reactor-IV accident at Chernobyl. *Journal of Radioanalytical and Nuclear Chemistry, Articles.* 116(2). pg. 291-306.

Date	Location	Media	Nuclide	Activity
May, 1986	Akita	30 day cumulative deposition	^{131}I	18,792 Bq/m^2
May, 1986	Akita	30 day cumulative deposition	^{137}Cs	414 Bq/m^2
May, 1986	Akita	30 day cumulative deposition	^{103}Ru	1,098 Bq/m^2

Aoyama, M., Hirose, K., Suzuki, Y., Inoue, H. and Sugimura, Y. (1986). High levels of radioactive nuclides in Japan in May. *Nature.* 321. pg. 819-820.

- The Chernobyl plume arrived in Japan in the form of an abrupt change of atmospheric radioactivity - with an increase of about a thousand fold in air concentration of ^{137}Cs.

Higuchi, H., Fukatsu, H., Hashimoto, T., Nonaka, N., Yoshimizu, K., Omine, M., Takano, N. and Abe, T. (1988). Radioactivity in surface air and precipitation in Japan after the Chernobyl accident. *J. Environ. Radioactivity.* 6. pg. 131-144.

- Twenty nuclides were observed in two kinds of plumes in early (1,500 m) and late May (6,300 m), with a highest level in Northwestern Japan; the average deposition of ^{137}Cs from May 1 to May 22 was 95 Bq/m^2, 550 times higher than the pre-Chernobyl value.

Hisamatsu, S., Takizawa, Y. and Abe, T. (1987). Reduction of ^{131}I content in leafy vegetables and seaweed by cooking. *J. Radiat. Res.* 28(1). pg. 135-140.

Date	Location	Media	Nuclide	Activity
May 1986	Akita	Ground deposition	^{131}I	2,500 Bq/m^2
May 1986	Akita	Edible seaweed	^{131}I	1,300 Bq/kg fresh weight

- "The decontamination ratio of ^{131}I content in leafy vegetables and edible wild grass samples boiled in water to that in washed samples was 0.51 +/- 0.21. The ratio of ^{131}I content in leafy vegetables and edible wild grass samples boiled in water to that in washed samples was 0.51 +/- 0.19 on an average." (pg. 135).

Imanaka, T. and Koide, H. (1986). Fallout in Japan from Chernobyl. *J. Environ. Radioactivity*. 4. pg. 149-153.

- Twenty nuclides were detected in air, the dominant species being ^{131}I, ^{103}Ru, ^{137}Cs and ^{134}Cs; maximum air concentration for ^{131}I was 0.8 Bq/m^3 (800,000 µBq/m^3) on May 5.

Ishida, J., Miyagawa, N., Watanabe, H., Asano, T. and Kitahara, Y. (1988). Environmental radioactivity around Tokai-Works after the reactor accident at Chernobyl. *J. Environ. Radioactivity*. 7. pg. 17-27.

Date	Location	Media	Nuclide	Activity
May, 1986	Tokai-mura, Japan	aerosol associated	^{131}I	1×10^{-1} Bq/m^3
May, 1986	Tokai-mura, Japan	gaseous in air	^{131}I	3×10^{-1} Bq/m^3
May, 1986	Tokai-mura, Japan	plants	^{131}I	2.1×10^2 Bq/kg

Kawamura, H., Sakurai, Y., Shiraishi, K. and Yanagisawa, K. (1988). Concentrations of ^{131}I in the urine of Japanese adults and children following the Chernobyl nuclear accident. *J. Environ. Radioactivity*. 6. pg. 185-189.

- ^{131}I peak concentrations to 3.3 Bq/dm^3 in a male adult.

Kusakabe, M. and Ku, T.L. (1988). Chernobyl radioactivity found in mid-water sediment interceptors in the N. Pacific and Bering Sea. *Geophysical Research Letters*. 15(1). pg. 44-47.

- Nuclides in fine air-borne particles have been incorporated into biogenic material formed in the surface ocean and transferred downward with velocities of ±100 meters per day.

Nishizawa, K., Takata, K., Hamada, N., Ogata, Y., Kojima, S., Yamashita, O., Ohshima, M. and Kayama, Y. (1986). ^{131}I in milk and rain after Chernobyl. *Nature*. 324. pg. 308.

- ^{131}I concentrations in milk to 400 pCi/l, 4-5 times higher than in rainwater.

Noguchi, H. and Murata, M. (1988). Physiochemical speciation of airborne ^{131}I in Japan from Chernobyl. *J. Environ. Radioactivity.* 7. pg. 65-74.

- "Physiochemical forms of ^{131}I were 19% particulate iodine, 5% I^2, 6% HOI and 70% organic iodides." (pg. 65).

Ooe, H., Seki, R. and Ikeda, N. (1988). Particle-size distribution of fission products in airborne dust collected at Tsukuba from April to June 1986. *J. Environ. Radioactivity.* 6. pg. 219-223.

- [137]Cs activity over a thousand times the usual level was detected in airborne dust from the end of April to the beginning of May. AMAD (activity median aerodynamic diameters) range was 0.25 to 0.70 μm.

Seki, R., Endo, K. and Ikeda, N. (1988). Determination of radioiodine species in rain water collected at Tsukuba near Tokyo. *J. Environ. Radioactivity.* 6. pg. 213-217.

- Iodide was the predominant species; no iodate was noted 1-6 May; iodate concentration was 1/3 that of iodide in mid-May sampling period; by June 1, ^{131}I species were not detectable.

Uchiyama, M. and Kobayashi, S. (1988). Consequences of the Chernobyl reactor accident on the ^{137}Cs internal dose to the Japanese population. *J. Environ. Radioactivity.* 8. pg. 119-127.

- Before the Chernobyl accident ^{137}Cs body burdens were about 30 Bq, rising the following year, 1986, to over 50 Bq with values still increasing in May of 1987. This compares to body burdens in England of 250-450 Bq. Differences in ^{137}Cs body burdens among individuals studied have become smaller with time.

Yoshida, S., Muramatsu, Y. and Ogawa, M. (1994). Radiocesium concentrations in mushrooms collected in Japan. *J. Environ. Radioactivity.* 22(2). pg. 141-154.

Date	Location	Media	Nuclide	Activity
1990	Akita	Mushrooms	[137]Cs	16,300 Bq/kg

- Mushroom ^{137}Cs activity varied widely; no other sample was greater than 3,110 Bq/kg.

- "The proportions of ^{137}Cs originated from the Chernobyl accident... and were calculated in the range of 7-60% (1989) and 10-23% (1990)." (pg. 143).

Monaco

Ballestra, S.B., Holm, E., Walton, A. and Whitehead, N.E. (1987). Fallout deposition at Monaco following the Chernobyl accident. *J. Environ. Radioactivity.* 5. pg. 391-400.

Date	Location	Media	Nuclide	Activity
May 4-5, 1986	Monaco	Deposition in rainfall	^{131}I	7,517 Bq/m^2
May 4-5, 1986	Monaco	Deposition in rainfall	^{103}Ru	2,350 Bq/m^2

- A total of 33 radioisotopes were detected; one week later, activities had fallen to about 1% of peak values.

Fowler, S.W., Buat-Menrad, P., Yokoyama, Y., Ballestra, S., Holm E. and van Nguyen, H. (1987). Rapid removal of Chernobyl fallout from Mediterranean surface waters by biological activity. *Nature.* 329. pg. 86-88.

- ^{141}Ce and ^{144}Ce were rapidly removed from surface waters and transported to 200 m in a few days primarily by zooplankton grazing (by rapidly sinking fecal pellets).

Whitehead, N.E., Ballestra, S., Holm, E. and Walton, A. (1988). Air radionuclide patterns observed at Monaco from the Chernobyl accident. *J. Environ. Radioactivity.* 7. pg. 249-264.

- Two pulses of Chernobyl-derived radioactivity were noted: the first on April 29-30; the second activity peak was May 4-5, allowing detection and differentiation of two different phases of the accident, even though Monaco and Chernobyl are about 1900 km apart. (pg. 260-61).
- Unusual radionuclides such as 105Ru, 111Ag, 125Sn and 126Sb were identified as being present in small amounts, out of a total of 34 nuclides observed.

Whitehead, N.E., Ballestra, S., Holm, E. and Huynk-Ngoc, L. (1988). Chernobyl radionuclides in shellfish. *J. Environ. Radioactivity.* 7. pg. 107-121.

- Chernobyl-derived radionuclides in mussels (Mytilus galloprovincialis) included 129Te: 2,203 Bq/kg; [129m]Te: 2,774 Bq/kg; [131]I: 5,387 Bq/kg; [132]Te: 3,168 Bq/kg (all soft parts, wet weight).
- Fifty days later, the concentration factor was 1/20 to 1/50 of the May 7 values.

Netherlands

Frissel, M.J., Stoutjesdijk, J.F., Koolwijk, A.C. and Koster, H.W. (1987). The Cs-137 contamination of soils in the Netherlands and its consequences for the contamination of crop products. *Netherlands Journal of Agricultural Science*. 35. pg. 339-346.

- Total ^{137}Cs deposition levels in the Netherlands in 1954 to 1982: 4,600 Bq/m^2, even distribution. Remaining weapons fallout deposition in soils in 1987, 1,500 Bq ^{137}Cs.
- Chernobyl-derived radiocesium deposition levels: estimated dry deposition in May, 1986 500 Bq/m^2, even distribution. Estimated wet deposition May, 1986, 0-5,500 Bq/m^2, extremely variable distribution.

Sloof, J.E. and Wolterbeek, B.T. (1992). Lichens as biomonitors for radiocesium following the Chernobyl accident. *J. Environ. Radioactivity*. 16(3). pg. 229-242.

Date	Location	Media	Nuclide	Activity
1986	Netherlands	Lichen (Armelia)	^{137}Cs	6,100 Bq/kg

- Activity concentration levels of Chernobyl-derived radiocesium in lichens closely followed ground deposition levels in the Netherlands (Activity per lichen dry weight vs. activity deposited per unit area (m^2) = approximately 1. (pg. 229).

van Dam, H. (1986). Silver from Chernobyl. *Nature*. 324. pg. 216.

- Chernobyl-derived radiosilver probably originated from 200 in core neutron detectors which had emitters composed of silver.

Voors, P.I. and van Weers, A.W. (1991). Transfer of Chernobyl radiocesium (^{134}Cs and ^{137}Cs) from grass silage to milk in dairy cows. *J. Environ. Radioactivity*. 13(2). pg. 125-140.

- 1986 radiocesium contaminated grass silage was used as feed in 1988 for dairy cows. (Average 1986 grass silage: ^{137}Cs 171.8 Bq/kg, dry weight.) Milk: to 4.7 Bq/l.
- 40-55% of the daily radiocesium intake was excreted daily.
- Average excretion rate of radiocesium for two cows was 4-5% in milk; 5-9% in urine and 32-43% in feces.

Norway

Online radiation fallout maps of Northern Europe; http://www.grida.no/db/maps/prod/europe/nc29_l.gif and http://www.mv.slu.se/ma/radio/radio/chern/cs-dep2.gif.

Blakar, I.A., Hongve, D. and Njastad, O. (1992). Chernobyl cesium in the sediments of Lake Hoysjoen, central Norway. *J. Environ. Radioactivity.* 17(1). pg. 49-58.

- Total radiocesium activity in soils in affected areas of Norway was in the range of 5,000-200,000 Bq m^{-2}.
- Study area deposition was about 50,000 Bq m^{-2}.
- Sediment concentration was similar to that in surrounding soil but higher accumulation was noted in a riverine plume due to runoff from frozen wetland after deposition and subsequent melting.
- Lake sediment contamination range "was rather homogeneous with activities between 25 and 75 kBq m^{-2} in most parts of the lake." (pg. 53).
- Stationary populations of brown trout and antic char contained about 1% of activity in the sediments." (pg. 53).

Brittain, J.E., Storruste, A. and Larsen, E. (1991). Radiocesium in brown trout (*Salmo trutta*) from a subalpine lake ecosystem after the Chernobyl reactor accident. *J. Environ. Radioactivity.* 14(3). pg. 181-192.

Date	Location	Media	Nuclide	Activity
Sep. 20	Ovre Heimdalsvatn	Brown trout	^{134,137}Cs	12,500 Bq/kg wet weight
Aug. 28	Ovre Heimdalsvatn	Lichen	^{134,137}Cs	60,000 Bq/kg wet weight

- 1986 average total cesium content rose to 7000 Bq/kg wet weight, falling to 4,000 Bq in 1988. No further declines were noted in 1988 and 1989, though there was considerable variation in content measured in individual fish.

- An observed half-life of 3.0 years for ^{137}Cs in trout was estimated.

Hongve, D., Blakar, I.A., Brittain, J.E. (1995). Radiocesium in the sediments of Ovre Heimdalsvatn, a Norwegian subalpine lake. *J. Environ. Radioactivity, 27*, 1-11.

Date	Location	Media	Nuclide	Activity
Autumn 1990	Norway	Stream substrate, organism	^{137}Cs	190,000 kBq/kg (190,000,000 Bq/kg)
Autumn 1990	Norway	Stream gravel	^{137}Cs	22,000 kBq/kg (22,000,000 Bq/kg)
Autumn 1990	Norway	Average sediment samples	^{137}Cs	500 kBq/m^2 (500,000 Bq/m^2)

- "One sediment sample with 1,846 kBq/m^2 (1,846,000 Bq/m^2) was discarded." (pg. 6).
- "The radiocesium concentration increased with increasing percentage of organic matter, water content, and with distance from the main inlet... the total radiocesium activity in the sediments may remain unchanged or even increase during the years to come." (pg. 1).

Pacyna, J.M., Semb, A. and Christenson, G.C. (1989). *Migration of ^{137}Cs from air to soil and plants in the Gulsvik area, Norway after the Chernobyl reactor accident.* Report No. NILU-TR-2/88. Lillestroem, Norsk Inst. for Luftforskning, Norway.

- "Uptake of ^{137}Cs from soil to plants through their root system is not a rapid process... the effect of the Chernobyl releases is not an acute but a long-term phenomenon." (Dialog abstract quote).

Pinglot, J. F., Pourchet, M., Lefauconnier, B., Hagen, J. O., Vaikmae, R., Punning, J. M., Watanabe, O., Takahashi, S. and Kameda, T., (1994). Natural and artificial radioactivity in the Svalbard glaciers. *J. Environ. Radioactivity*. 25. pg. 161-176.

- "Natural and artificial radioactivity in the snow of 10 Svalbard glaciers has been measured from 31 ice core samples, drilled between 1981 and 1993. Of these ice cores seven exhibit the well-known level arising from the fallout of the 1961-62 atmospheric thermonuclear tests. The second level, due to the Chernobyl accident

(26 April 1986), has been detected in all the studied glaciers; the maximum ^{137}Cs fallout reaches 22 Bq kg^{-1} and shows a high variability." (pg. 161).

- "The Chernobyl fallout is about 20Bq m^{-2} per annum, similar to the value found in Greenland, and smaller than that found in the French Alps (400 Bq m^{-2} per annum)." (pg. 174).
- "The ^{137}Cs fallout from the atmospheric thermonuclear tests reaches about 10 to 20 times the Chernobyl fallout, at 200-540 Bq m^{-2}, at the deposition date." (pg. 174).
- "The natural radioactivity (mainly ^{210}Pb) is higher than the artificial deposits... The fallout of ^{210}Pb since 1986 is about 5 times the deposit of ^{137}Cs from Chernobyl." (pg. 174).

Pourchet, M., Lefauconnier, B., Pinglot, J. F. and Hagen, J. O. (1995). Mean net accumulation of ten glacier basins in Svalbard estimated from detection of radioactive layers in shallow ice cores. *Zeitschrift fur Gletscherkunde und Glazialgeologie*. 31. pg. 73 - 84.

- "Present and recent past mean net accumulations are estimated from the detection of dated radioactive layers in 31 shallow ice-cores over ten glaciers in Svalbard. The reference layer due to fall-out of radioactive elements after the Chernobyl accident has been detected in 25 ice cores and gives a clear mark of the 1986 ablation surface. A layer corresponding to fall-out from atmospheric nuclear tests conducted in 1961 and 1962 has also been detected in eight ice cores." (pg. 73).

Solem, J.O. and Gaare, E. (1992). Radiocesium in aquatic invertebrates from Dovrefjell, Norway, 1986 to 1989, after the Chernobyl fall-out. *J. Environ. Radioactivity*. 17(1). pg. 1-12.

Date	Location	Media	Nuclide	Activity
1985	Dovrefjell, Norway	Reindeer Lichen, 5 samples	^{137}Cs	350-450 Bq/kg
1986	Dovrefjell, Norway	Reindeer Lichen, 6 samples	^{137}Cs	10,000 to 25,000 Bq/kg
1986	Dovrefjell, Norway	Ground deposition	^{137}Cs	to 80,000 Bq/m^2
1986	Dovrefjell, Norway	Plankton feeders	^{137}Cs	9,855 Bq/kg dry weight

- Contamination in aquatic invertebrates varied widely due to patchy rainfall associated deposition.

- Some samples showed an increase between 1988 and 1989.

Staaland, H., Garmo, T.H., Hove, K. and Pedersen, O. (1995). Feed selection and radiocesium intake by reindeer, sheep and goats grazing alpine summer habitats in southern Norway. *J. Environ. Radioactivity.* 29(1). pg. 39-56.

Date	Location	Media	Nuclide	Activity
1987	Jotunheim, Norway	Lichen	^{137}Cs	40,040 Bq/kg dry weight
1988	Jotunheim, Norway	Moss	^{137}Cs	20,290 Bq/kg dry weight
1987	Jotunheim, Norway	Moss	^{137}Cs	40,180 Bq/kg dry weight

- Mean value of ground deposition in all grazing areas equaled 54,000 Bq m^{-2}.
- Depending on the type of vegetation in the grazed areas, the transfer of radiocesium from soil to grazed vegetation (Bq kg^{-1} dry extrusa/Bq m^{-2} soil) was estimated to 0.02-0.04 in sheep, 0.02-0.05 in goats and 0.02-0.43 in reindeer for 1987. (pg. 39).

Steinnes, E. and Njastad, O. (1993). Use of mosses and lichens for regional mapping of ^{137}Cs fallout from the Chernobyl accident. *J. Environ. Radioactivity.* 21(1). pg. 65-74.

Date	Location	Media	Nuclide	Activity
1986	Norway	Lichen (Cladonia stellaris)	^{134,137}Cs	60,000 Bq/kg d.w.
1986	Norway	Moss (Hylocomium splendens)	^{134,137}Cs	40,000 Bq/kg d.w.

- Concentration of cesium was much less in another lichen, hylocomium physodes, and was less in spruce forest due to absorption of cesium by needles prior to deposition on lichen.

Strand, T. (1987). *Doses to the Norwegian population from naturally occurring radiation and from the Chernobyl fallout.* Doctoral Dissertation. National Institute of Radiation Hygiene, Oslo.

Date	Location	Media	Nuclide	Activity
1986	Norway	Reindeer meat	^{134,137}Cs	100,000 Bq/kg
1986	Norway	Freshwater fish	^{134,137}Cs	55,000 Bq/kg
1986	Norway	Goat's milk products	^{134,137}Cs	4,200 Bq/kg
1986	Norway	Mutton	^{134,137}Cs	15,000 Bq/kg

- The Norwegian survey used approximately 30,000 samples on foodstuffs in the first year after the Chernobyl accident; the above data are peak concentrations.
- The estimated annual intake of radiocesium following the Chernobyl accident for the average Norwegian consumer was 10,500 Bq.

Poland

Broda, R. (1987). Gamma spectroscopy analysis of hot particles from the Chernobyl fallout. *Acta Physica Polica.* B18. pg. 935-950.

Example of radioactive fallout composition in Krakow - Poland. A soil sample of 0.5 cm layer was measured 1st of May.	
^{132}Te - 29.3 kBq/m^2	^{140}Ba - 2.5 kBq/m^2
^{132}I - 25.7 kBq/m^2	^{140}La - 2.4 kBq/m^2
^{131}I - 23.6 kBq/m^2	^{99}Mo - 1.7 kBq/m^2
^{129m}Te - 8.0 kBq/m^2	^{106}Ru - 1.3 kBq/m^2
^{103}Ru - 6.1 kBq/m^2	^{127}Sb - 0.8 kBq/m^2
^{137}Cs - 5.2 kBq/m^2	^{136}Cs - 0.7 kBq/m^2
^{134}Cs - 2.7 kBq/m^2	

- Ground level activity up to 360,000 Bq/m^2 in Kracow area; the above sample is from a less contaminated area of Kracow.
- Numerous hot particles were found in both the Kracow and Masurian Lakes regions. See the summary of the same citation in Hot Particles above.

Jaworowski, Z. and Kownacka, L. (1988). Tropospheric and stratospheric distributions of radioactive iodine and cesium after the Chernobyl accident. *J. Environ. Radioactivity.* 6. pg. 145-150.

- On April 30, 1986, Warsaw ground level air concentration of ^{131}I was 11.5 Bq/m^3 (11,500,000 µBq).

Mietelski, J. W., Broda, R. and Sieniawski, J. (1988). Long lived isotopes in the Chernobyl radioactive cloud at Kracow. *J. Radioanal. Nucl. Chem., Letters.* 127(5). pg. 367-378.

- Peak gross air contamination in Kracow reached 250 Bq/m^3 on April 30, 1986. Little nuclide specific data available. Ruthenium hot particles were observed.

Piasecki, E. (1987). Spatial distribution of radioactive fallout in Poland. *J. Radioanal. Nucl. Chem, Letters.* 118(5). pg. 369-372.

- Very large spatial and temporal variability of radioactive fallout in Poland. Total beta measurements were made only with crude instrumentation.

Rich, V. (1986). Fallout pattern puzzles Poles. *Nature.* 322(28). pg. 765.

- Many hot spots were noted measuring tens to hundreds of meters across and exhibiting levels radioactivity ten times higher than the surrounding area. Ruthenium was the predominant hot spot nuclide although a few hot spots also had barium or lanthanum as the predominant nuclide.

Rich, V. (1986). Polish fallout underestimated. *Nature.* 324. (number and page unavailable).

- Early official reports from Poland were very inaccurate.

Robbins, J.A. and Jasinski, A.W. (1995). Chernobyl fallout radionuclides in Lake Sniardwy, Poland. *J. Environ. Radioactivity.* 26(2). pg. 157-184.

- Total ^{137}Cs loading of Lake Sniardwy estimated at an average of 6,100 Bq m^{-2}.
- Approximate particles settling rate one meter per day.
- "The average Activity with ^{137}Cs in the flesh of bream increased up to 120 times that of pre Chernobyl levels measured in 1985. (Bream to 493 mBq/g)" (pg. 157-158).

274

Seaward, M.R.D., Heslop, J.A., Green, D. and Bylinska, E. A. (1988). Recent levels of radionuclides in lichens from southwest Poland with particular reference to [134]Cs and [137]Cs. *J. Environ. Radioactivity.* 7. pg. 123-129.

Date	Location	Media	Nuclide	Activity
August 1986	SW Poland	Lichen	[95]Nb	8,114 Bq/kg
August 1986	SW Poland	Lichen	[103]Ru	2,065 Bq/kg
August 1986	SW Poland	Lichen	[106]Ru/[106]Rh	16,570 Bq/kg
August 1986	SW Poland	Lichen	[134]Cs	18,263 Bq/kg
August 1986	SW Poland	Lichen	[137]Cs	36,630 Bq/kg
August 1986	SW Poland	Lichen	[144]Ce	18,500 Bq/kg

- August 1986 samples of *Umbilicaria* collected in SW Poland showed significant increases in radioactivity over 1978-79 levels. Lichen is a highly efficient accumulator of radionuclides.

Skwarzec, B. and Bojanowski, R. (1992). Distribution of plutonium in selected components of the Baltic ecosystem within the Polish economic zone. *J. Environ. Radioactivity.* 15(3). pg. 249-264.

- Average values of ^{239,240}Pu in seawater: 5 mBq m^{-3} of which 70% constituted filterable forms (<0.45 μm).
- Concentration levels of plutonium were one order of magnitude greater per kilogram of dried specimens vs. wet weight concentration.
- Total plutonium deposition ranged from 30 Bqm^{-2} to 98 Bq m^{-2} at three sampling locations.
- The highest concentration of Pu in sediment were probably due to the Vistula river, which delivered 192 MBq ^{239,240}Pu in 1989 to the Baltic Sea.
- Little impact from Chernobyl plutonium, unlike Chernobyl cesium.
- The concentration factor range was 600-27,100 in benthic animals and sea plants.

Portugal

Carvalho, F.P. (1986). *Radioactivity fallout in Portugal following the Chernobyl accident*. Laboratorio Nacional de Engenharia e Tecnologia Industrial, Saracem, Portugal.

- Almost none of the Chernobyl fallout reached Portugal compared to the rest of Europe. Within the Azore Islands, peak values of ^{131}I in milk 1.5 Bq/l, and for ^{137}Cs 19 Bq/l.

Romania

Pourchet, M., Melieres, M. A., Silvestru, E., Rajka, G., Candaudap, F. and Carbonnel, J. P. (May 26-30, 1996). Radionuclides in a cave sediment core from Ghetarul de sub Zgurasti (Romania). *14th Intern. Symposium on Theoretical and Applied Karstology.* Baile Herculane- Romania.

- "The transmission of pollution from the earth surface to remote areas like underground cave is very effective. To our knowledge it is the first time that such substantial contamination is revealed." (pg. 1).
- "The ^{137}Cs inventory relative to nuclear tests (estimated from the ^{90}Sr deposit) has been measured at Vienna (Austria) and is of 2.95 $kBq.m^{-2}.y^{-1}$. The Chernobyl ^{137}Cs deposition in Romania has been estimated at Cluj to be 4 $kBq.m^{-2}$, but shows in Romania a wide range of values from 0.2 to 8.5 $kBq.m^{-8}$." (pg. 2).

Russia and Former USSR

Abaturov, Y.D., Abaturov, A.V., Bykov, A.V., et al. (1996). *The effect of ionizing irradiation on the pine forests in the nearest zone of the Chernobyl nuclear power plant.* ISBN 5-02-001918-6. M. Nauka. pp. 140.

- "Materials of researches in 1987-1992 over the territory adjacent to Chernobyl nuclear power plant are generalized in the monograph. A three-dimensional picture of area of irradiation damage of pine along the western trace of radioactive fall-outs is presented. The most important elements of irradiation situation during an active phase of the incident are reconstructed." (abstract).

276

AUIAR. *Cs-137 and Cs-134 contamination map of Byelorussia based on data from 1989*. Byelorussian Department of the All Union Institute for Agricultural Radiology, Gomel. (unpublished).

Cooper, E.L., Zeiller, E., Ghods-Esphahani, A., Makarewicz, M., Schelenz, R., Frindik, O., Heilgeist, M. and Kalus, W. (1992). Radioactivity in food and total diet samples collected in selected settlements in the USSR. *J. Environ. Radioactivity.* 17(2+3). pg. 147-158.

Date	Location	Media	Nuclide	Activity
1987	Novozybkov	Wild blueberries	^{137}Cs	1240 Bq/kg wet weight
1987	Savici	Mushroom	^{137}Cs	131,000 Bq/kg wet weight
1987	Daleta	Total dietary intake per kg of food	^{137}Cs	6,370 Bq/kg total diet average
1987	Daleta	Total dietary intake	Total alpha	1300 µBq/g (1.3 Bq/kg) dry weight

- Daleta, an area with the "lowest" surface contamination, had ^{137}Cs deposition levels of 1-5 Ci km^2 (37,000 to 185,000 Bq/M^2).

Egorov, Nikolai N., Novikov, Vladimir M., Parker, Frank L. and Popov, Victor K., Eds. (June 2000). *The radiation legacy of the Soviet nuclear complex*. Earthscan Publications Ltd., United Kingdom.

- The annotations of this text are in the RADNET Section on Russian nuclear power plants.

Fleishman, D.G., Nikiforov, V.A., Saulus, A.A. and Komov, V.T. (1992). ^{137}Cs in fish of some lakes and rivers of the Bryansk region and north-west Russia in 1990-1992. *J. Environ. Radioactivity.* 24(2). pg. 145-158.

- The highest levels of ^{137}Cs ranged up to 15,000-21,000 Bq/kg wet weight. (pg. 145).
- "Fish from lakes generally contained more ^{137}Cs (up to about ten times) than those from rivers, both in the heavily contaminated zone and in the lightly contaminated regions." (pg. 145).

- "Between 1990 and 1992 there was no significant general decrease of ^{137}Cs content in fish." (pg. 145).

Holliday, B., Binns, K.C. and Stewart, S.P. (1986). Monitoring Minsk and Kiev students after Chernobyl. *Nature.* 321. pg. 820-821.

- "Using published NRPB information, an acceptance level with 30 Bq/cm^2 (300,000 Bq/m^2)... was adopted as the clearance level for clothing contamination on returning students." (pg. 820).
- "Most of the clothing showed positive indications of contamination but only 2% of the clothing monitored was in excess of the 'clearance level.'" (pg. 821).
- Approximate thyroid radioactivity of 99 students reentering to the United Kingdom ranged from 800 Bq to 6,900 Bq with a maximum calculated intake on 26 April of 53,000 Bq.

IAEA (1990). *The radiological consequences in the USSR from the Chernobyl accident: Assessment of health and environmental effects and evaluation of protective measures.* Project description and scheme of implementation as approved by the International Advisory Committee, International Atomic Energy Agency report, and appendices. IAEA, Vienna.

Kliashtorn, A.L., Tikhomirov, F.A. and Shcheglov, A.I. (1984). Lysimetrical study of radionuclides in the forests around the Chernobyl nuclear power plant. *J. Environ. Radioactivity.* 24. pg. 81-90.

- "Sum total accumulation deposition of the radionuclides (Ce-144; Cs-134,137; Ru-106; Sr-90) ranged from 90 MBq/m^2 to 0.5 Mbq/m^2 (90,000,000 Bq/m^2 - 500,000 Bq/m^2)." (pg. 81).
- "It was shown that 0.01-0.6% of the total amount of radionuclides was lost from the 20-30 cm layer of soil every year depending on the type of radionuclide, type of ecosystem and the location of study plot." (pg. 207).

Knatko, V.A., Skomorokhov, A.G., Asimova, V.D., Strakh, L.I., Bogdanov, A.P. and Mironov, V.P. (1996). Characteristics of ^{90}Sr, ^{137}Cs and ^{239,240}Pu migration in undisturbed soils of southern Belarus after the Chernobyl accident. *J. Environ. Radioactivity.* 30(2). pg. 185-196.

Date	Location	Media	Nuclide	Activity
1994	Southern Belarus	Total activity; soil horizon 0-20 cm.	^{137}Cs	6,700,000 Bq/m^2
1994	Southern Belarus	Total activity; soil horizon 0-20 cm.	^{239,240}Pu	37,700 Bq/m^2

- "These values are 10^2-10^3 times higher than the global fallout." (pg. 189).

Knatko, V.A., Gurkov, V.V., Asimova, V.D., Shpakovskaya, E.B. and Shimanovich, E.A. (1994). Soil-milk transfer of ^{137}Cs in an area of Byelorussia after the Chernobyl accident. *J. Environ. Radioactivity.* 22(3). pg. 269-278.

- "Some hundreds of kilometers from Chernobyl the level of soil contamination varies over several orders of magnitude and is sometimes higher than 1,000,000 Bq m^{-2}." (pg. 269).
- "TF (transfer) values decrease with increasing soil contamination levels, except a constant TF value is achieved in highly contaminated soil." (pg. 277).

Kryshev, I.I. (1995). Radioactive contamination of aquatic ecosystems following the Chernobyl accident. *J. Environ. Radioactivity.* 27(3). pg. 207-220.

Date	Location	Media	Nuclide	Activity
May 1986	Chernobyl	Cooling pond sediments	^{95}Zr	54,000,000 Bq/m^2
May 1986	Chernobyl	Cooling pond sediments	^{95}Nb	50,000,000 Bq/m^2
May 1986	Chernobyl	Cooling pond sediments	^{137}Cs	5,000,000 Bq/m^2
May 1986	Chernobyl	Cooling pond sediments	^{144}Ce	40,000,000 Bq/m^2
May 1986	Chernobyl	Algae	^{137}Cs	90,000 Bq/m^2
1987	Chernobyl	Pike muscle	^{137}Cs	420,000 Bq/kg fresh weight
1989	Kiev reservoir	Mollusc	^{90}Sr	1,200 Bq/kg fresh weight

- "The CNPP cooling pond... was formed by cutting off part of the Pripyat River plain with a dike." (pg. 209).
- "Reduction in the ^{137}Cs concentration proceeded slowly in most of the aquatic ecosystems." (pg. 207).

Likhtarev, I.A., Kovgan, L.N., Vavilov, S.E., Gluvchinsky, R.R., Perevoznikov, O.N., Litvinets, L.N., Anspaugh, L.R., Kercher, J.R. and Bouville, A. (1996). Internal exposure from the ingestion of foods contaminated by ^{137}Cs after the Chernobyl accident. Report 1. General model: Ingestion doses and countermeasure effectiveness for the adults of Rovno oblast of Ukraine. *Health Physics.* 70(3). pg. 297-317.

- "Fallout of ^{137}Cs in these regions [Rovno Oblast] of Ukraine was lower than in other regions of Ukraine. However, the transfer of ^{137}Cs from soil to milk in the region considered is high (up to 20 Bq L^{-1} per kBq m^{-2}" (pg. 297).

Poiarkov, V.A., Nazarov, A.N. and Kaletnik, N.N. (1995). Post-Chernobyl radiomonitoring of Ukrainian forest ecosystems. *J. Environ. Radioactivity.* 26. pg. 259-271.

- ^{137}Cs activities now measured in Ci km^{-2} in litter, and per 0-2 and 5 cm units of soil depth.
- Peak concentration in litter: 12.3 Ci km^{-2} (1 Ci/km^2=37,000 Bq/m^2).
- "The bulk of the activity occurs in the top 15 cm of the soil profile... Soil Profiles indicate no substantial differences for 1991 and 1992 suggesting that ^{137}Cs has a limited mobility." (pg. 259).

Polikarpov, G.G., Kulebakina, L.G., Timoshchuk, V.I. and Stokozov, N.A. (1991). ^{90}Sr and ^{137}Cs in surface waters of the Dnieper River, the Black Sea and the Aegean Sea in 1987 and 1988. *J. Environ. Radioactivity* 13(1). pg. 25-38.

- Concentrations of ^{90}Sr in the Dnieper River, which drains the Pripjat River near Chernobyl, were up to ten times greater than cesium concentrations, "probably due to the ability of cesium to become absorbed on suspended particles... and the high solubility of strontium in fresh water." (pg. 36).

Robertson, D.E., Perkins, R.W., Lepel, E.L. and Thomas, C.W. (1992). Radionuclide concentrations in environmental samples collected around Chernobyl during the International Chernobyl Project - analyses conducted by Battelle, Pacific Northwest Laboratory. *J. Environ. Radioactivity.* 17(2+3). pg. 159-182.

Date	Location	Media	Nuclide	Activity
1990	Svjatsk	Creek sediment	^{137}Cs	4.53 Bq/g
1990	Soboli	Rock moss scrapings	^{137}Cs	552 Bq/g (552,000 Bq/kg)
1990	Bragin	Tombstone moss scrapings	^{239,240}Pu	0.979 Bq/g (979 Bq/kg)
1990	Bragin	Soil core	^{129}I	5.11 x 10^{-6}Bq/g

- ^{129}I 1/2T = 15.9 million years.
- "Moss species appear to be efficient collectors of the Chernobyl fallout in this region." (pg. 159).
- "Analyses of undisturbed soil cores indicated that the Chernobyl fallout was still concentrated in the upper several centimeters of soil as of August, 1990. Cesium-137 is the least mobile of all of the radionuclides measured.' (pg. 159).

Salbu, B., Oughton, D.H., Ratnikov, A.V., Zhigareva, T.L., Kruglov, S.V., Petrov, K.V., Grebenshakikova, N.V., Firsakova, S.K., Astasheva, N.P., Loshchilov, N.A., Hove, K. and Strand, P. (1994). The mobility of ^{137}Cs and ^{90}Sr in agricultural soils in the Ukraine, Belarus, and Russia, 1991. *Health Physics.* 67(5). pg. 518-528.

- "The sites, representing plowed and natural pastures, were located at varying distances between 50 and 650 km and varying directions from the Chernobyl reactor site." (pg. 518).
- "The ^{137}Cs activity concentrations in the upper 0-5 cm soil layer ranged from 25-1,000 kBqm^{-2} and were higher in natural pastures as compared to plowed pastures." (pg. 518).
- "For ^{90}Sr, activity levels ranged from 1.4-40 kBq m^{-2}" (pg. 518).

Steinhausler, F. (1992). Uncertainties associated with the corroboration of official USSR environmental data in areas contaminated by Chernobyl fallout. *J. Environ. Radioactivity.* 17(2+3). pg. 211-232.

Date	Location	Media	Nuclide	Activity
1990	Bragin	Lichen (Parmelia sulcata)	^{137}Cs	326,000 Bq/kg

- "The results show that the uncertainty associated with official data on ^{137}Cs is relatively low.' (pg. 211).

Tracy, B.L., Kramer, G.H. and Gamarnik, K. (1993). Radiocesium in children from Belarus. *Health Physics.* 66(4). pg. 439-443.

- "The body burdens of ^{137}Cs were measured in 74 children from Belrus who had been exposed to fallout from the Chernobyl accident [who were] visiting the Ottawa area during the summers of 1991 and 1992." (pg. 439).
- "The body burdens were related to some extent, with recorded fallout levels in the region of origin but did not appear to change significantly from 1991 to 1992." (pg. 439).
- "During their stay in Canada, radiocesium was being cleared from the children's bodies with a mean half-time of 33 d (range = 12-77 d)." (pg. 439).

USSR State Committee on Hydrometeorology. (1991). *Radiation maps in the territory of the European part of the USSR as of December 1990. Densities of area contamination by cesium-137, strontium-90, and plutonium-239, 240.* SCH, Minsk.

van den Berg, G.J., Tyssen, T.P.M., Ammerlaan, M.J.J., Volkers, K.J., Woroniecka, U.D., de Bruin, M. and Wolterbeek, H. Th. (1992). *J. Environ. Radioactivity.* 17(2+3). pg. 115-128.

Date	Location	Media	Nuclide	Activity
1986	Bragin	Lichen	^{137}Cs	1,630,000 Bq/kg

- "The lichen ^{137}Cs levels are one to two orders of magnitude higher than levels determined in 1986 in Poland, Greece or the Netherlands." (pg. 115).
- "Averaged lichen ^{137}Cs levels (kBq kg^{-1}) are proportional to soil surface deposition (kBq m^{-2}), except for higher soil ^{137}Cs activity classes (>500 kBqm^{-2}), where lichen data may lead to underestimates of actual deposition." (pg. 128).

Scotland

Also see the listings under United Kingdom.

(1998). *The Scottish Environmental Statistics - 1998.* Economic Advice and Statistics, The Scottish Office, Edinburgh, UK.

- Online maps of radiation in Scotland; http://www.scotland.gov.uk/library/stat-ses/sesm7-5.htm.

Martin, C.J., Heaton, B. and Robb, J.D. (1988). Studies of ^{131}I, ^{137}Cs and ^{103}Ru in milk, meat and vegetables in north east Scotland following the Chernobyl accident. *J. Environ. Radioactivity.* 6. pg. 247-259.

- "A heavy rainstorm occurred between 2100 and 2300 hours on the evening of 3 May and contained summed concentrations of the nuclides $^{132}Te/^{132}I$, ^{131}I, ^{103}Ru, ^{137}Cs, ^{134}Cs and $^{140}Ba/^{140}La$ totaling 7,000 Bq/l ... corresponding to activity levels of 41,000 Bq/m^2." (pg. 249).

Martin, C.J. et. al. (1989). Cesium-137, Cs-134 and Ag-110 in lambs grazing pasture in NE Scotland contaminated by Chernobyl fallout. *Health Physics.* 56(4). pg. 459-464.

Martin, C.J. and Heaton, B. (1989). The impact of Chernobyl on the marine environment in Northern Scotland. *J. Environ. Radioactivity.* 9. pg. 209-221.

- The biological half-lives of Cs, Ru, and Ag radionuclides in *Fucus vesiculosis* were 57, 80 and 210 days respectively.
- Radionuclide levels in sea spume were several thousand times greater than in sea water in June of 1986.
- Cesium quickly migrated to the sediments; ruthenium and radioactive silver lingered in the spume.

Scottish Development Department. (1988). *Chernobyl accident, monitoring for radioactivity in Scotland.* Statistical Bulletin No 1(E). Scottish Office, Edinburgh.

Date	Location	Media	Nuclide	Activity
May 6, 1986	Strathelyde	Ground deposition	Gross beta	88,425 Bq/m^2
May 11, 1986	Galloway	Evaporated milk	$^{134,137}Cs$	689 Bq/kg
May 12, 1986	Highland	Goats milk	^{131}I	1,460 Bq/L
June 3, 1986	Strathelyde	Free range hen	^{131}I	23,330 Bq/kg
June 3, 1986	Strathelyde	Free range hen	$^{134,137}Cs$	6,600 Bq/kg
July 1986	Borders	Mutton	$^{134,137}Cs$	4,218 Bq/kg

- A difficult to read but very detailed record of pervasive Chernobyl-derived radioactive contamination in all areas of Scotland, with increasing levels of radiocesium in mutton throughout the summer.

Watson, W.S. (1986). Human ^{134}Cs / ^{137}Cs levels in Scotland after Chernobyl. *Nature.* 323. pg. 763-764.

- Mean activity in 18 adults: ^{134}Cs: 172 Bq; ^{137}Cs: 363 Bq; 40K: 4,430 Bq. Peak concentrations of ^{134}Cs: 285 Bq; ^{137}Cs: 663 Bq.

Spain

Baeza, A., del Rio, M., Miro, C., Moreno, A., Navarro, E., Paniagua, J.M. and Peris, M.A. (1991). Radiocesium and radiostrontium levels in song-thrushes (*Turdus philomelos*) captured in two regions of Spain. *J. Environ. Radioactivity.* 13(1). pg. 13-24.

- Peak mean concentration in a group of 14 song thrushes captured at Valencia, ^{137}Cs: 208 Bq/kg dry weight for this group.

Sweden

Online radiation fallout maps of Northern Europe;
http://www.grida.no/db/maps/prod/europe/nc29_1.gif and
http://www.mv.slu.se/ma/radio/radio/chern/cs-dep2.gif.

Maps of Sweden; http://www.gr.is/nsfs/finck.htm.

Ahman, B. and Ahman, G. (1994). Radiocesium in Swedish reindeer after the Chernobyl fallout: Seasonal variations and long-term decline. *Health Physics.* 66(5). pg. 506.

Date	Location	Media	Nuclide	Activity
1986-1987	Vilhelmina Sodra	Ground deposition	^{137}Cs	80,000 Bq/m^2
1986-1987	Vilhelmina Sodra	Reindeer meat, average of 29 samples	^{137}Cs	44,800 Bq/kg

- "^{137}Cs activity concentrations in reindeer during winter exceed those found during summer by about 20 times." (pg. 503).
- "Activity concentrations of ^{137}Cs in reindeer were fairly well correlated to ground deposition ." (pg. 503).
- "The ratio between ^{137}Cs in reindeer (kBq kg^{-1} wet weight) and ground deposition (kBq m^{-2}) was calculated to be 0.76 m^{2}kg^{-1} for the winter period, January-April, in 1987." (pg. 503).

Carbol, P., Ittner, T. and Skalberg, M. (1988). Radionuclide deposition and migration of the Chernobyl fallout in Sweden. *Radiochimica Acta.* 44/45. pg. 207-212.

- "Surface activity varies between 14 and 300,000 Bq (300 kBq) ^{137}Cs per square meter at seven sampling points." (pg. 171).
- "Two different forms of ruthenium are observed, one which is insoluble and observed in hot spots, and one form which is transported by water." (pg. 171).

Carlson, L. and Holm, E. (1992). Radioactivity in *Fucus vesiculosis L.* from the Baltic Sea following the Chernobyl accident. *J. Environ. Radioactivity.* 15(3). pg. 231-248.

Date	Location	Media	Nuclide	Activity
July 1986	Baltic Sea	Fucus vesiculosis	^{239,240}Pu	268 mBq/kg dry weight
July 1986	Baltic Sea	Fucus vesiculosis	^{137}Cs	600 Bq/kg dry weight
July 1986	Baltic Sea	Fucus vesiculosis	^{99}Tc	108 Bq/kg dry weight

- In Aug-Sept. 1987, the activity concentration of radiocesium level increased by a factor of 2 to 3 at most localities off the Swedish coast.
- There was no increase in plutonium or americium between 1986 and 1987.
- In another article (SSI Project P 393-86 University of Lund, Sweden), Carlson and Holm found that Fucus vesiculosis preferentially accumulates radionuclides in the following order: technicium -> americium -> plutonium -> cesium.

Dahl, C. and Grimas, U. (1987). Report of radionuclides in *Aedes communis* pupae from central Sweden, 1986. *Journal of American Mosquito Control Association.* 3(2). pg. 328-331.

Date	Location	Media	Nuclide	Activity
May 18, 1986	Central Sweden	aquatic insect larvae	[131]I	5,000 Bq/kg dry weight
1986	Central Sweden	aquatic insect larvae	[103]Ru	1,060 Bq/kg d.w.
May 4, 1986	Central Sweden	diatoms	[131]I	31,000 Bq/kg d.w.
1986	Central Sweden	diatoms	[103]Ru	6,400 Bq/kg d.w.

Danell, K., Nelin, P. and Wickman, G. (1989). [137]Caesium in Northern Swedish moose: The first year after the Chernobyl accident. *Ambio.* 18(2). pg. 108-111.

Date	Location	Media	Nuclide	Activity
April 1986	Sweden	Ground deposition	[137]Cs	60,000 Bq/m^2
April 1986	Sweden	Moose meat	[137]Cs	665 Bq/kg (mean)

- "Concentration of cesium in moose muscle correlated positively with ground deposition."
- "3,661 moose: Average was 470 Bq/kg fresh weight for calves and 300 Bq/kg for older animals." (pg. 108).
- The average level before accident was 33 Bq/kg.

Devell, L., Aarkrog, A., Blomquist, L, Magnusson, S. and Tveten, U. (1986). How the fallout from Chernobyl was detected and measured in the Nordic countries. *Nuclear Europe.* 11. pg. 16-17.

- "Actual impact upon agriculture seemed modest in the beginning." (pg. 16).
- Peak concentrations in Norway and Sweden were soon noted in excess of 100,000 Bq/m^2 in some locations.

Devell, L., Tovedal, H., Bergstrom, U., Appelgren, A., Chyssler, J. and Andersson, L. (1986). Initial observations of fallout from the reactor accident at Chernobyl. *Nature.* 321. pg. 192-193.

- A hot particle of nearly pure ruthenium was noted (the diameter is ~ 1μm); [103]Ru: 10,000 Bq; [106]Ru: 2,800 Bq; 51Cr: 1,700 Bq; [99]Mo, 99Mtc: 1,600 Bq.

Erlandsson, B. and Mattsson, S. (1988). Uptake of dry-deposited radionuclides in *Fucus* - a field study after the Chernobyl accident. *J. Environ. Radioactivity.* 6. pg. 271-281.

- In an area of very low deposition (11.6 Bq/m^2) the activity concentration ratio of cesium-137 m^2/kg of *Fucus* ranged from 0.66-1.3, where fallout was primarily dry deposition of large and not particularly soluble particles.

Finck, Robert. (August 1996). Local reference measurements of gamma radiation in Sweden. Experiences from seven years of measurements. *Proceedings Nordic Society for Radiation Protection.* Swedish Radiation Protection Institute, Stockholm, Sweden. http://www.gr.is/nsfs/finck.htm.

Fox, B. (1988). Porous minerals soak up Chernobyl's fallout. *New Scientist.* 2. pg. 36.

- "Boiling contaminated reindeer meat (15,000 Bq/kg) in 1/2 liter water containing 10 grams mordenite reduced contamination to 1,200 Bq/kg."

Hakanson, L., Andersson, T. and Nilsson, A. Radioactive cesium in fish in Swedish lakes 1986-1988 - General pattern related to fallout and lake characteristics. *J. Environ. Radioactivity.* 15(3). pg. 207-230.

Date	Location	Media	Nuclide	Activity
1986	Sweden	Perch	^{134,137}Cs	3,585 (mean) Bq/kg wet weight
1988	Sweden	Perch	^{134,137}Cs	6,042 (mean) Bq/kg wet weight

- "A register was compiled containing a broad set of data from 644 Swedish lakes. The median ^{137}Cs concentration in fish increased between 1986 and 1987 by between 13% (trout) and 240% (pike)." (pg. 207).
- "The increase between 1987 and 1988 has stagnated for most species, but not for pike where the concentration increased 82% in 'the median lake'." (pg. 207).
- "About 14,000 lakes in Sweden had fish ('100g perch') with ^{137}Cs concentrations above 1500 Bq/kg (wet weight)... during the autumn of 1987." (pg. 207).

Hakansson, L. (1991). Radioactive cesium in fish in Swedish lakes after Chernobyl - geographical distributions, trends, models and remedial measures. In : *The Chernobyl fallout in Sweden.* Moberg, L., Ed. Stockholm. pg. 239-281.

Hardy, E., Krey, P., Klusek, C., Miller, K., Helfer, I., Sanderson, C. and Rivera, W. (1986). Observations and sampling by EML in Sweden, with preliminary gamma-ray spectrometric data. In: *Environmental Measurements Laboratory: A compendium of the Environmental Measurements Laboratory's research projects related to the Chernobyl nuclear accident: October 1, 1986.* Report No. EML-460. New York, US Department of Energy, New York, NY. pg. 224-243.

Date	Location	Media	Nuclide	Activity
May 5, 1986	Skutskar, Sweden	Ground Deposition	^{95}Zr	14,000 Bq/m^2
May 5, 1986	Skutskar	G. Deposition	^{95}Nb	14,000 Bq/m^2
May 5, 1986	Skutskar	G. Deposition	^{103}Ru	7,800 Bq/m^2
May 5, 1986	Skutskar	G. Deposition	^{131}I	890,000 Bq/m^2
May 5, 1986	Skutskar	G. Deposition	^{132}Te, ^{132}I	260,000 Bq/m^2
May 5, 1986	Skutskar	G. Deposition	^{134}Cs	150,000 Bq/m^2
May 5, 1986	Skutskar	G. Deposition	^{136}Cs	44,000 Bq/m^2
May 5, 1986	Skutskar	G. Deposition	^{137}Cs	240,000 Bq/m^2
May 5, 1986	Skutskar	G. Deposition	^{140}Ba	180,000 Bq/m^2
May 5, 1986	Skutskar	G. Deposition	^{140}La	210,000 Bq/m^2
May 5, 1986	Skutskar	G. Deposition	^{141}Ce	10,000 Bq/m^2

- Prior to the Chernobyl accident, ^{137}Cs deposition at Skutskar was estimated to be 1,500 Bq.
- In Skutskar, background radiation rose to 900 μRh^{-1}; in Stockholm, it was 30 μRh^{-1}.
- This data is an example of when "the dose and subsequent health risks to the population of Western Europe are minimal." (pg. 259).

Holmberg, M., Edvarson, K. and Finck, R. (1988). Radiation doses in Sweden resulting from the Chernobyl fallout: a review. *Int. J. Radiat. Biol.* 54(2). pg. 151-166.

- "Equivalent surface deposition as an average for the whole country: about 3,000 Bq/m^2 of ^{134}Cs and 5,000 Bq/m^2 of ^{137}Cs." (pg.152).

- "Accumulated cesium deposition from nuclear weapons fallout 1955-1980: about 4,000 Bq/m^2 of ^{137}Cs (Cs-134 was only present in small amounts)." (pg.152).
- "Using a risk factor of 0.02 fatal cancers for man-Sievert the Chernobyl fallout over Sweden might cause 100 to 200 fatal cancers." (pg. 152).

Johanson, K.J., Bergstrom, R., Eriksson, O. and Erixon, A. (1994). Activity concentrations of ^{137}Cs in moose and their forage plants in mid-Sweden. *J. Environ. Radioactivity.* 22(3). pg. 251-267.

Date	Location	Media	Nuclide	Activity
May 5, 1986	Skutskar	Deposition in wet soil	^{131}I	2,139,000 Bq/m^2
May 5, 1986	Skutskar	Deposition in wet soil	^{132}Te	2,094,000 Bq/m^2
May 5, 1986	Skutskar	Deposition in wet soil	^{137}Cs	289,000 Bq/m^2
May 5, 1986	Skutskar	Deposition in wet soil	^{140}Ba	362,000 Bq/m^2
1989	Kramfors	Heather	^{134,137}Cs	15,129 Bq/kg dry weight
1989	Kramfors	Waterlily	^{134,137}Cs	14,745 Bq/kg dry weight

- "The mean ^{137}Cs activity concentration in moose muscle samples from the three areas with a ground deposition from 20 to 60 kBq m^{-2} varied between 540 and 915 Bq kg^{-1} [N= 1,119 moose]." (pg. 251).

Kresten, P. and Chyssler, J. (1989). The Chernobyl fallout: Surface soil deposition in Sweden. *Geologiska Foreningens i Stockholm Forhandlingar.* 111(2). pg. 181-185.

Date	Location	Media	Nuclide	Activity
1 May 1986	Skatan, Sweden	Surface soil	^{137}Cs	200,000 Bq/kg
1 May 1986	Skatan, Sweden	Surface soil	^{134}Cs	117,000 Bq/kg
1 May 1986	Skatan, Sweden	Surface soil	^{140}La	80,000 Bq/kg
1 May 1986	Skatan, Sweden	Surface soil	^{131}I	669,000 Bq/kg
1 May 1986	Skatan, Sweden	Surface soil	^{103}Ru	51,200 Bq/kg

- Areas of wet deposition exhibited the highest levels of fallout.
- "Components in the fallout show wide variations...^{137}Cs dominated the coast of southern Norrlan; ^{131}I to the north and south, with the central Uppland area characterized by ^{132}Te dominating over ^{131}I." (pg. 182).

Krey, P.W., Klusek, C.S., Sanderson, C., Miller, K. and Helfer, I. (1986). Radiochemical characterization of Chernobyl fallout in Europe. In: *Environmental Measurements Laboratory: A compendium of the Environmental Measurements Laboratory's research projects related to the Chernobyl nuclear accident: October 1, 1986.* Report No. EML-460, US Department of Energy, New York, NY. pg. 155-213.

Date	Location	Media	Nuclide	Activity
May 5, 1986	Skutskar	Deposition in wet soil	^{131}I	2,139,000 Bq/m^2
May 5, 1986	Skutskar	Deposition in wet soil	^{132}Te	2,094,000 Bq/m^2
May 5, 1986	Skutskar	Deposition in wet soil	^{137}Cs	289,000 Bq/m^2
May 5, 1986	Skutskar	Deposition in wet soil	^{140}Ba	362,000 Bq/m^2

Lindahl, Patric, Roos, Per, Eriksson, Mats and Holm, Elis. (2004). Distribution of Np and Pu in Swedish lichen samples (*Cladonia stellaris*). *Journal of Environmental Radioactivity.* 73(1). pg. 73-85.

Malmgren, L. and Jansson, M. (February 1996). The fate of Chernobyl radiocesium in the River Ore catchment, northern Sweden. *Oceanographic Literature Review.* 43(2). pg. 196.

Mascanzoni, D. (December 1987). Chernobyl's challenge to the environment: A report from Sweden. *Science of the Total Environment.* 67(2-3). pg. 133-48.

Mattson, S. and Vesanen, R. (1988). Patterns of Chernobyl fallout in relation to local weather conditions. *Environmental International.* 14. pg. 177-180.

- "99% of Chernobyl-derived radionuclides were deposited in one single period of rain on May 8, 1986; dry deposition was 1% of the remaining." (pg.177).

Mellander, H. (1987). Early measurements of the Chernobyl fallout in Sweden. *IEEE Transactions on Nuclear Science.* NS-34(1). pg. 590-594.

- Peak air concentrations of ^{137}Cs to 1.7 Bq/m^3 April 28; peak weapons fallout was 6.7 mBq/m^3 in 1959.

Nelin, P. (1995). Radiocesium uptake in moose in relation to home range and habitat composition. *J. Environ. Radioactivity.* 26. pg. 189-203.

- 1991 mean concentration levels of ^{137}Cs in moose (range: 105-1060 Bq/kg fresh weight) from 8 moose home ranges showed no significant correlation with ground deposition during the summer home range, and a sample of 119 moose from a larger area of Sweden showed only a partial correlation with ground deposition.

Persson, C., Henning, R. and De Geer, L.E. (1987). The Chernobyl accident - A meteorological analysis of how radionuclides reached and were deposited in Sweden. *Ambio.* 16(1). pg. 20-31.

- Deposition of cesium mainly occurred through wet deposition; ^{95}Zr and ^{239}Np mainly occurred as dry deposition.

Petersen, R.C., Landner, L. and Blanck, H. (1986). Assessment of the impact of the Chernobyl reactor accident on the biota of Swedish streams and lakes. *Ambio.* 15. pg. 327.

- The total radioisotopic activity in lake algae was 296,000 Bq/kg, 2,000 Bq of which was naturally occurring 40K; the majority of the remaining activity was Chernobyl-derived ^{134,137}Cs, ^{103,106}Ru, ^{140}Ba, ^{95}Nb and ^{110}Ag.

Reizenstein, P. (1987). Carcinogenicity of radiation doses caused by the Chernobyl fall-out in Sweden, and prevention of possible tumors. *Med. Oncol. and Tumor Pharmacother.* 4(1). pg. 1-5.

- External gamma radiation levels in Sweden 2-500 µRh^{-1} above the 10-15 µRh^{-1} background, with peak values at 1,000 µRh^{-1}.
- ^{131}I deposition to 170,000 Bq/m^2; initial activities in pure rainwater to 500,000 Bq/l; raw farm milk to 2,900 Bq/l.

Rosen, K., Andersson, I. and Lonsjo, H. (1995). Transfer of radiocesium from soil to vegetation and to grazing lambs in a mountain area in northern Sweden. *J. Environ. Radioactivity.* 26. pg. 237-257.

- "Activity analyses of soil samples... showed a mean deposition of ^{137}Cs of 15.7 (range 14.1-17.6) kBq/m^2." (pg. 237).
- "^{137}Cs concentration of the herbage cut at the various sites decreased with time from 1,175 to 900 Bq/kg dry weight." (pg. 237).
- "The average ^{137}Cs concentration in the abdomen wall muscle of lamb carcasses was 1,087, 668, 513 and 597 Bq/kg wet weight in the years 1990-1993 respectively." (pg. 237).
- "All carcasses exceeded the intervention level applied in Sweden, 300 Bq ^{137}Cs/kg, and were thus discarded for human consumption." (pg. 237).

Snoeijs, P. and Notter, M. (1992). Benthic diatoms as monitoring organisms for radionuclides in a brackish-water coastal environment. *J. Environ. Radioactivity.* 18(1). pg. 23-52.

Date	Location	Media	Nuclide	Activity
May 6, 1986	N. Baltic Sea	Diatoms	^{140}Ba	439,000 Bq/kg dry weight
May 6, 1986	Brackish lagoon	Diatoms	^{103}Ru	540,000 Bq/kg dry weight
May 6, 1986	Brackish lagoon	Diatoms	^{144}Ce	612,000 Bq/kg dry weight
May 6, 1986	Brackish lagoon	Diatoms	^{141}Ce	825,000 Bq/kg dry weight
May 6, 1986	Brackish lagoon	Diatoms	^{95}Zr	864,000 Bq/kg dry weight
May 6, 1986	Brackish lagoon	Diatoms	^{95}Nb	1,022,000 Bq/kg dry weight

Switzerland

Baltensperger, U., Gaggeler, H.W. and Jost, D.T. (1987). Chernobyl radioactivity in size-fractionated aerosol. *J. Aerosol. Sci.* 18. pg. 685-688.

- "Highest activity of ^{131}I was found in size fractions smaller than those for the nuclides ^{137}Cs, ^{132}Te and ^{103}Ru." (pg. 685).
- ^{242}Cm was also noted along with hot particles containing ^{95}Zr, ^{95}Nb and ^{144}Ce. (Please see additional citations on hot particles above.)

Haeberli, W. and Schotterer, U. (1988). The signal from the Chernobyl accident in high-altitude firn areas of the Swiss Alps. *Annals of Glaciology.* 10. pg. 48-51.

- ^{137}Cs deposition to above 43,000 Bq/m^2 noted in SW Switzerland.

Jost, D.T., Gaggeler, H.W., Baltensperger, U., Zinder, B. and Haller, P. (1986). Chernobyl fallout in size-fractioned aerosol. *Nature*. 324. pg. 22-23.

- "Maximum ambient air concentration occurred May 1, peaking at 2 Bq/m^3 for ^{137}Cs (2,000,000 µBq/m^3), one half that of Studsvik, Sweden." (pg. 22).
- "^{131}I mainly present in the gas phase... ^{137}Cs, 131Te and ^{103}Ru were presumably ejected as particles or attached very early to aerosols and grew by coagulation with other particles during transport." (pg.22).

Rybach, Ladislaus, Schwarz, Georg F. and Medici, Fausto. (date unknown). *Construction of radioelement and dose-rate baseline maps by combining ground and airborne radiometric data*. Swiss Federal Nuclear Safety Inspectorate and Swiss Federal Institute of Technology, Zurich.

- Online radiation maps of Switzerland are part of this document. http://www.hsk.psi.ch/pub_eng/gmap96.html.

Swiss Nuclear Safety Inspectorate. (1987). *Chernobyl nuclear power plant accident: Radiological situation in Switzerland and corresponding response*. Report No. CH-5303. Federal Office of Public Health, Berne, Switzerland.

Date	Location	Media	Nuclide	Activity
May 1986	Switzerland	Milk	^{131}I	2,000 Bq/kg
May 1986	Switzerland	Goat's milk	^{131}I	10,000 Bq/kg
May 1986	Switzerland	Milk	^{137}Cs	650 Bq/kg
May 1986	Switzerland	Mutton	^{137}Cs	4,000 Bq/kg
May 1986	Switzerland	Ground deposition	^{137}Cs	26,000 Bq/m^2

- Fallout patterns were correlated with erratic rainfall; all areas received some dry deposition.
- External dose rate to 150 µRh^{-1}.

Tobler, L., Bajo, S. and Wyttenback, A. (1988). Deposition of ^{134,137}Cs from Chernobyl fallout on Norway spruce and forest soil and its incorporation into spruce twigs. *J. Environ. Radioactivity.* 6. pg. 225-245.

Date	Location	Media	Nuclide	Activity
May 1986	Switzerland	Ground deposition from Chernobyl	^{134,137}Cs	6,200 Bq/m^2
1950-1985	Switzerland	Nuclear weapons fallout	^{134,137}Cs	2,600 Bq/m^2

- "^{137}Cs on the surface of the Needles was found to be water insoluble; ... activity in twigs is one half that in soil." (pg. 225-226).

Turkey

Gedikoglu, A. and Sipahi, B.L. (1989). Chernobyl radioactivity in Turkish tea. *Health Physics.* 56(1). pg. 97-101.

- 90% of the activity measured was due to ^{134,137}Cs; activity range in dry tea was 1,064 Bq/kg to 44,000 Bq/kg (1,196,800 pCi/kg). The ratio of activity transferred to brewed tea was estimated at 65%.
- None of the tea samples were below the USFDA action level for contaminated foodstuffs.
- A paradigm for the impact of Chernobyl over vast areas of Turkey, Iran, Iraq and Afghanistan, where voiceless millions have no access to radiological surveillance data.

Unlu, M.Y., Topcuoglu, S., Kucukcezzar, R., Varinlioglu, A., Gungor, N., Bulut, A.M. and Gungor, E. (1995). *Health Physics.* 68(1). pg. 94-99.

Date	Location	Media	Nuclide	Activity
May 1986	Turkey	Tea leaves	^{137}Cs	25,000 Bq/kg

United Kingdom

Also see the listings under Scotland.

Online fallout maps of the United Kingdom;
http://www.infoukes.com/history/chornobyl/gregorovich/thumb06.gif and
http://www.antenna.nl/wise/349-50/eng.gif.

Cambray, R.S., Cawse, P.A., Garland, J.A., Gibson, J.A.B., Johnson, P., Lewis, G.N.J.,
Newton, D., Salmon, L. and Wade, B.O. (1987). Observations on radioactivity from the
Chernobyl accident. *Nucl. Energy.* 26(2). pg. 77-101.

Date	Location	Media	Nuclide	Activity
1-6 May 1986	Lerwick, Shetland	Ground deposition	^{131}I	26,000 Bq/m^2
1-6 May 1986	Holmrook, Cumbria	Ground deposition	^{131}I	41,000 Bq/m^2

- Maximum levels of Chernobyl-derived ^{137}Cs are reported as only 10,000 Bq/m^2, also at Holmrook, Cumbria, much less than peak concentrations noted in later reports.

Camplin, W.C., Leonard, D.R.P., Tipple, J.R. and Duckett, L. (1989). Radioactivity in
freshwater systems in Cumbria (UK) following the Chernobyl accident. *Fish. Res. Data
Rep.*, MAFF Direct. Fish Res. Lowestoft. 18. pg. 1-90.

Camplin, W.C., Mitchell, N.T., Leonard, D.R.P. and Jefferies, D.F. (1986).
*Radioactivity in surface and coastal waters of the British Isles. Monitoring of fallout
from the Chernobyl reactor accident.* Aquatic Environment Monitoring Report No. 15.
Ministry of Agriculture, Fisheries and Food, Lowestoft.

- This report shows only a light pulse of Chernobyl radioactivity in the surface and coastal waters; many Chernobyl-derived fission products are recorded in sediments, but data concerning contamination in other media is limited by the number of samples taken and the fact that Chernobyl-derived radioactivity did not have enough time to bioaccumulate in significant quantities in sentinel organisms.
- Pre-Chernobyl concentrations of ^{137}Cs (Bq/kg wet weight) were noted as codfish 52 Bq/kg; shrimp 42 Bq/kg; winkles 80 Bq/kg; cockles 22 Bq/kg; mussels 13 Bq/kg, and in fresh water, brown trout at 339 Bq/kg.

Clark, M.J. (1986). Fallout from Chernobyl. *J. Soc. Radiol. Prot.* 6(4). pg. 157-166.

Date	Location	Media	Nuclide	Activity
1986	UK	Milk	^{131}I	500 Bq/l
1986	UK	Milk	^{137}Cs	500 Bq/l
1986	UK	Grass	^{131}I	15,000 Bq/m^2
1986	UK	Grass	^{137}Cs	10,000 Bq/m^2

Clark, M.J. and Smith, F.B. (1988). Wet and dry deposition of Chernobyl releases. *Nature*. 332. pg. 245-249.

- ^{137}Cs was present in the atmosphere, mostly as particulate species with wet deposition mechanisms dominating... ^{131}I was present as particulate and vapor phase material... both wet and dry deposition mechanisms were important.

Copplestone, D., Jackson, D., Hartnoll, R.G., Johnson, M.S., McDonald, P. and Wood, N. (2004). Seasonal variations in activity concentrations of ^{99}Tc and ^{137}Cs in the edible meat fraction of crabs and lobsters from the central Irish Sea. *Journal of Environmental Radioactivity*. 73(1). pg. 29-48.

Fry, F.A., Clarke, R.H. and O'Riordan, M.C. (1986). Early estimates of UK radiation doses from the Chernobyl reactor. *Nature*. 321. pg. 193-195.

- "Mechanisms of human exposure to the contaminants from Chernobyl... (in sequence) external irradiation by the passing cloud; inhalation of radioactive material in the cloud; beta ray contamination of the skin; external irradiation by material deposited on the ground; ingestion of contaminated foods." (pg. 194).

Fulker, M.J. (1987). Aspects of environmental monitoring by British Nuclear Fuels plc following the Chernobyl reactor accident. *J. Environ. Radioactivity*. 5. pg. 235-244.

- ^{137}Cs measured at 15,000 Bq/m^2 of which 7,400 Bq/m^2 was attributed to Chernobyl fallout.

Hamilton, E.I., Zou, B. and Clifton, R.J. (1986). The Chernobyl accident - radionuclide fallout in S.W. England. *The Science of the Total Environment*. 57. pg. 231-251.

Date	Location	Media	Nuclide	Activity
8th May	SW England	Lichen	^{131}I	1,260 Bq/kg
7th June	SW England	Lichen	^{137}Cs	260 Bq/kg

- SW England received low fallout levels from Chernobyl.

Her Majesty's Stationery Office. (1986). *Levels of radioactivity in the UK from the accident at Chernobyl USSR, on 26 April 1986: A compilation of the results of environmental measurements in the U.K.* Her Majesty's Stationery Office, London.

- This early report grossly underestimates the amount of Chernobyl-derived fallout and its radiological impact on the United Kingdom.

- While much of the emphasis is on air concentration, it is obvious that measurements of air concentration alone are insufficient to provide a comprehensive understanding of the radiological impact of a nuclear accident such as Chernobyl. Later and more comprehensive surveys of biological media indicated the impact of Chernobyl was much more significant than this initial report indicates. Particularly misleading are the early reports (32 samples) for radionuclide concentrations in meats (pg. 169) which, while showing substantial contamination in Cumbria, failed to document the full impact of the Chernobyl accident in the northern sections of the United Kingdom.

- This report does document substantial contamination of milk and milk products throughout the United Kingdom, with particularly high readings in Cumbria. Peak concentrations of radiocesium are reported as high as 380 Bq/l. Extensive contamination with radioiodine is also noted. The data in this report illustrate the fact that the fallout patterns from an accident such as Chernobyl are highly erratic, affecting one area while skipping another.

Hill, C.R., Adam, I., Anderson, W., Ott, R.J. and Sowby, F.D. (1986). Iodine-131 in human thyroids in Britain following Chernobyl. *Nature*. 321. pg. 655.

- ^{131}I activity range in the neck region: adults 8-33 Bq; children 2-16 Bq.

Howard, B.J., Beresford, N.A., Mayes, R.W. and Lamb, C.S. (1993). Transfer of ^{131}I to sheep milk from vegetation contaminated by Chernobyl fallout. *J. Environ. Radioactivity*. 19(2). pg. 155-162.

- "The daily proportion of ^{131}I intake which was secreted in milk was 56%. This is an order of magnitude higher than for cattle." (pg. 155).

Hunt, G.J. (1988). Radioactivity in surface and coastal waters of the British Isles, 1987. In: *Aquatic Environment Monitoring Report No. 19*. Ministry of Agriculture, Fisheries and Food (MAFF), Lowestoft.

- This report documents a relatively light pulse of Chernobyl-derived radiocesium in the marine food chain and gives no hint of much higher levels of radiocesium in the terrestrial food chain. The highest levels of radioactivity are reported in freshwater fish, with mean radioactive concentration levels reported as high as 2,300 Bq/kg of ^{137}Cs in pike in Scotland. Radiocesium-134,137 levels in many samples of freshwater fish are frequently reported in the ±1000 Bq/kg range.

Jackson, D., Jones, S.R., Fulkner, M.J. and Coverdale, N.G.M. (1987). Environmental monitoring in the vicinity of Sellafield following deposition of radioactivity from the Chernobyl accident. *J. Soc. Radiol. Prot.* 7(2). pg. 75-87.

- "Peak concentrations reached in May 1986 in milk were... 500 to 1000 times the mean values reported for 1985 for ^{131}I and ^{137}Cs. Levels of ^{90}Sr were elevated by a factor of no more than 3 to 4." (pg. 81).
- Total cesium to 63,000 Bq/m^2, 60% of which can be attributed to Chernobyl.

Johnston, K. (1987). UK upland grazing still contaminated. *Nature.* 326. pg. 821.

- "Initial mathematical model predictions of a quick decline in cesium levels proved incorrect... acidic peat soils with low clay content do not bind the cesium... over 270,000 sheep plus their lambs are (still) under restrictions compared to 4.2 million initially." (pg. 821).

Johnston, K. (1987). British sheep still contaminated by Chernobyl fallout. *Nature.* 328. pg. 661.

- Radiocesium more accessible to plant roots than expected due to lower than expected binding with clay or loam; restrictions on sheep extended to farms previously designated as safe (August, 1987).

Jones, G.D., Forsyth, P.D. and Appleby, P.G. (1986). Observation of ^{110m}Ag in Chernobyl fallout. *Nature.* 322. pg. 313.

- ^{110m}Ag peak concentration in beef and lamb liver: 74 Bq/kg.

Leonard, D.R.P., Camplin, W.C. and Tipple, J.R. (1990). The variability of radiocaesium concentrations in freshwater fish caught in the UK following the Chernobyl nuclear reactor accident: An assessment of potential doses to critical group consumers. In: *Proc. Int. Symp. on Environmental Contamination Following a Major Nuclear Accident*. IAEA-SM-306/15. IAEA, Vienna.

Livens, F.R., Fowler, D. and Horrill, A.D. (1992). Wet and dry deposition of ^{131}I, [134]Cs and ^{137}Cs at an upland site in northern England. *J. Environ. Radioactivity.* 16(3). pg. 243-254.

- Study was done in an area of very light Chernobyl fallout. Pu: 1,230 Bq/m^2 at the summit of test area (elevation 847 m).
- "Only 20% of the ^{131}I is wet deposited, compared with almost all of the [103]Ru and cesium isotopes." (pg. 252).
- Enhanced cesium activity was noted at the summit, suggesting "turbulent cloud water deposition" as the clouds encountered the summit top.

Mason, C.F. and MacDonald, S.M. (1988). Radioactivity in otter scats in Britain following the Chernobyl reactor accident. *Water, Air and Soil Pollution.* 37. pg. 131-137.

Date	Location	Media	Nuclide	Activity
July 1986	Galloway, Scotland	Otter scats	total activity	79,500 Bq/kg dry weight

- Pre-Chernobyl peak concentration was 7,400 Bq/kg, with a geometric mean of 640 Bq/kg for 52 samples taken in Wales in January 1985.

McAuley, I.R. and Moran, D. (1989). Radiocesium fallout in Ireland from the Chernobyl accident. *J. Radiol. Prot.* 9(1). pg. 29-32.

- The initial Chernobyl fallout ^{137}Cs to [134]Cs ratio was 1.90; mean deposition level was 3,200 Bq/m^2; peak concentration was 14,200 Bq/m^2.
- The overall mean pre-Chernobyl figure for 111 sites was 600-800 Bq/m^2 to a depth of 30 mm, excluding ^{137}Cs which had been mechanically redistributed to deeper soil.

Mitchell, N.T. and Steele, A.K. (1988). The marine impact of Cesium-134 and -137 from the Chernobyl reactor accident. *J. Environ. Radioactivity.* 6. pg. 163-175.

- "Chernobyl derived radiocesium... largely masked the activity of fuel reprocessing (BNF-Sellafield) origin previously found in the North Sea." (pg. 173).

Mondon, K.J. and Walters, C.B. (1990). Measurements of radiocaesium, radiostrontium and plutonium in whole diets following deposition of radioactivity in the UK originating from the Chernobyl power plant accident. *Food Additives and Contaminants.* 7(6). pg. 837-848.

Nair, S. and Darley, P.J. (1986). A preliminary assessment of individual doses in the environs of Berkeley, Gloucestershire, following the Chernobyl nuclear reactor accident. *J. Soc. Radiol. Prot.* 6(3). pg. 101-108.

- In an area of very low Chernobyl fallout, ^{132}Te in fresh grass: 460 Bq/kg; ^{103}Ru in fresh spinach: 140 Bq/kg; initial estimate of ^{137}Cs ground deposition: 630 Bq/m^2. This is probably an underestimation of the actual deposition in this area.

National Radiological Protection Board. (1986). *Levels of radioactivity in Wales from the accident at Chernobyl, USSR on 26 April 1986: A compilation of the results of environmental measurements in Wales*. Welsh Office, National Radiological Protection Board, London.

Date	Location	Media	Nuclide	Activity
to May 23, 1986	Wales	Milk	^{131}I	190 Bq/l
to May 23, 1986	Wales	Milk	^{137}Cs	443 Bq/l

Nicholson, K.W. and Hedgecock, J.B. (1991). Behavior of radioactivity from Chernobyl - weathering from buildings. *J. Environ. Radioactivity.* 14(3). pg. 225-232.

- In an area of relatively low dry deposition, both concrete and clay tiles showed negligible weathering of radiocesium and, in some cases, an increase in radiocesium levels that cannot be explained.

Rafferty, B., McGee, E.J., Colgan, P.A. and Synnott, H.J. (1993). Dietary intake of radiocesium by free ranging mountain sheep. *J. Environ. Radioactivity.* 21(1). pg. 33-46.

Date	Location	Media	Nuclide	Activity
1986	Ireland	Ground deposition	^{137}Cs	15,000 Bq/m^2

- "Feces radiocesium activity was shown to be more appropriate than vegetation radiocesium activity as a predictor of in-vivo radiocesium activity in free ranging mountain sheep." (pg. 33).

Sandalls, F.J. and Gaudern, S.L. (1988). Radiocesium on urban surfaces in West Cumbria five months after Chernobyl. *J. Environ. Radioactivity.* 7. pg. 87-91.

- Pre-Chernobyl ^{137}Cs: 3,640 Bq/m^2; Chernobyl-derived ^{134,137}Cs: 15,169 Bq/m^2.

Sanderson, D.C.W. and Scott, E.M. (1989). *Aerial radiometric survey in West Cumbria 1988: Project N611*. Ministry of Agriculture and Fisheries and Foods (MAFF), London.

300

- This is a landmark aerial radiometric survey and follows an initial survey by the Scottish Universities' Reactor Research Centre (see Wynne, 1989) which showed Chernobyl-derived ^{137}Cs deposition in Cumbria up to forty times higher than originally reported by the MAFF Institute for Terrestrial Ecology ground survey.

- This aerial survey showed deposition levels exceeding 60,000 Bq/m^2 in areas of Cumbria. Follow-up investigations located isolated areas of contamination up to 300,000 Bq/m^2 near the Sellafield fuel reprocessing facility, at least half of which have been attributed to the Chernobyl accident (see Wynne, 1989).

- The detail and the quality of the radiometric map which resulted from this survey marks a new era in radiological surveillance. The technology for such a survey has long been available and has often been used in the past in the search for oil-bearing geological deposits and their naturally occurring radioactive signals. This is the first time such a survey of a nuclear accident has been made available to the public, and the detailed polychromatic map of the Chernobyl fallout patterns raises the question of why such surveys are not available for many other source points of radioactive contamination.

- Following the publication of the Cumbria aerial radiometric survey, a second aerial radiometric survey has been made public by the Swedish University of Agricultural Sciences. To view the Deposition Map, just click on this link (http://www.mv.slu.se/ma/radio/radio/chern/cs-dep2.gif) or contact Sverker Forsberg by email at (sverker.forsberg@mv.slu.se). The Nordic Deposition Map characterizes slightly more intense and equally erratic levels of Chernobyl-derived contamination in Southern Finland, Central Sweden, and Norway.

- Radiometric surveys of Chernobyl fallout within Russia have recently been issued and will be cited as soon as we can obtain copies or more information about the maps.

Sherlock, J., Andrews, D., Dunderdale, J., Lally, A. and Shaw, P. (1988). The *in vivo* measurement of radiocesium activity in lambs. *J. Environ. Radioactivity.* 7. pg. 215-220.

Date	Location	Media	Nuclide	Activity
June 1986	Cumbria	Lamb	^{134,137}Cs	3,898 Bq/kg

Walling, D.E. and Bradley, S.B. (1988). Transport and redistribution of Chernobyl fallout radionuclides by fluvial processes: some preliminary evidence. *Environmental Geochemistry and Health.* 10(2). pg. 35-39.

- ^{137}Cs loading in floodplains soils up to 100 times greater than in adjacent areas above the floodplain.
- ITE (Institute of Terrestrial Ecology) vegetation survey results in gross underestimation of Chernobyl-derived radiocesium (see Wynne, B., 1989).

Walters, B. (1988). Chernobyl derived activity in sheep: variation within a single flock and with time. *J. Environ. Radioactivity.* 7. pg. 99-106.

- In vivo monitoring of 100 live sheep introduced to a grazing area containing up to 2,000 Bq/kg of radiocesium-134,137 in herbage resulted in peak average activity levels of 1,300 Bq/kg in the grazing sheep.

Watson, W.S. (1987). Total body potassium measurement - the effect of fallout from Chernobyl. *Clin. Phys. Physiol. Meas.* 8(4). pg. 337-341.

- Eighteen healthy adults, group mean ^{134}Cs level: 174 Bq.

Welsh Office. (1998). *More areas freed from Chernobyl sheep restrictions.* W98018-Ag. Welsh Office Press Release, Cardiff.

Wynne, B. (1989). Sheep farming after Chernobyl. *Environment.* 31(2). pg. 10-39.

- Initial surveys of ^{137}Cs deposition on vegetation by the Institute for Terrestrial Ecology (ITE) showed peak concentrations below 5,000 Bq/m^2.
- In the fall of 1988, an aerial grid survey sponsored by Scottish Universities' Reactor Research Centre showed some areas of Scotland with up to 40 times the official MAFF (ITE) figures. Peak ground deposition concentrations of Chernobyl-derived radiocesium (134/137) were actually 100,000 Bq/m^2.
- Peak concentrations of radiocesium in the Sellafield vicinity were 300,000 Bq/m^2. While Sellafield and NRPB (National Radiological Protection Board) officials had previously denied any significant Sellafield-derived aerial deposition of radiocesium, it is unlikely that Chernobyl-derived radiocesium was greater than 100,000 Bq/m^2 in this vicinity, the Chernobyl accident having opened a Pandora's box of issues in a closet full of nuclear skeletons.
- A new and updated MAFF report: Aerial and Radiometric Surveys in West Cumbria, 1988, previously cited above, was issued after the first aerial survey and remains the first official comprehensive aerial radiometric survey of a specific, if limited, area available to the general public in radiological surveillance literature.

Yugoslavia

Byrne, A.R. (1988). Radioactivity in fungi in Slovenia, Yugoslavia, following the Chernobyl accident. *J. Environ. Radioactivity.* 6. pg. 177-183.

- "The median concentration of ^{137,134}Cs in R. caperata from over 40 sampling sites was about 22,000 Bq/kg dry weight." (pg. 177).
- "Radiocesium levels in certain species will probably increase further next year, and subsequently as Cs migrates down the soil profile." (pg. 177).
- "^{110m}Ag was found in concentrations of up to 500 Bq/kg dry weight in certain species known to be Ag accumulators." (pg. 177).

Juznic, K. and Fedina, S. (1987). Short Communication: Distribution of ^{89}Sr and ^{90}Sr in Slovenia, Yugoslavia, after the Chernobyl accident. *J. Environ. Radioactivity.* 5. pg. 159-163.

Date	Location	Media	Nuclide	Activity
May 1986	Slovenia	Ground deposition	Total radiostrontium	5,700 Bq/m^2
May 1986	Slovenia	Ground deposition	^{90}Sr	420 Bq/m^2
May 1986	Slovenia	Ground deposition	^{131}I	140,000 Bq/m^2
May 1986	Slovenia	Ground deposition	^{134}Cs	12,000 Bq/m^2
May 1986	Slovenia	Ground deposition	^{137}Cs	26,000 Bq/m^2

- "The maximum radiostrontium concentration in rain appeared during the period 3-5 May ... most of the activity in plants resulted from direct absorption through the leaves." (pg. 159).

USA

Juzdan, Z.J., Helfer, I.K., Miller, K.M., Rivera, W., Sanderson, C.G. and Silvestri, S. (1986). Deposition of radionuclides in the northern hemisphere following the Chernobyl accident. In: *Environmental Measurements Laboratory: A compendium of the Environmental Measurements Laboratory's research projects related to the Chernobyl nuclear accident: October 1, 1986.* Report No. EML-460. US Department of Energy, New York, NY.

Date	Location	Media	Nuclide	Activity
May 5, 1986	Forks, WA	Wet and dry deposition	^{137}Cs	301.55 Bq/m^2
May 5, 1986	Forks, WA	Wet and dry deposition	^{131}I	1200 Bq/m^2
May 5, 1986	Forks, WA	Wet and dry deposition	^{103}Ru	134.68 Bq/m^2
May 5, 1986	Forks, WA	Wet and dry deposition	^{134}Cs	72.89 Bq/m^2

Larsen, R.J., Sanderson, C.G., Rivera, W. and Zamichieli, M. (1986). The characterization of radionuclides in North American and Hawaiian surface air and deposition following the Chernobyl accident. In: *Environmental Measurements Laboratory: A compendium of the Environmental Measurements Laboratory's research projects related to the Chernobyl nuclear accident: October 1, 1986.* Report No. EML-460. US Department of Energy, New York, NY. pg. 1-104.

Date	Location	Media	Nuclide	Activity
May 11, 1986	Rexburg, Idaho	Air concentration	^{131}I	11,390 µBq/m^3
May 11, 1986	New York City	Air concentration	^{131}I	20,720 µBq/m^3
May 11, 1986	New York City	Air concentration	^{137}Cs	9,720 µBq/m^3
May 1986	Rexburg, Idaho	Total ground deposition	^{131}I	707 Bq/m^2
May 1986	Chester, NJ	Total ground deposition	^{131}I	168.4 Bq/m^2
May 1986	Chester, NJ	Total ground deposition	^{137}Cs	68.5 Bq/m^2

- Chernobyl debris appeared in both eastern and western sites of the US in similar magnitudes. The mean activity ratio of cesium-137 to cesium-134 was 1.9 at all sites.
- This is a detailed survey of air concentration radioactivity levels at Barrow, Alaska; Moosonee, Canada; Beaverton, Oregon; Rexburg, Idaho; Chester, NJ; NY, NY; Biscayne, Florida; Miami, Florida; Mauna Loa, Hawaii.
- Small quantities of numerous Chernobyl-derived radionuclides were noted at most stations including ^{103,106}Ru, ^{140}Ba, ^{140}La, ^{95}Zr, ^{95}Mo, ^{141}Ce, ^{144}Ce.

Larsen, R.J., Haagenson, P.L. and Reiss, N.M. (1989). Transport processes associated with the initial elevated concentrations of Chernobyl radioactivity in surface air in the United States. *J. Environ. Radioactivity.* 10. pg. 1-18.

- "The nearly simultaneous arrival of radioactive debris at widely separated locations resulted from different paths being taken by the debris released at different times during the course of the accident." The plume pathways crossed the Arctic within the lower troposphere, and the Pacific Ocean within the mid-troposphere. (pg. 1).

304

- Peak concentrations of gross beta: 2.0 pCi/m^3 (74,000 µBq/m^3 on May 10 in South-central Idaho).
- This report contains radiometric data (isopleths) of gross beta concentration which are unusually uniform given the erratic patterns of Chernobyl fallout noted in other countries and do not match the data contained in some other reports.

Larsen, R., Juzdan, Z.R. (1986). Radioactivity at Barrow and Mauna Loa following the Chernobyl accident. In: *Geophysical Monitoring for Climatic Change, No. 14, Summary Report 1985*. US Department of Commerce, Washington.

Date	Location	Media	Nuclide	Activity
May 1986	Barrow, Alaska	Surface air concentration	^{131}I	218.7 fCi/m^3
May 1986	Barrow, Alaska	Surface air concentration	^{134}Cs	18.6 fCi/m^3
May 1986	Barrow, Alaska	Surface air concentration	^{137}Cs	27.6 fCi/m^3
May 1986	Mauna Loa, Hawaii	Surface air concentration	^{131}I	28.5 fCi/m^3
May 1986	Mauna Loa, Hawaii	Surface air concentration	^{134}Cs	11.2 fCi/m^3
May 1986	Mauna Loa, Hawaii	Surface air concentration	^{137}Cs	22.9 fCi/m^3

- Air concentration is usually reported in picocuries, or in Europe, in microbecquerels (µBq); a femtocurie (10^{-15}) is three orders of magnitude less than a picocurie (10^{-12}). The minimum detectable level in air is 0.01 pCi/m^3 (10 femtocuries). Most US reports of radioactivity in air are not nuclide specific, but rather summarize gross beta activity in airborne particulates.
- These data show a minimal impact from the Chernobyl accident but make interesting points of comparison with the real time air concentrations of Chernobyl-derived radionuclides recorded in Finland, as well as in other sections of North America (see Canada).

United States Environmental Protection Agency. (1986). *Environmental radiation data: Report 46: April 1986-June 1986*. Report No. EPA520/5-87-004. US EPA, Washington, D.C.

Date	Location	Media	Nuclide	Activity
May 11, 1986	Montpelier, VT	Precipitation	Gross beta	6.26 nCi/m^2
May 12, 1986	Spokane, WA	Precipitation	^{131}I	6,620 pCi/l
May 13, 1986	Cheyenne, WY	Precipitation	^{137}Cs	710 pCi/l
May 16, 1986	Cheyenne, WY	Precipitation	Gross beta	710 pCi/l
May 1986	Lincoln, NE	Air particulate	Gross beta	14.3 pCi/m^3
June 4, 1986	Seattle, WA	Milk	^{137}Cs	66 pCi/l

- A wide variety of gamma emitting nuclides noted throughout the US in May, '86 including ^{103}Ru, ^{106}Ru, ^{134}Cs, ^{136}Cs, ^{137}Cs, ^{140}Ba, ^{140}La, ^{132}I and ^{95}Zr/

305

- Extensive ^{131}I in precipitation in May throughout US in all stations, with 15 reports above 1000 pCi/l.
- 14.3 pCi/m^3 is in excess of 500,000 µBq/m^3, one of the highest levels of air particulate activity in the US since the termination of nuclear weapons testing.
- Ground deposition activities are not included in EPA reports.

Maine

Toppan, C. (Memo released May 8, 1986). *Iodine detected in rain water by DHS.* Manager, Department of Human Services, Augusta, ME.

Date	Location	Media	Nuclide	Activity
May 7-8, 1986	Augusta	Wet deposition	^{131}I	1,561 pCi/m^2
May 9, 1986	Augusta	Rainfall	^{131}I	110 pCi/l

Toppan, C. (Memos released May 9-13, 1986). *Update on DHS radiation monitoring.* Manager, Department of Human Services, Augusta, ME.

Date	Location	Media	Nuclide	Activity
May 12, 1986	Augusta	Air concentration	^{131}I	0.80 pCi/m^3

Schell, R. (Memo released July 9, 1986). *Quarterly radiation monitoring report.* Department of Human Services, Augusta, ME.

Date	Location	Media	Nuclide	Activity
May 16, 1986	Maine	Milk	^{131}I	52.5 pCi/l
May 1986	Wiscasset, ME	Seaweed	^{131}I	176 pCi/kg
June 1986	Maine	Milk	^{137}Cs	20.29 pCi/l
June 1986	Maine	Milk	^{134}Cs	9.68 pCi/l

Maryland

Dibb, J.E. and Rice, D.L. (1988). Short communication: Chernobyl fallout in the Chesapeake Bay region. *J. Environ. Radioactivity.* 7. pg. 193-196.

Date	Location	Media	Nuclide	Activity
May 8- June 20, 1986	Solomons MD	Ground deposition	^{137}Cs	4,250 Bq/m^2
May 8- June 20, 1986	Solomons MD	Ground deposition	^{134}Cs	2,000 Bq/m^2
May 8- June 20, 1986	Solomons MD	Ground deposition	^{103}Ru	22,000 Bq/m^2

- The highest levels of fallout were on May 22 and were associated with rainfall events.
- ^{134}Cs/^{137}Cs isotopic ratios were the same as those measured in Sweden, Paris, Japan and Tennessee.
- The high levels of Chernobyl fallout reported in Maryland contradict EPA and other federal reports indicating little or no Chernobyl fallout in the Maryland area.
- Ground deposition for ^{137}Cs at 4,250 comes close to equaling or exceeding total weapons testing cesium fallout at this location.

New Jersey

Dreicer, M. and Klusek, C.S. (1988). Transport of ^{131}I through the grass-cow-milk pathway at a Northeast US dairy following the Chernobyl Accident. *J. Environ. Radioactivity.* 7. pg. 201-207.

Date	Location	Media	Nuclide	Activity
May 15, 1986	Chester, NJ	Grass	^{131}I	72.2 Bq/kg dry weight
May 17, 1986	Chester, NJ	Milk	^{131}I	1.47 Bq/l

- Several EML reports compiled at the EML laboratory in Chester, New Jersey are reported under the New York (metropolitan area) section heading.

Dreicer, M., Helfer, I.K. and Miller, K.M. (1986). Measurement of Chernobyl fallout activity in grass and soil at Chester, New Jersey. In: *Environmental Measurements Laboratory: A compendium of the Environmental Measurements Laboratory's research projects related to the Chernobyl nuclear accident: October 1, 1986.* Report No. EML-460. Department of Energy, New York, NY. pg. 265-284.

Date	Location	Media	Nuclide	Activity
May 17, 1986	Chester, NJ	Deposition on grass	^{137}Cs	9.40 Bq/m^2
May 23, 1986	Chester, NJ	Deposition in soil	^{131}I	47.2 Bq/m^2
June 3, 1986	Chester, NJ	Deposition in soil	^{103}Ru	18.46 Bq/m^2

New York

Feely, H.W., Helfer, I.K., Juzdan, Z.R., Klusek, C.S., Larsen, R.J., Leifer, R. and Sanderson, C.G. (1988). Fallout in the New York metropolitan area following the Chernobyl accident. *J. Environ. Radioactivity.* 7. pg. 177-191.

Date	Location	Media	Nuclide	Activity
May 1986	NY City	Air/gaseous	^{131}I	23 mBq/m^3
May 1986	NY City	Air/aerosol	^{131}I	20 mBq/m^3
May 1986	NY City	Air/aerosol	^{137}Cs	9.5 mBq/m^3
May 1986	NY City	Milk	^{137}Cs	1.5 Bq/l

- "The total deposition of Chernobyl ^{137}Cs deposition was <1% of that already present in the soil from fallout from past nuclear weapon tests." (pg. 177).
- The report of a peak concentration of 130Cs in grass at 8 Bq/m^2 is at variance with numerous other reports of much higher levels of cesium deposition at most other reporting stations in the US (see Dibbs, 1988; Maryland). [http://courses.lib.odu.edu/ocen/mmulholl/boynton_article.pdf gives that reference as: DIBBS, J. E. 1988. The dynamics of Beryllium-7 in Chesapeake Bay. Ph.D. Dissertation, State University of New York, Binghamton, New York.]
- The total iodine air activity of 43 mBq/m^3 (43,000 μBq/m^3) is a distinctly elevated level of environmental radioactivity.

Klusek, C.S., Sanderson, C.G. and Rivera, W. (1986). Concentrations of ^{131}I and ^{134}Cs and ^{137}Cs in milk in the New York metropolitan area following the Chernobyl reactor accident. In: *Environmental Measurements Laboratory: A compendium of the Environmental Measurements Laboratory's research projects related to the Chernobyl nuclear accident: October 1, 1986.* Report No. EML-460. Department of Energy, New York, NY. pg. 308-326.

- On May 12, 1986, in the New York metropolitan area, ^{131}I levels in milk were 40 pCi/l and ^{137}Cs 80 pCi/l for fresh milk. Pasteurized milk showed a delayed ^{131}I peak concentration on May 28 of 82 pCi/l.

Leifer, R., Helfer, I., Miller, K. and Silvestri, S. (1986). Concentrations of gaseous ^{131}I in New York City air following the Chernobyl accident. In: *Environmental Measurements Laboratory: A compendium of the Environmental Measurements Laboratory's research projects related to the Chernobyl nuclear accident: October 1, 1986.* Report No. EML-460. Department of Energy, New York, NY. pg. 301-307.

- The air concentration of gaseous ^{131}I in New York City on May 10-12, 1986, was measured at 23 mBq/m^3.

Miller, K.M. and Gedulig, J. (1986). Measurements of the external radiation field in the New York metropolitan area. In: *Environmental Measurements Laboratory: A compendium of the Environmental Measurements Laboratory's research projects related to the Chernobyl nuclear accident: October 1, 1986.* Report No. EML-460. Department of Energy, New York, NY. pg. 284-290.

Date	Location	Media	Nuclide	Activity
May 23,1986	Chester, NJ	Ground deposition	[131]I	30 Bq/m^2
May 23,1986	Chester, NJ	Ground deposition	[103]Ru	15 Bq/m^2

Oregon

Gebbie, K.M. and Paris, R.D. (1986). *Chernobyl: Oregon's response.* Radiation Control Section, Office of Environment and Health Systems, Health Division, Oregon Department of Human Resources, Portland, Oregon.

Date	Location	Media	Nuclide	Activity
May 9 1986	Portland	Rain	Gross beta	481 pCi/l
May 5-10	Portland	Air activity	Gross beta	1.031 pCi/m^3
May 11	Portland	Ground deposition	[131]I	9157 pCi/m^2
May 11-12	Portland	Total air	[131]I	2.9 pCi/m^3
May 12	Willamette Valley	Milk	[131]I	167 pCi/l
May 19	Willamette Valley	Milk	[137]Cs	97 pCi/l

- This is a more detailed report than that available from the state of Washington and provides excellent documentation of the relatively moderate impact of Chernobyl fallout in Oregon.

Tennesse

Bondietti, E.A., Brantley, J.N. and Rangarajan, C. (1988). Size distributions and growth of natural and Chernobyl-derived submicron aerosols in Tennessee. *J. Environ. Radioactivity.* 6. pg. 99-120.

- Carrier aerosols provide the key to the transport of volatile radioactive aerosols released during the Chernobyl accident. A linear growth rate of ~ 0.013 μm/day was observed in the lower troposphere after the median diameter had reached ~ 0.4 μm.
- Two pulses of Chernobyl-derived radioactivity were noted. The first arrived May 10, and a second peak on May 20-23 contained high levels of ^{103}Ru.
- The change in the fission product spectrum illustrated a second phase of the Chernobyl accident (May 5) in which much of the cesium was released.

Bondietti, E.A. and Brantley, J.N. (1986). Characteristics of Chernobyl radioactivity in Tennessee. *Nature*. 322. pg. 313-314.

- "The Chernobyl reactor accident was first detected in air samples at Oak Ridge, Tennessee on 10 May 1986... Two distinct phases in airborne radioactivity were evident in our measurements. The first phase lasted from 10 to 17 May and was characterized by a ^{137}Cs/^{103}Ru activity ratio of ~1.5...The second phase began on 18 May, when precipitation from a convective storm yielded much higher ^{103}Ru and ^{140}Ba activities, relative to ^{137}Cs, than previously found in air or precipitation." (pg. 313).

Vermont

Vermont State Department of Health. *Vermont state environmental radiation surveillance program*. (1986). Division of Occupational and Radiological Health, Vermont State Department of Health, Montpelier, VT.

Date	Location	Media	Nuclide	Activity
May 1986	Vermont	Air particulates	Gross beta	0.113 pCi/m^3
May 1986	Vermont	Milk	^{131}I	88 pCi/l
June 1986	Vermont	Fiddleheads	^{137}Cs	328 pCi/kg
June 1986	Vermont	Fiddleheads	^{103}Ru	261 pCi/kg
September 1986	Vermont	Wild boletus	^{137}Cs	3,750 pCi/kg
September 1986	Vermont	Wild russla	^{137}Cs	3,660 pCi/kg

- The EPA report number 46 (Environmental Radiation Data, EPA 520-5-8704) records ^{131}I wet deposition as 12,300 pCi/m^2 and with rain containing 1,660 pCi/l on May 11 at Montpelier.
- Information detailing the extent of ground deposition of radiocesium is sorely lacking not only in Vermont but also in most US reporting locations.

310

Washington

Pickett, B. (1987). *DSHS activities relating to the Chernobyl Nuclear Accident.* Department of Social and Health Services, Olympia, WA.

Date	Location	Media	Nuclide	Activity
May 5, 1986	Spokane	Rainwater	^{131}I	6,620 pCi/l
May 5, 1986	Redland	Milk	^{131}I	560 pCi/l
May 5, 1986	Spokane	Gross beta	^{131}I	2.2 pCi/l
May 5, 1986	E. Washington	Food	^{134,137}Cs	1250 pCi/kg

- This report is far superior to the EPA quarterly reports in terms of the quality and clarity of the graphic representations, although containing much less data.

Picket, B.D. (1987). *Assessment of Chernobyl fallout in the state of Washington: 32nd annual meeting of the Health Physics Society.* Department of Social and Health Services, Olympia, WA.

Date	Location	Media	Nuclide	Activity
May 12, 1986	Portland	Rain	^{131}I	6000 pCi/l
May 12, 1986	Portland	Rain	Gross Beta	2.2 pCi/m^3

- This report contains limited data, some of which is included in the previous report.

IX. Nuclear Dada: The Maine Yankee Atomic Power Plant

Introduction

The Maine Yankee Atomic Power Plant (MYAPC) operated in Wiscasset, Maine, from 1973 to 1997. The plant closed as a result of fuel cladding failures within the reactor vessel, a possible scenario in any aging nuclear reactor. Unlike the subprime General Electric boiling reactor design utilized at the Pilgrim Nuclear Generating Station, Plymouth, Massachusetts, and the Fukushima Daiichi complex of six reactors in Japan, the Maine Yankee facility had a more robust pressure vessel and a less vulnerable spent fuel pool located adjacent to the reactor building rather than within it.

The most common type of accident at operational nuclear power plants involves leakage from reactor water systems. Leakage of cooling water contaminated with tritium and small amounts of other fission products are ubiquitous, well documented, routine events, which eventually characterize most reactor operations. The fuel cladding failure accident, which resulted in the closing of Maine Yankee, is a much less frequent occurrence at US nuclear reactors (±8%?) Most fuel cladding failure events are not made public either by the utilities operating the reactor, the NRC, or local and regional media outlets, which report on routine reactor operations. In the case of Maine Yankee, the licensee made documentation of fuel cladding failures available to Ray Shadis and Friends of the Coast, Maine's foremost environmental organization dedicated to questioning the efficacy of nuclear power in Maine. Maine media, including the Maine Public Broadcasting Network, which received substantial funding from Maine Yankee, never explored the role fuel assembly failures had in closing Maine Yankee a decade before it otherwise would have ceased operations under normal conditions. The documents and letters that follow allow the review of now forgotten issues that resulted in the premature closing of a nuclear power facility due to the high costs and difficulties of replacing the damaged fuel assemblies and associated cleanups. The importance of these documents lies in the light they may shed on fuel assembly-related problems at other nuclear reactors, of which there are now 442 operating in many nations. Most fuel assembly failures occur at undamaged reactor vessels with intact and functional cooling capabilities. The ongoing fuel assembly melting events at the Fukushima Daiichi complex are, in contrast, the result of tsunami-derived total loss of cooling capabilities, which resulted in massive hydrogen gas explosions, which then destroyed both the secondary reactor vessel containment buildings and the roofing and some of the containment associated with the adjacent spent fuel pools. The accident at the Fukushima Daiichi complex is the first of its kind, a fuel assembly melting accident involving seven sites in one reactor complex that was totally unexpected. Routine fuel assembly failure in the context of robust and normally functioning reactor cooling

312

systems, such as those at Maine Yankee, is a possibility at any operating reactor. The probability of further fuel cladding failures, including at less carefully managed reactors, should make all operating nuclear power plants the subject of additional scrutiny in the post-Fukushima Daiichi era.

While being a notable case study for fuel cladding failures that could occur at any reactor, the Maine Yankee facility was unique in one respect: its rapid decommissioning and low budget disposal in a landfill in South Carolina. The probable envy of all future decommissioning efforts, the burial of a reactor vessel containing large quantities of fuel assembly-derived fission products was a unique landmark in America's nuclear industrial history. A onetime event, the burial of the Maine Yankee reactor vessel included both greater than class C activation products (see below) and a substantial quantity of fuel assembly-derived fuel pellets, possibly in excess of a million curies. The South Carolina site had a low-level waste disposal limit of 50,000 curies. The licensee joined ratepayers, NRC, Maine state regulators, the South Carolina waste site operators, the state of South Carolina, and all media outlets in conveniently ignoring the high levels of fission products that were being disposed of in a primitive landfill. The taxpayers and ratepayers of Maine owe the State of South Carolina a debt of gratitude for saving them the hundreds of millions of dollars that would now have to be spent for the less hazardous disposal of fuel-cladding-failure-derived nuclear waste.

Also see the 1998 Union of Concerned Scientists' report on fuel cladding failures at NRC supervised nuclear reactors.

Historical Overview

The history of the Maine Yankee Atomic Power Company (MYAPC) is littered with clues that suggest significant loss of radiological controls occurred during "normal" reactor operations. As early as 1976, Charles Hess' post-operational radiological survey noted the existence of an activation products hot particle discovered in one of only 50 samples in his survey. This was the first hint that "normal" reactor operations included discharges of significant amounts of radioisotopes and/or radiologically contaminated materials. During the early years of operation numerous other studies were made under the auspices of the Maine Sea Grant program to determine the viability of growing oysters and other marine shellfish in the heated effluents discharged from the MYAPC reactor into Montsweag Bay. All of these studies, without exception, documented alarming quantities of reactor-derived activation and fission products in marine pathways (see Brack 1986, *A Review of Radiological Surveillance Reports*). Among the most controversial of these reports was Vaughn Bowen's 1981 study of *Transuranium Nuclides Released from Water Cooled Nuclear Power Plants*. When Bowen noted trace amounts of ^{239}Pu in Maine Yankee batch discharges, his three year contract (Woods

Hole Oceanographic Institute) was discontinued. In fact, all secondary as well as federally sponsored research reports on MYAPC operations and discharges and their environmental impact were discontinued; the last report in the literature other than the GTS Duratek *Site Characterization Report* was S. Murray's 1982 *Oceanography* article on Retention of Co-60 by the Sediments of Montsweag Bay, Maine.

The Maine Yankee proprietary inventory of its spent fuel pool issued on April 16, 1998, well after the reactor shutdown in December 1997, clearly documents at least two episodes of fuel cladding failure. While pinhole leaks in fuel assemblies are a relatively normal component of light water reactor operations, discharge of fuel pellets from the fuel assemblies into any environment on the grounds of a nuclear reactor indicates that a nuclear accident has occurred. Unfortunately, fuel cladding failure is one of the most common types of nuclear accident that afflicts operating nuclear reactors. Appendix A, reprinted below, summarizes the radioactive material stored in the Maine Yankee spent fuel pool. Some items, such as control element assemblies, are highly radioactive components common to all nuclear power plant operations.

Table A.7 (Appendix A), filtered waste stored in MYAPC spent fuel pool (SFP) constitutes a smoking gun with respect to prior fuel cladding failures at MYAPC. As the table notes, the filters date to a cleanup campaign in 1992; fuel cladding failures, probably involving axial cracking on the fuel assemblies, were severe enough so that loose fuel pellets were vacuumed up from the floor of the spent fuel pool. The vacuum filters are so radioactive that they must be sited as high-level waste, and will eventually end up in MYAPC dry casks in its new onsite high-level waste storage facility, the ISFSI (Independent spent fuel storage installation).

Unfortunately, the spent fuel pool is not the only location in which fuel pellets from failed fuel assemblies were discharged. Prior to removal to the spent fuel pool the damaged fuel assemblies also discharged fuel pellets within the reactor vessel itself. MYAPC is a pressurized water reactor. In this type of reactor heat is transferred from the core to a heat exchanger via water kept under high pressure. Fuel pellets from axial cracks on the fuel assemblies were discharged into this reactor water system prior to removal of the damaged fuel to the spent fuel pool. These fuel pellets entered an environment in which the water was moving at a high velocity, under high pressure and at high temperatures. Under these conditions these fuel pellets, which at this point are essentially spent nuclear fuel wastes, would be broken up into minute particles and spread throughout the reactor water systems. The high levels of radioactivity contained in the reactor water storage tank leak of 1984 resulted from this contamination of the plant environment. The GTS Duratek Site Characterization is a landmark document in the history of nuclear power in the United States as it clearly shows the impact of fuel cladding failure in the form of a highly contaminated reactor water system and some of the highest levels of environmental contamination ever documented at a US nuclear facility.

David Lockbaum at the Union of Concerned Scientists has expressed the opinion that only a handful of US nuclear power plants would have experienced fuel cladding failure via axial cracking to a significant enough degree to necessitate vacuuming the spent fuel pool to retrieve spent fuel pellets. The October 31, 1974, fuel cladding failure at the Dresden, Illinois, reactor is the most severe known incident of loss of radiological controls at an NRC facility other than the Three Mile Island accident. Ironically, the Dresden fuel failures occurred within a few months of the first of Maine Yankee's fuel assembly failures. The fuel failure at MYAPC in the early years of its operation was not, however, the only incident of its kind at MYAPC. Table A.12 not only lists highly radioactive filters collected during a second 1995 cleanup, but also references filters (BT17) derived from a 6/11/97 vacuuming of the top of fuel assemblies as well as the reactor vessel cavity prior to removal of spent fuel from the reactor vessel after the 1997 final shutdown. As is well known in certain circles, the direct cause of the MYAPC reactor shutdown was a second round of fuel assembly failures that resulted in additional contamination of reactor water systems and environs. A complete inventory of the contents of the Maine Yankee spent fuel pool awaits further study by the licensee -- even they don't know what's in the spent fuel pool.

The following confidential documents represent only the tip of an iceberg of licensee generated data and reports that document the size and the extent of the fuel cladding failures at MYAPC. We do not have copies of Table A.13 Non-fuel special nuclear material or Table A.14 fuel cycle (core history).

Other incidents of known fuel cladding failures have been recorded at Haddam Neck in Connecticut (apparently more severe than that at MYAPC), Oyster Creek, Seabrook in New Hampshire, and most recently, at Vermont Yankee.

Maine Yankee Reactor Vessel Inventory (TLG Engineering Report, 1987)

Readers, please note that this essay was written in 1999, prior to the burial of the reactor vessel in Barnwell, SC. The GTCC (greater-than-class C) wastes at Maine Yankee are similar to those at all other operating nuclear power plants in the US and in other countries. In the US, it is now unlikely that GTCC wastes in other reactor vessels, which are essentially high level wastes, will be disposed of in a shallow land fill as was the case at Barnwell. The burial of the Maine Yankee reactor vessel also included a large inventory of spent fuel-derived fission products.

Due to the recent evolution of MYAPC decommissioning plans away from the likely use of the Texas compact for low-level waste disposal (no greater than class C wastes allowed: see entry for October 3) and the anticipated availability of the Barnwell, South Carolina "low-level waste" facility for uncontained burial of an intact reactor vessel (706.4 tons) including all GTCC reactor vessel components (±4,170,000 curies at two years cooling), the 1987 TLG Reactor Vessel Inventory is reproduced below. Recent conversations with the editors and some news staff at the State Newspaper in Columbia, South Carolina indicate that the Barnwell, SC, low-level waste facility is now selling space in this facility to utilities seeking nuclear waste disposal capacity in the future. While no public announcement has been made in Maine or in Connecticut, presumably both MYAPC and Connecticut Yankee would be the nuclear utilities most interested in purchasing these space allotments. Neither South Carolina authorities nor Barnwell management appear to have any objections to the "new paradigm" of the NRC for GTCC reactor internals disposal: average the greater than class C reactor vessel components with class A reactor vessel parts and site the whole package in one unit as class C wastes in an uncontained land burial site at Barnwell, SC.

The following Maine Yankee Reactor Vessel Inventory was first published in a restricted TLG report to the Maine Yankee Atomic Power Company in 1987 as a component of TLG's decommissioning cost estimate. (TLG, a contractor for estimating decommissioning costs, estimated decommissioning costs in 1987 as $178,097,900.) In 1992, Uldis Vanags inadvertently republished this confidential information in "A Study of Radioactive Wastes" (Maine State Planning Office). The editor of RADNET reproduced this inventory in *Legacy for Our Children: The Unfunded Costs of Decommissioning the Maine Yankee Atomic Power Station* (Brack 1993, 116).

A complete copy of the most relevant components of the Maine Yankee Reactor Vessel Inventory is reprinted below. This includes the greater than class C reactor vessel internals which are symbolized by >C in the lower parts of the third column, which reference the radioactivity in the lower core support barrel, the core shroud, and the lower core support plate. The recent development, courtesy of more liberal NRC waste disposal standards, of allowing greater than class C wastes to be disposed of as class C low-level waste by averaging GTCC wastes with class A wastes (equals class C low-level wastes) in one large package (706.4 tons) is an important development for any person or organization concerned about what actually goes into a low-level waste facility. A key question for the evolution of "low-level waste" facilities in Ward Valley, California, Sierra Blanca, Texas or in other locations in Nevada or Utah is, Will these facilities be as willing to receive GTCC wastes in the form of "hot C wastes" as is the Barnwell, SC facility?

MAINE YANKEE REACTOR VESSEL INVENTORY

Output Tuesday, September 29, 1987 at 14:10

Maine Yankee Atomic Power Station -- Prompt Reactor Vessel Removal (w/2 yr decay) -- Last revised on Tuesday September 29, 1987

(This is a partial list of relevant data in the TLG report)

COMPONENT	Total Wt. Lbs.	Sp Activ. Ci/Lb	10 CFR 61 [A/C/>C]	Activity Ci	Payload K/hr	No. of Shipments
*** Reactor Vessel ***	933,270.00					
Upper Head	197,400.00	3.555E-05	A	7.02		1
Vessel Flange	69,268.94	3.555E-05	A	2.46		*
Upper Head Pkg Closure Plate	16,785.53	0.000E+00	n/a	0.00		*
	------------			------------		------------
Total Upper Head Package	283,454.47		A	9.48	0.01	1.00
High Activation Zone Vessel	254,588.79	2.250E-02	C	5,728.25		14.14
High Activation Zone Clad	8,753.48	1.830E-01	C	1,601.89		1.09
	------------			------------		------------
Total High Activation Zone	263,342.27		C	7,330.13	24.40	15.24
Upper Nozzle Zone Base Metal	173,759.85	3.555E-05	C	6.18		9.65
Upper Nozzle Zone Clad	5,171.36	2.891E-04	C	1.50		0.29
Outlet Nozzles	65,583.00	3.555E-05	C	2.33		3.64
Inlet Nozzles	60,261.00	3.555E-05	C	2.14		3.35
	------------			------------		------------

MAINE YANKEE REACTOR VESSEL INVENTORY

Output Tuesday, September 29, 1987 at 14:10

Maine Yankee Atomic Power Station -- Prompt Reactor Vessel Removal (w/2 yr decay) -- Last revised on Tuesday September 29, 1987

(This is a partial list of relevant data in the TLG report)

COMPONENT	Total Wt. Lbs.	Sp Activ. Ci/Lb	10 CFR 61 [A/C/>C]	Activity Ci	Payload K/hr	No. of Shipments
Total Upper Nozzle Zone	304,775.21		C	12.15	0.05	16.93
Reactor Vessel Insulation	2,000.00	1.762E-03	C	3.52	0.10	0.11
Neutron Shield Tank (inner wall)	41,563.00	1.167E-04	C	9.00	0.10	2.31
Neutron Shield Tank (remaining structure)	48,147.00	3.555E-05	A	1.71	0.05	2.67
Lower Head & Flow Skirt	98,483.58	8.089E-05	A	7.97		1
Lower Head Pkg Closure Plate	8,029.30	0.000E+00	n/a	0.00		*
	------------			------------		------------
Lower Head Package	106,512.88		A	7.97	0.01	1.00
	=========			=========		
Vessel Totals	1,049,794.84 lbs.			7,373.97 curies		
*** Internals ***						
UGS Support Plate	70,762.00	2.232E-04	C	15.79	0.10	3.93
Control Element Shroud Assys	10,883.00	2.415E-03	C	26.29	0.49	0.60
Fuel Assy Alignment Plate	10,200.00	1.410E+00	C	14,382.00	488.00	1.28

MAINE YANKEE REACTOR VESSEL INVENTORY

Output Tuesday, September 29, 1987 at 14:10

Maine Yankee Atomic Power Station -- Prompt Reactor Vessel Removal (w/2 yr decay) -- Last revised on Tuesday September 29, 1987

(This is a partial list of relevant data in the TLG report)

COMPONENT	Total Wt. Lbs.	Sp Activ. Ci/Lb	10 CFR 61 [A/C/>C]	Activity Ci	Payload K/hr	No. of Shipments
Upper Core Support Barrel	39,221.00	1.210E-01	C	4,745.74	48.80	4.90
Thermal Shield	63,813.00	1.460E+00	C	93,166.98	976.00	7.98
Lower Core Support Barrel	69,304.00	7.960E+00	>C	551,659.84	2,928.00	13.79
Core Shroud	37,873.00	8.370E+01	>C	3,169,970.10	2,928.00	79.25
Lower Core Support Plate	8,700.00	3.750E+01	>C	326,250.00	2,928.00	8.16
Core Support Columns	10,200.00	8.960E-01	C	9,139.20	97.70	1.28
Lower Core Support Beam Assys	42,082.00	2.060E-02	C	866.89	24.40	2.34
	------------			------------		
Total Internals	363,038.00 lbs.			4,170,222.83 curies		[123]

Maine Yankee Atomic Power Plant Proprietary Information

Maine Yankee Atomic Power Company. (April 16, 1998). *Appendix A: Spent fuel and other radioactive material stored in the Maine Yankee spent fuel pool.* MYPS-101, Rev. 0. Maine Yankee Atomic Power Company, Wiscasset, ME.

This inventory contains the following information: Out of 1,434 spent fuel assemblies now contained in the spent fuel pool, at least 15% are nonstandard and include the following:

· 66 fuel assemblies with "confirmed failure"
· 10 are consolidated or otherwise damaged

· 50 exhibit "physical damage"

· 18 have fuel rods replaced

· 80 have hollow rods (the present location of the spent fuel originally in these rods is not indicated in this inventory)

· 1 fuel rod is cemented into a pipe

· 1 fuel rod is stuck in a conduit

· 1 fuel rod was removed during a "1992 disposal campaign"

· In addition to this inventory of fuel assemblies in the spent fuel pool, this report indicates five filters "determined to be unsuitable for shipment due to dose rates" are in the spent fuel pool and contain loose fuel pellets derived from vacuuming up the debris released during the failure of the 66 fuel assemblies

· Numerous other filters are within the fuel pool and also contain fuel pellets but as of 1998 the number of these is not known

· 61 additional filters are contained in trash baskets in the spent fuel pool which also contain solid wastes from thermal shield positioning pin repairs; these filters may also contain spent fuel pellets derived from the fuel cladding failure accidents at MYAPC

Summary:

Minor grid-to-rod fretting fuel assembly damage and pinhole leaks are a normal component of reactor operations, as the NRC has made clear to the Union of Concerned Scientists (UCS) in response to its petitions submitted to the NRC on September 25, 1998 and November 9, 1998, re: failed fuel assemblies at the River Bend, LA and Perry, OH, reactors. The NRC response to the UCS petitions is dated April 18, 1999, and should be reviewed by all persons concerned with fuel failure. The NRC is emphatic in insisting that fuel "leakage" is a normal part of reactor operations, but the NRC response to the UCS is also a model of equivocation and sophistry in that it does not differentiate between "leakage," a word repeated dozens of times in the NRC response, and "failure," the concern of the UCS petition.

In the case of MYAPC, 66 *failed* fuel assemblies with spilled spent fuel pellets throughout the containment; this is a radically different issue than the 50 leaky damaged fuel assemblies, where no fuel pellets may have escaped and where emissions are limited to noble gasses, tritium and ^{131}I. Each fuel assembly contains an excess of 100,000 curies (Ci) of long-lived spent fuel waste. If 10% of this radioactivity was released during fuel cladding failure accidents, 660,000 Ci of spent fuel waste would have the following potential destinations:

320

· Vacuuming and water system filters now in the spent fuel pool as noted above
· Filters and resin beads shipped to Barnwell, SC as class B low-level waste
· Spent fuel-derived pellets still lodged in inaccessible areas of the containment
· Fuel pellets and pellet-derived fission products released outside of the reactor containment pressure boundary, but still within reactor water systems (piping, reactor water storage tank, other water system tanks, etc. as indicated by elevated fission product activity in the cursory GTS Duratek site characterization smear sample spectroanalyses)
· Fuel cladding failure-derived radioactivity released through stack emissions (e.g. tritium, cesium iodide? etc.) and in water systems leaks 1984, 1988, etc.
· Free released radioactivity during normal plant discharges to Montsweag Bay, during decommissioning reactor containment decontamination and water systems tank discharges

The actual amount of radioactivity released during fuel cladding failure accidents at MYAPC is unknown. The NRC has demonstrated gross negligence in the documentation of the source term (quantities of accident-derived effluents) of fuel cladding failures at MYAPC including all of the above potential destinations in the rush to minimize MYAPC owner and stockholder liability for the costs of the radioactive legacy of the MYAPC reactor. The NRC and the licensee have skipped the historical site assessment phase of reactor decommissioning in the hope that no one will request documentation of the accidents of the past. The release of this proprietary information by MYAPC and the Community Advisory Panel is a giant step in confronting and documenting fuel cladding failure accidents at all US nuclear reactors.

APPENDIX A

SPENT FUEL AND OTHER RADIOACTIVE MATERIAL
STORED IN THE
MAINE YANKEE SPENT FUEL POOL

3.0   Table A.9 and Figure A.2 provide information on 168 Control Element
       Assemblies (CEA) stored in the spent fuel pool.

4.0   Figure A.3 provides information on 8 CEA plugs stored in the spent fuel pool.

5.0   Figure A.4 provides information on 5 neutron sources (3 Pu-Be, 2 Sb-Be) stored
       in the spent fuel pool.

6.0   Table A.10 and Figure A.5 provide information on the 138 incore instrument
       thimbles stored in the spent fuel pool.

7.0   Table A.11 and Figure A.6 provide information on in-core instruments stored in
       the spent fuel pool.

8.0   Figure A.7 provides information on surveillance capsules stored in the spent fuel
       pool. Maine Yankee reserves the right to redirect some or all of the surveillance
       capsules as required to support decommissioning

9.0   Table A.12 provides the inventory of the trash baskets stored in the spent fuel
       pool.

10.0   Table A.13 provides information on nonfuel special nuclear material stored in the
       spent fuel pool.

11.0   Table A.14 provides information on each fuel cycle (core history).

MAINE YANKEE PROPRIETARY

CONFIDENTIAL

MYPS-101
April 16, 1998
Rev: 0
Page: A.63

TABLE A.7

FILTER WASTE STORED IN SFP

The following filters were utilized with a TriNuke vacuum system at Maine Yankee. The filters were determined to be unsuitable for shipment due to dose rates during a disposal campaign in late 1992. Each filter is housed in a stainless mesh container. Efforts may be possible to consolidate the contents or repackage the filters for storage.

Filter Number	Maximum Dose Reading	Waste Stream and Comments
1	2000R	SFP floor clean up; may contain loose fuel pellets
2	100R	SFP floor clean up; may contain loose fuel pellets
3	1000R	SFP floor clean up; may contain loose fuel pellets
4	200R	SFP floor clean up; may contain loose fuel pellets
5	200R	Machining debris from thermal shield repair
6	25R	SFP floor clean up; may contain loose fuel pellets

Four (4) specially designed filters were used to capture debris from Electro Discharge Machining (EDM) during repair of the thermal shield positioning pins. Dose rates on these filters were unavailable at the release of this document. These filters were designed to be stored in the SFP storage racks. Other filters containing radioactive material may also exist in the spent fuel pool.

MAINE YANKEE PROPRIETARY

MYPS-101
April 16, 1998
Rev: 0
Page: A.69

CONFIDENTIAL

TABLE A.12

INVENTORY OF TRASH BASKETS IN MAINE YANKEE SPENT FUEL POOL

TRI-NUKE FILTERS, STORED IN SPENT FUEL POOL

Trash can in SFP	Source of Filters
BT2	5 filters on November 1995 or cut up ICI's
BT4	5 filters from vacuuming upender pit and thermal shield repair Cycle 14 refueling
BT9	5 filters on November 1995 or cut up ICI's (One cover bolt is stripped)
BT10	5 filters on November 1995
BT13	5 filters on November 1995
BT14	5 filters on November 1995
BT15	5 filters on November 1995
BT17	4 filters cycle 16 cavity purification and vacuuming on top of fuel (6/11/97)
BT19	5 filters on November 1995
BT20	4 filters on November 1995
BT21	4 filters on November 1995, and debris bucket with 60 R/hr piece of metal
BT23	4 filters (1 from vacuuming, others from cavity upender pit vacuuming)
BT25	5 filters on November 1995

A Summary of Safety Concerns

Re: Maine Yankee-NRC Meeting of September 14, 1995 Steam Generator Sleeving Update

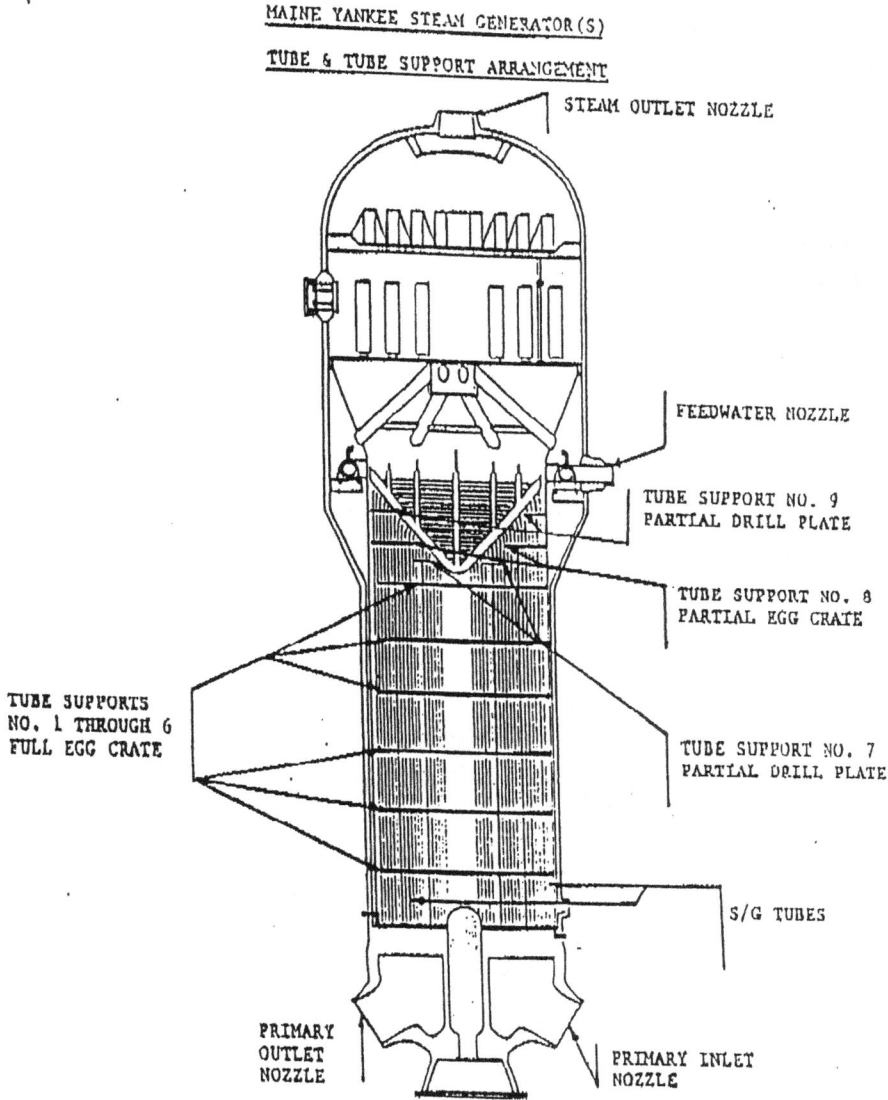

Figure 12. Maine Yankee steam generator(s) tube and tube support arrangement (NRC Figure 1).

"Laboratory testing of sleeving joints (between alloy 600 and 690) consistently produces cracking in the weld-heat affected zone of the parent tube." (Nuclear Regulatory Commission 1995b)

1. Circumferential crack growth in the Maine Yankee steam generator tubes results primarily from primary water stress corrosion cracking (PWSCC) and cannot be predicted from simulated testing. Other processes contributing to the damage in the tubes include "outside diameter stress corrosion cracking (ODSCC), intergranular attack (IGA), pitting, denting, and vibration induced wear" (U. S. Nuclear Regulatory Commission 1995c).

2. Inserting and welding the new sleeves to repair the circumferential cracks in the 16,000 steam generator tubes in the Maine Yankee steam generator creates residual stresses in the sleeve-tube transition area which then must be heat treated to repair stress damage.

3. Both the sleeve insertion and the heat treatment to repair damage caused by the sleeve insertion create additional "far field post-weld stress in the parent tube structure" (U. S. Nuclear Regulatory Commission 1995c).

4. Hard rolling the sleeve edges after insertion is an additional source of residual stress damage. To minimize this damage, hard rolling the sleeve edge is done after heat treatment of the damaged weld, but remains a source of both post-weld transition area stress and far field stress damage.

5. Extensive sludge deposits have been documented by the NRC and MYAPC throughout the steam generator. The largest deposits are at the base of the steam tubes at the junction of the tube sheet. Other sludge deposits are associated with the nine horizontal ("eggcrate" and drill plate) supports. The new sleeves (12, 20 and 30 inch) are located at the intersection of the steam tubes and the tube sheet, well below areas of sludge deposits and stress which have accumulated at these horizontals. The areas of sludge deposits and stress located at these horizontals are not subject to repair in the sleeving process. Not all parent tubing subject to circumferential cracking or other corrosion will be sleeved.

6. MYAPC locked support mock-up tests indicate that sludge deposits combine with copper scaling on the steam tubes to create a locking effect at the horizontal "eggcrate" supports which results in both severe bowing and tube displacement off the vertical during the post-weld heat treatment process (see NRC Figure 2 & Figure 3). The MYAPC report indicates that tube bowing and lateral displacement (deformations) occur early in the stress relief process.

7. Shorter 12-inch and 20-inch sleeves are frequently used in areas with the least sludge deposits to lower the frequency of post-weld induced deformations. The sludge deposits, tube scaling, and other corrosive products result in the non-uniform distribution of pressure and temperature which cannot be quantified in pre-test mock-up trials. "Sleeve joint life will be maximized if the joints are made above the sludge pile" (Nuclear Regulatory Commission 1995b).

8. Extensive copper scaling tube deposits project above the sludge deposits, sometimes overlapping into the weld transition areas and complicating the distribution of operational, weld, and post-weld heat treatment stress damage (see NRC Figure 2).

9. Additional mechanisms causing microstructural degradation of the older parent tubes are not fully understood and cannot be empirically reproduced in mock-up tests. "Recent experience at operating plants has emphasized the sensitivity of the Alloy 600 parent steam generator tube material to stress corrosion cracking when unfavorable residual stresses are introduced by processes such as sleeving" (MYAPC 1995).

10. The formation of both sludge deposits and copper scaling tube deposits result from corrosion processes which are also not fully comprehended. Use of the alloy 690 in the new sleeves, which has twice as much chromium as the parent tube alloy 600, doubles corrosion resistance of the new sleeve. The older less corrosion resistant parent tube continues to be the weakest link in a steam generator being subject to all the degradation processes noted in Section 1, as well as the residual stresses from the sleeving process.

11. Welding during the sleeve insertion results in copper scale tube flaking, further increasing sludge deposits and complicating residual stress distribution.

12. Maine Yankee and Westinghouse steam corrosion tests "indicate a properly stress relieved weld has four to five times the expected life of a comparable weld joint" (Nuclear Regulatory Commission 1995c). All 16,000 welds therefore must be carefully and successfully heat treated to complete a safe sleeving.

13. Ultra sonic testing to analyze tube sleeve welds results in extensive "attenuated backwall signals" due to the copper scaling tube deposits and other corrosive products outside the tube, making it impossible to verify the success of the welds. "Effectiveness of post-weld heat treatments in relieving residual stresses created by the welding process is unknown" (Nuclear Regulatory Commission 1995b).

14. The MYAPC report concludes that "tube OD (outside diameter) deposits can attenuate backwall signals in structurally acceptable welds." Since successful

welds can therefore not be verified with the ultra sonic tests, there is no way to locate defective welds.

15. Defective welds can only be verified by a tube-by-tube preoperational safety analysis. Such a detailed safety check is too time consuming and thus extremely expensive for the licensee, Maine Yankee, in that it would significantly delay the reopening of the plant. The verification of the weld efficacy will only be made as the steam generator is returned to service, or after the first cycle of operation through the use of ECT (eddy current testing).

16. Only full operation of the steam generator will verify the extent of weld deficiencies, far field stress damage, and other structural degradations, including bowing and lateral deformations, as well as additional operational corrosion damage. The confirmation of such damage will only come in the form of leaks or breaks in the steam generator tubes, some types of which could lead to a serious nuclear accident.

The NRC issued the following reassurances in the September 14th report:

A. Tube bowing and tilting (deformations) resulting from the locking effect of the sludge deposits "might result in tube contact, but there are no indications such contacts will be hazardous" (Nuclear Regulatory Commission 1995c). This is an extraordinary admission of steam tube degradation.

B. "Local deformations have equal or better corrosion resistance than the stress relieved joints" (Nuclear Regulatory Commission 1995c). In view of the superior corrosion resistance of the new sleeve, the sleeve tube transition area remains that component of the steam generator most vulnerable to a catastrophic failure.

C. "Sleeving induced deformations and resulting stresses in the Maine Yankee steam generator tubes, should a tube be locked at a horizontal position [by sludge deposits], will be low enough to insure a satisfactory sleeve performance through Maine Yankee's life" (Nuclear Regulatory Commission 1995c). This controversial assertion, which is clearly contradicted in the MYAPC Locked Support Testing Program report, can never be verified empirically: safe operation of this damaged generator through the year 2008 would be an extraordinary stroke of good luck.

Safety Issue Summary: The NRC and MYAPC locked tube test program clearly documents a grossly degraded steam generator with extensive steam tube circumferential cracking due to primary water stress corrosion cracking (PWSCC) and widespread accumulations of sludge deposits (corrosion products) throughout the steam generator. The steam generator suffers additional system wide welding-induced stress damage during the sleeving process, which must be successfully repaired with a post-weld heat treatment process which in itself either reinforces pre-existing far field stress damage in the parent tube or extends the damage further down the parent tube. The

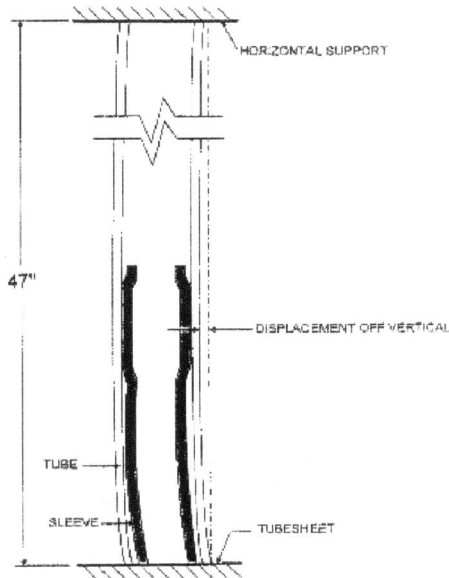

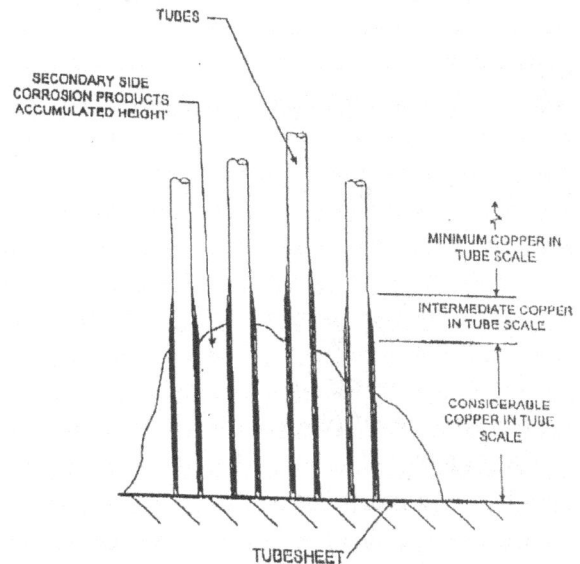

Figure 13. Fuel assembly deformation (NRC Figure 3 [left] and Figure 2 [right]).

successful operation of the repaired steam generator is further undermined by steam tube deformations (bowing and vertical displacement) which occur in the repair process due to extensive preexisting corrosion-derived sludge and copper scale tube deposits. These deformations cannot be repaired. The sludge deposits and scaling are located throughout the steam generator, with a particularly large accumulation of corrosion products (see NRC figure 2) at the tube sheet intersection and smaller deposits at the horizontal supports. These widespread sludge deposits combined with tube scaling render tests to verify weld efficacy useless.

The sleeving process, which includes the plugging of some steam tubes that cannot be successfully welded, results in a reduction in the reactor coolant flow rate. Future plant operation may approach, equal, or exceed the 9 percent reduction in maximum power which is the cutoff point for the safe operation of the Maine Yankee plant. This reduced reactor coolant flow, combined with possible weld deficiencies, sludge deposits, and microstructural degradation, provide the basis for Maine Yankee's recent controversial application for a reduced power license. The unprecedented restart of such a degraded and decrepit facility is a dangerous experiment which poses a clear and present danger to the public safety in the State of Maine.

The return of this degraded equipment to service in lieu of replacement with a new steam generator greatly increases the chance of a substantial nuclear accident and constitutes an act of criminal negligence by the NRC, the supervisory agency for this obsolete and dangerous nuclear energy dinosaur.

H.G. Brack
Director of Center for Biological Monitoring
Hulls Cove, Maine
cbm@davistownmuseum.org

REFERENCES

- Maine Yankee Atomic Power Company. (1995). *Maine Yankee Steam Generator Tube Sleeving Locked Support Testing Program. Safety evaluation presented to the Office of Nuclear Reactor Regulation.* Maine Yankee Atomic Power Station, Wiscasset, Maine. Docket No. 50-309. MN-95-108, GDW-95-44.
- Nuclear Regulatory Commission. (1995b). Undated NRC reply to questions listed in a letter written by Dr. Ravender Chona, Texas A & M University to Pat Dostie, Maine State Nuclear Safety Inspector, dated July 20, 1995.
- Nuclear Regulatory Commission. (1995c). *Steam generator sleeving update.* Presented at Maine Yankee/NRC Meeting, September 14, 1995.

Whistleblower's Letter

Anonymous letter released to the public pertaining to falsified computer data, deficiencies in the emergency core cooling system and misrepresentation of the reactor vessel pressurization capabilities

The whistleblower's letter issued in early December, 1995, and sent to Robert Pollard of the Union of Concerned Scientists prior to its release to the general public is the single most important document pertaining to the twilight of the nuclear era among all the reports, journal articles and research papers cited and annotated within RADNET. In one short year, this revealing fragment of information has had a vast impact on public awareness about the policies and practices of the nuclear industry. The person who blew the whistle on MYAPC, YAEC, and the NRC opened a vast Pandora's box of safety issues, the ramifications of which will continue for generations. The full copy of this letter is preceded by an introductory letter by the Union of Concerned Scientists.

USCS Cover Letter

UNION OF CONCERNED SCIENTISTS

December 1, 1995

Mr. Uldis Vanags
State Nuclear Safety Advisor
Maine State Planning Office
184 State Street, Station #38
Augusta, ME 04333

Dear Mr. Vanags:

I am writing to bring to your attention a matter that is of utmost importance in determining whether operation of the Maine Yankee nuclear power plant will pose an unacceptable risk to the health and safety of the public.

I have received documentation, purportedly from a longtime employee of the Yankee Atomic Electric company, indicating that the management of Maine Yankee deliberately falsified reports to the US Nuclear Regulatory Commission in order to receive approval of an increase in the reactor's maximum allowable power level. Specifically, the individual asserts that management officials manipulated computer calculations to avoid disclosing that the emergency core cooling systems at the Maine Yankee plant are inadequate to prevent overheating of the reactor fuel following a small break loss-of-coolant accident. The individual also asserts that the Yankee Atomic Electric Company fraudulently modified its analysis of the reactor containment building to avoid disclosing that a large break loss-of-coolant accident will pressurize the building above the pressure that it was designed to withstand.

It is apparent that this information was provided by someone who is knowledgeable of the subject matter and has access to documents that are not publicly available. It is also apparent that the person knows that it is the responsibility of the US Nuclear Regulatory Commission to ensure public safety, but has concluded, as I have, that the NRC fails to fulfill that responsibility. I assume that this information was provided to the Union of Concerned Scientists because the individual wishes it to be made public. Therefore, I am distributing this letter and the individual's three-page letter to the public.

A copy of that unsigned, undated letter and copies of the other five documents that I received earlier this week are enclosed. The handwritten notations were on the documents before I received them. I trust that you will make these documents available to the US Nuclear Regulatory Commission and the citizens of Maine.

I urge you to recommend that the State of Maine take the position that the Maine Yankee plant should not be permitted to resume operation until a thorough, factual investigation of the individual's allegations is completed and made available for public scrutiny. I am convinced, based on 26 years of experience, that the NRC will not conduct such an investigation unless the State of Maine demands it.

Sincerely,

Robert D. Pollard Nuclear Safety Engineer

Washington Office: 1616 P Street NW Suite 310. Washington, DC 20036 (202)332-900 FAX: (202)332-0905
Cambridge Headquarters Two Brattle Square Cambridge, MA 02238 (617)547-5552 FAX:
(617)864-9405 California Office 2397 Shattuck Avenue Suite 203 Berkeley, CA 94704 (510)843-1872. FAX (510)843-3785

Anonymous Report of Safety Violations at Maine Yankee

Dear Sir,

I must report to you some of the flagrant violations of NRC regulations by Yankee Atomic Electric Company (YAEC). I have worked at YAEC for several years, with each passing year a belief that NRC is a nuisance as an organization and its staff technically incompetent, has become stronger at YAEC. Surely, YAEC's management has actively supported this belief and jeopardized public safety on several occasions. The disregard for public safety is manifested by the temerity with which Maine Yankee Power Plant's rated power was increased from 2630 MWth to 2700 MWth in 1989. YAEC's management knew that the Emergency Core Cooling System (ECCS) and the containment system of Maine Yankee (MY) did not meet the licensing requirements even for the pre-1989 power rating of 2630 MWth, never the less they made misrepresentations to NRC and obtained the license to operate MY at 2700 MWth. The deficiencies in ECCS and containment have still not been rectified. To ensure public safety, NRC should immediately derate the plant to 2400 MWth, its original power, and fine Maine Yankee.

Deficiencies in ECCS: As a consequence of the Three Mile Island Accident (TMI), NRC issued a set of requirements for the nuclear power plant licensees in its report NUREG 0737 (Reference 1). Item II.K.3.30 of this report required all licensees to upgrade their method (computer code) for analyzing the Small Break Loss-Of-Coolant-Accidents (SBLOCA's), and Item II.K.3.3 1 required the licensees to use the new method to assess their ECCS's performance during SBLOCA's.

To meet the requirement II.K.3.30, YAEC spent several years (1980 to 1983) to develop the RELAP5YA(PWR) computer code (Reference 2). This code was able to predict the LOFT SEMIS CALE and other experiments reasonably well. However, preliminary SBLOCA analysis of the Maine Yankee plant with this code showed that the plant's ECCS is grossly inadequate, i.e., calculated peak clad temperatures (PCTs) were higher than 2200 °F. MY management refused to even discuss the possibility of upgrading the ECCS. Hence YAEC did not submit the code for NRC review. Between 1983 and 1987, YAEC analyzed and re-analyzed these MY accidents, made modifications to the computer code, but with any reasonable code modification and input parameters the results showed that Maine Yankee ECCS is inadequate, i.e. the fuel rod cladding temperature was calculated to exceed 2200 0F during the LOCAs. As a last resort, in 1987, YAEC considered scrapping the code and approached Combustion Engineering (CE) to perform the analysis with its (CE's) new method to show adequacy of the MY's ECCS. Alter some preliminary analysis CE turned down the offer. At this point, under pressure from NRC to close out Item II.K.3.30, YAEC submitted to NRC the RELAP5YA(PWR) in 1987.

Consider the ethical bankruptcy: Knowing that once the new method is approved, it will have to be applied. The new method, at 2630 MWth, will give MY, at best, very limited margins in PCT. This will eliminate the possibility of MY ever applying for power up-rate. Hence, while the new method was under review, despite the knowledge of the inadequacy of the ECCS, MY and YAEC management decided to apply for a power up-rate for MY in 1988 (Reference 3). To support this power up-rate application they used the small break analysis performed by CE in 1973, and told NRC that they were working on a new analysis, with 2700 MWth to meet the post TMI NRC requirements. YAEC staff was aware of the fact that applying for power up-rate while knowing the inadequacy of the ECCS, was dishonest. However, it was thought by the management that YAEC should get the approval for power up-rate before Mr. Pat. Sears, NRC project manager for MY, moved to a different position in January 1989. Mr. Sears was considered to be a particularly lenient person, therefore YAEC wanted to get the approval before he left. YAEC wanted to apply between thanksgiving and Christmas, when NRC staff is least vigilant. Open discussion of these considerations is indicative of a disregard for public safety. They applied for the power up-rate and got it, as planned.

In 1990, under pressure from NRC, YAEC decided to fulfill its commitment to perform a new small break analysis according to the post-TMI rules. This analysis, as expected, showed inadequacy of the MY ECCS. At this point, a new scheme was devised by Mr. R.K. Sundaram: we will do the break spectrum analysis with the Best Estimate (BE) assumptions, and perform an Evaluation Model (EM) analysis of the limiting break from the BE break spectrum analysis. Since the limiting break in the BE break spectrum analysis will not be the limiting break in the EM break spectrum, we will be analyzing a non limiting break and showing a lower PCT. The scheme was approved and put into action. It was decided that the scheme will be justified to NRC by stating that the BE analyses are useful for operator training etc., therefore, to conserve resources, the break spectrum analysis is done with BE assumptions and only the limiting break is analyzed with EM assumptions. In reality, making input changes from BE input to EM input and running the code did not take much. However, the results of this "limited EM" analysis gave PCT higher than 2200 °F!

At this point, the conservatism in the decay heat and the break flow calculations were removed from the EM input deck. The decay heat was calculated by the un-approved (by NRC) 1979 ANS standard and the break flow was calculated with the RELAP5 critical flow model (not the licensing Moody Model). In calculations with these

334

fraudulent models, decay heat was under estimated by the decay heat model, and the combination of non-licensing break flow model with the licensing assumption of one ECCS train assured that we were analyzing a non-limiting break. In fact we assured that we did not even analyze a realistic accident scenario.

The results of analysis with the above non-conservatism's were presented as 95% confidence level results. This is fraudulent, RELAPS was approved by NRC only as a licensing code (with several stipulations, indicating lack of confidence in the code). Also, the method of performing BE LOCA analysis to obtain results that are considered as 95/95, is completely different. YAEC management, specifically R.K Sundaram, clearly defrauded NRC in this regard. After completing this analysis Mr. Sundaram and other YAEC officials reported to NRC that MY ECCS performance was satisfactory, and all post-TMI and licensing requirements have been met. NRC simply acknowledged this report. The Maine Yankee plant is operating on the basis of this fraudulent analysis at 2700 MW$_{th}$. I hope an occasion to use ECCS does not arise.

Alter the TMI accident, nuclear industry declared that it had learned its lesson from the accident and will use the experience to improve the public safety. In case of YAEC, it was doing every thing to cover up, rather than repair, the deficiencies in the safety systems exposed by TMI.

Deficiencies in Containment System: The containment design analysis for Maine Yankee was performed by Stone and Webster Co. for a design power of approximately 2430 MW$_{th}$ (1970). For this analysis it was assumed that a hot leg LOCA would result in maximum possible containment pressure, and the maximum pressure from such a break was calculated to be less than 55 psi. Hence MY containment was designed for 55 psi.

In the 1970s MY applied for two power up-rates, from 2430 MW$_{th}$ to 2550 MW$_{th}$ and then to 2630 MW$_{th}$. For these power up-rates, a containment analysis was performed with the help of Combustion Engineering (CE). This analysis showed that during a cold leg guillotine break the containment pressure would exceed the design pressure (55 psi). Specifically, the mass and energy released to the containment during the reflood period of the LOCAs caused the containment pressure to increase beyond the design pressure. During the reflood period a significant source of energy is the hot water contained in the secondary side of the steam generators. YAEC decided to fraudulently exclude from the calculations this energy. Additionally, the containment free volume was assumed to be highest of the estimates (lower bound, best estimate and upper bound) given by Stone and Webster Co. These tricks in the safety analysis produced acceptable results and the plant was up-rated to 2630 MW$_{th}$.

In 1985, 86, 87, preliminary analyses (performed by L. Schor) had shown that the MY containment could not safely contain the mass and energy released during a LOCA from a power level of 2630 MWth. This did not deter the YAEC management from applying for the power up-rate in 1988. The YAEC management indicated to NRC that during operation at 2700 MWth the average temperature of primary coolant was going to be maintained at the same value as it was for operation at 2630 MWth (Reference 3). This implied that the energy content of the primary coolant was not going to change, hence the containment response to LOCA from 2700 MWth was going to be the same as that from 2630 MWth. Since the containment analysis was considered acceptable for 2630 MWth it would also be considered acceptable for 2700 MWth. This would be a fair argument, if the fluid mass on the hot side of the primary system was equal to that on the cold side, and if there was some margin in the existing containment analysis. However, the public safety concerns were put aside and power up-rate was gotten.

I think these violations of NRC regulations are serious enough to derate the MY plant and to levy fines against YAEC and MY. Also, the management, particularly Mr. Sundaram who used these activities for self promotion, should be seriously reprimanded.

References:
1. NUREGO737
2. 11RELAP5YA, *A Computer Program for Light Water Reactor System Thermal-Hydraulic Analysis* YAEC 1 300P.
3. Maine Yankee Power Uprate Application, December 1988.

**The Peter Atherton Letter of November 15, 1996
to the Center for Biological Monitoring**

A friend provided me with your RADNET nuclear information from the Internet, which I personally don't yet have access to.

For your info, I blew the whistle in 3/78 on Maine Yankee within the executive branch of the federal govt. to the White House. My evaluation covered fire protection and raised safety concerns thruout the entire plant while I worked for the US Nuclear Reg. Comm'n. I suggested solutions.

As a GS-13 engineer I was subjugated, my mental health was both threatened and challenged, and the evaluation never made it to the public document room after I was fired in 5/78. I checked after the 1991 fire. Maine Yankee is not my idea of a model nuclear power station. But they are not alone. If you are truly interested in nuclear safety, I could help. If you are not or you are a facade, pass this letter to a concerned group who is interested in nuclear safety. I sacrificed my job & ultimately my family for nuclear safety.

P J A

Nuclear Accident-in-Progress at the Maine Yankee Atomic Power Company

Spent fuel cladding failure
A nuclear accident in slow motion, 1973-???

There is growing evidence of the existence of a nuclear accident at the Maine Yankee Atomic Power Company (MYAPC) in Wiscasset, Maine. The release phase of this accident, which has occurred in spurts over a long time frame (1973 to the present), will continue as the decommissioning process remobilizes fission and activation product contamination throughout the plant.

Item: MYAPC released information about the contents of the spent fuel pool at the last Community Advisory Panel (CAP) meeting (February 18, 1999) in Wiscasset. Of the 1,436 fuel assemblies, up to 298 are "nonstandard." This is a deceptive way of acknowledging that over 20% of MYAPC has undergone fuel cladding failure at some time in the past. Such failure releases substantial amounts of fission product into the primary cooling circulation water where it can spread to other areas of the plant.

The spent fuel pool also contains a variety of other equipment and items which may be further evidence of the nuclear accident that occurred at MYAPC. These include control element assemblies and plugs, incore instrument thimbles, filter cartridges, neutron sources and, of particular interest, 22 storage baskets, the contents of which have not yet been made public.

Item: On March 30, 1984, during a snowstorm, the reactor water storage tank (RWST) sprung a leak and released 7,000 gallons of radioactive water onto frozen ground. The Licensee Event Report (84-004-00) indicates the water went into a *storm drain* and was diluted in the Forebay with *no detectable radioactivity* released. The 1998 *Duratek Site Characterization Report* contradicts the NRC licensee event report by documenting significant soil contamination along the west side of MYAPC (Volume 6, Supplemental Survey 2501). In a 10,000 sq. ft. area, which was frozen at the time of the release, Duratek site characterization data confirms fuel-rod-derived residual ^{137}Cs contamination of at least 3 million nanocuries. There is no other reactor source for this isotope. The maximum permissible concentration of ^{137}Cs in water released in the environment is 1 nanocurie/liter (10 CFR Part 20 Appendix B). Seven thousand gallons of water = 26,600 liters; the MYAPC release limit for this quantity of contaminated water is therefore 26,600 nanocuries, far below residual contamination levels documented in the frozen soil in the 1984 accident. The contamination documented in Supplemental Survey 2501 is unlikely to be the only radiation released by this component of the accident. (Portions of these reports have now been scanned and can be seen at this link: http://www.davistownmuseum.org/cbm/Rad9e2.html.)

Item: The *Duratek Site Characterization Report* also documented high levels of radiocesium adjacent to and outside of the reactor containment and bioshield, with contamination as high as 156,000 pCi/kg ^{137}Cs in grid 103. This is further evidence of systemic failure of fission product barriers at MYAPC. (RADNET Volume 6, Survey package R0100).

Item: The Duratek report documents substantial activation product contamination outside of plant systems and equipment, with samples as high as 33,600,000 pCi/kg of ^{60}Co in a hot particle (CRUD) on Bailey Point (RADNET Volume 6, Survey package R0500).

Item: Gamma spectroanalysis of smear samples (*Duratek Site Characterization Report,* Volume 3) document significant fission and activation product contamination throughout the pipes, drains, valves, steam generators and internal components of the plant outside of the containment and spent fuel pool. Very limited Duratek smear sample spectroanalysis (48 smears in 15 systems) of removal contamination indicates:

^{60}Co contamination exceeding 2,000,000 pCi/kg in 15 of 48 smears. Contamination peak value 86,400,000 pCi/kg.

^{137}Cs contamination exceeding 1,000,000 pCi/kg in 18 of 48 smears with peak values (PV) of 88,300,000 pCi/kg.

Evidence of fission product contamination in all smear samples from drains, valves, piping and other water system components.

Item: The reactor water storage tank (RWST), location of the March, 1984, accident, appears to be one of the many repositories of significant amounts of fission product contamination. Meager site characterization of Bailey Cove and Montsweag Bay, which show elevated levels of fission product contamination 2-5 times background, was executed prior to the free release (unfiltered) of 290,000 gallons of radioactive water into lobster fishing areas adjacent to recently reopened clam flats.

Item: Pre-free release analysis of water in the reactor water storage tank (RWST) was taken as a non-representative grab sample of quiescent water and provides no information about fission and activation product contamination in tank sludge, sediments and scaling. The NRC, the licensee and the state of Maine failed to provide CBM with the tank release data (including the neutron shield and other tank release data) when we requested it in the spring of 1998. No specific public notification was provided by the NRC, the licensee or the CAP to area fishermen that the free release of the RWST was underway.

Item: MYAPC did not release the last 10,000 gallons of water in the RWST during the time of this free release. Data on fission product contamination of the remaining contents of the RWST is not contained in any NRC or licensee documents and has not been made available despite requests made by the Center for Biological Monitoring.

Item: Significant fission product contamination remains within the reactor containment vessel. The *Duratek Site Characterization Report* contains a complete activation product analysis of reactor internals (+3,000,000 Ci) but no information whatsoever about fission and activation product hot particles, "fuel fleas," and other contaminants in the reactor containment. Underwater segmentation of the reactor vessel internals during the decommissioning process is a potential pathway for the release of spent-fuel-derived contamination within the reactor containment.

Item: The Duratek site characterization documents high radiation fields remaining in and around the containment (removal of the spent fuel has been completed), as well as around the base and floor adjacent to the RWST, and in many other locations. The question remains unanswered as to what extent these radiation fields derive from normally occurring shine from irradiated reactor internals versus radiation fields created by breach of the fission product barriers. Contamination derived from the reactor water storage tank produced ground shine along the west side fence line which the licensee has always attributed to shine from normal reactor operations.

Spent fuel failure at MYAPC is an accident-in-progress. The release of fission products in the 1984 RWST leak is one recently documented pulse of fission product contamination which has been followed by the undocumented free release of much of the remaining content of the RWST. Decommissioning decontamination activities for the purpose of constructing a high-level waste repository at Wiscasset will result in the remobilization of fission product contamination which has already escaped the fission product barriers (containment bioshield, spent fuel pool) at MYAPC.

The NRC, the licensee, the CAP, the state of Maine and Maine media have either lied or provided misinformation about the status of the MYAPC facility. The site is not clean as reported by MPBN, etc. None of the above have had the integrity to report that a nuclear accident has occurred at MYAPC in the form of spent fuel cladding failure and that this accident has spread radioactive contamination throughout the environs of the MYAPC facility. The fact that the release phase of this accident may resume with major

339

decommissioning activities has been evaded. The cover up of this relatively small nuclear accident involves numerous violations of federal law and NRC regulations by NRC and MYAPC staff. We have requested the assistance of the Office of the US Attorney for Maine in reviewing the situation and intervening to prevent further releases of fission and activation products and hot particle contamination during the decommissioning process until a careful review and recharacterization of plant facilities and the environs clarifies the extent of spent-fuel-derived contamination. It is important that the NRC provide an independent contractor, such as the GTS Duratek Co., to recharacterize spent-fuel-derived residual radioactivity and that the staff of MYAPC and the NRC have no further involvement with this process. Public safety considerations mandate that a thorough recharacterization be completed before any further decommissioning activities begin.

This situation is particularly unfortunate because there is evidence for a similar and perhaps more serious nuclear accident, also involving fuel cladding failure, at the Haddam Neck, Connecticut, reactor.

Checklist of Steam Generator Repair Safety Issues at Maine Yankee

The ancient history of safety issues at Maine Yankee is particularly pertinent to the continued operation of 104 aging US nuclear reactors and their upcoming "decommissioning."

12/1/95

1. Circumferential cracking in the steam generator.

2. Ongoing corrosion and deformation processes.

3. Accumulated residual stress damage from steam generator fabrication and operation.

4. Weld-induced stress damage at the sleeve-tube transition.

5. Post-weld heat treatment efficacy for repair of weld-induced damage.

6. "Far-field" stress damage in parent tube resulting from sleeve welds and post-weld heat treatment.

7. Steam-tube sludge deposits and corrosion scaling.

8. Tube deformations (tube bowing and lateral displacement) induced by sleeve insertion, welding, and sludge deposits.

9. Sludge deposit, tube plugging, and deformation-induced changes in pressure gradient and reactor coolant flow.

10. Increased likelihood of steam tube contact during reactor operation.

11. Non-uniform sleeve length fails to cover all parent tubing.

12. Stress resistance and corrosion resistance differentials between new sleeves and old parent tube and tube welds.

13. Reduced effectiveness of ultra-sonic weld inspections due to attenuation of signals by the extensive sludge deposits and tube scaling.

14. The dilapidated condition of the steam generator is a paradigm for the condition of the rest of the reactor at Maine Yankee (embrittled reactor vessel, obsolete spent-fuel storage facility, etc.)

Concerns about Maine Yankee Safety and the Current Traffic in Nuclear Waste

- Reactor re-start creates a public safety emergency.
- Existing emergency planning is obsolete.
- Biological monitoring database necessary for interpreting accidents, decommissioning, etc. is non-existent.
- The steam generator sludge deposits, plugged tubes, corrosion damage and weld-induced deformations are the tip of an iceberg of safety issues (e.g. reactor vessel embrittlement, spent fuel storage, health effects of plant releases, decommissioning scenarios, etc.) at Maine Yankee.
- The failure to fund disposal of Maine Yankee nuclear wastes is a very lucrative criminal activity.

Overview of the GTS Characterization Survey Report

Maine Yankee Atomic Power Company. (April 1998). *GTS Duratek characterization survey report for the Maine Yankee Atomic Power Plant, revision 1.* Nine volume report prepared by GTS Duratek, Inc. for the Maine Yankee Atomic Power Plant, Wiscasset, ME.

GTS DURATEK

CHARACTERIZATION SURVEY REPORT

for the

MAINE YANKEE ATOMIC POWER PLANT

APRIL 1998

REVISION 1

VOLUME 1: CHARACTERIZATION SURVEY DESCRIPTION

Prepared by: _____ GTS Duratek, Inc. _____ Date 4-28-98

Reviewed by: _____ Signature on File _____ Date 4-28-98
        Dave Lovett
        Project Manager

Reviewed by: _____ Date 4-28-98
        Dave Hall, CHP
        Manager, RE&DS Technical Department

Approved by: _____ Date 4-28-98
        Harvey F. Spray
        Director, Radiological Engineering
        and Decommissioning Services

Prepared By:

**GTS Duratek
628 Gallaher Road
Kingston, TN 37763**

342

The project team did not make *a priori* determinations of scan sensitivities. Instead, a retrospective assessment of scan sensitivities compared the quantitative soil sample analysis results to the findings documented during the scans. One comparison used biased soil sample analysis results from locations identified during the scanning process as having elevated levels of radioactive material. A second comparison used soil sample analysis results associated with random soil samples collected from areas where the scanning process did not identify elevated levels of radioactive material.

Gamma scans performed with the VRM-1X detector identified 24 locations as having elevated levels of radioactive material. Soil samples from two of the 24 locations showed evidence of licensed radionuclides:

- A small area containing Co-60 in the dry cask storage area

- An area where a discrete Co-60 particle was discovered on Bailey Point

For areas where the scanning process, using the VRM-1X detector, did not identify elevated levels of radioactive material, the soil sample analysis results did not indicate activity concentrations in excess of 2.0 pCi/g of either Co-60 or Cs-137.

Gamma scans performed with the 44-2 detector identified one area as having elevated levels of radioactive material. However, the follow-up investigation did not find evidence of licensed radionuclides. Of the 24 areas identified during scans performed with the VRM-1X detector as having elevated levels of radioactive material, only the area on Bailey Point could be identified during follow-up investigations using the 44-2 detector. For areas where the scanning process, using the 44-2 detector, did not identify elevated levels of radioactive material, the soil sample analysis results did not indicate activity concentrations in excess of 2.0 pCi/g of either Co-60 or Cs-137.

Although gamma scanning sensitivities depend on many parameters that are difficult to define, the survey team collected sufficient empirical data during the characterization survey to directly compare the relative sensitivities of the two different detectors used to perform gamma scans of open land areas.

The following Characterization Summary: Survey Package Number: R2501 documents the largest incident of soil contamination in the public records of the Nuclear Regulatory Commission. This survey was kept secret by the reactor licensee, Maine Yankee Atomic Power Company (MYAPC), its contractor, GTS Duratek, and the Nuclear Regulatory Commission. No mention was made of its existence in the executive summary of the nine volume *Characterization Survey Report*. Only one copy of this report, in the Maine State Library at Augusta, is currently available for perusal by members of the public. The secret survey R2501 is contained at the very end of volume six. Its significance lies in the documentation of soil contamination in an area in excess of 10,000 square feet with levels of the indicator radioisotope ^{137}Cs far above 2,000 pCi/kg (2 pCi/gm). Contamination is documented by this special survey reaching tens of thousands of picocuries per kilogram with some samples showing soil contamination levels of ^{137}Cs higher than any other samples in surveys of NRC licensees.

The soil contamination noted in special survey R2501 is a result of a reactor water tank leak on March 30, 1984 (LER 84-004-00). The water leak released ±8,000 gallons of water to surface soil before being allegedly recaptured and directed into the seal pit forebay area. The key question with respect to the 1984 leak is: why was there so much radiocesium in the reactor water storage tank that the leaking water left behind such excessive levels of soil contamination? The answer to this question, long a puzzle, was answered by documents released by the licensee pertaining to damaged fuel assemblies in the spent fuel pool. These damaged fuel assemblies discharged some of their spent fuel into reactor water systems; the reactor water storage tank became a repository for some of these fission products that were released as early as 1973.

Subsequent to the 1984 reactor water tank leak, the Maine Yankee Atomic Power Company suffered an additional episode of fuel cladding failure, which led to its closing. As noted in many sections of RADNET, the important unanswered question is: what is the source term of the fuel cladding failures at MYAPC? That is, what quantity of spent fuel fission products leaked out of the damaged spent fuel assemblies, what were their pathways and what is the current location(s) of this contamination?

Readers of the section of the *Characterization Summary* printed in this file please note the following two pieces of information.

In the following characterization survey description a previous drive over scan of the Maine Yankee site is described, with only two areas of contamination from MYAPC activities noted. This summary is highly deceptive, because the preliminary drive over scan failed to document a highly contaminated area of soil on the west side fence line of the plant, part of which extended off over the boundaries of the plant facility. The licensee probably knew about this contaminated area prior to the final site characterization survey, which clearly documented its existence as is noted in the characterization summary for R2501. Needless to say, the existence of this separate survey was not mentioned in any press releases or in the executive summary of the 8 volume report.

The characterization survey recording unit is pCi/gm, always a tip-off that something is amiss, as the standard reporting unit for radioactive contamination is per kilogram of soil. Use of the smaller reporting unit makes contamination levels sound and look much less significant and makes it much more difficult to evaluate long-term changes in contamination levels.

344

EXAMPLE: SAMPLE LOCATION 00022 REPORTS ^{137}Cs ACTIVITY OF 66.40 pCi/gm. THIS TRANSLATES TO 66,400 pCi/kg OF ACTIVITY IN A SAMPLE OF 1.716 kg OF SOIL. This is the highest level of ^{137}Cs soil contamination ever documented in the public records of a Nuclear Regulatory licensed facility. Maximum average levels of ^{137}Cs from weapons testing fallout in the Maine Yankee area soils and sediments are ±600 pCi/kg (0.60 pCi/gm). In contrast, the Yankee Rowe facility, which had no known episodes of fuel cladding failure, had maximum levels of ^{137}Cs contamination in the sediments of the effluent outfall basin, where the maximum contamination is to be expected, of not more than ±10,000 pCi/kg of ^{137}Cs. The maximum soil contamination at MYAPC was well away from the effluent outfall basin. No data is available for samples taken within the confines of the effluent outfall basin at MYAPC.

Please note pages 4 - 26 of the 36 page supplemental survey are duplicated below. Soil contamination levels in the remaining samples are generally at or near the minimum level of delectability, 0.05 to 0.08 pCi/gm (50 to 80 pCi/kg). Due to the decay of weapons testing ^{137}Cs and its mobility in surface soil and marine sediments, there is no uniform level of cesium contamination in soils unaffected by source points such as MYAPC. The maximum average levels noted above for contamination not derived from MYAPC operations takes into consideration that erosion patterns and silting will often concentrate weapons testing-derived radiocesium in certain locations, especially areas where soils and silt collect after rainfall events.

Maine Yankee Atomic Power Plant Site Characterization

Radiological Engineering & Field Services

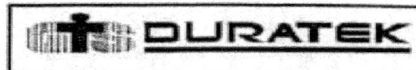

Maine Yankee Atomic Power Plant Site Characterization

04/28/98

CHARACTERIZATION SUMMARY

SURVEY PACKAGE NUMBER R2501   ENVIRONS

PACKAGE DESCRIPTION
Follow-up Surveys

SURVEY AREA DESCRIPTION
Follow-up Sampling at Elevated Soil Sample Locations (north of Forebay and Proposed Dry Cask Storage Area)

GENERAL HISTORICAL INFORMATION (Operational history, etc.)

Package R2501 consists of follow-up evaluations of three locations that have exhibited plant derived radioactive materials in soil samples taken from these locations. Two of these locations were initially identified from gamma spectroscopic results of surface soil samples taken from grids #130 and #122 in survey packages R0900 and R1000, respectively. Limited follow-up soil samples were obtained under these two packages to confirm the initial analytical results and to expand the sampling area somewhat. Because of the findings of these follow-up surveys, i.e., the contamination was found to be more widespread than just at grid stakes #130 and #122, the evaluation of these two locations was broadened to include larger areas and this additional work was performed under Package R2501.

The third location, which is in the Proposed Dry Cask Storage Area, was initially identified by the drive-over gamma scanning surveys using a large plastic scintillator detector. Actually, two elevated areas were initially identified in this manner, but after follow-up surveys of these two flagged areas (areas #7 and 8 on Figure 3) were performed under Package R2500, only flagged location #8 required further investigation. Note that this location #8 corresponds to area #2 on Attachment R2500-6 in Package R2500. Because the evaluation of location #8 was to be broadened to encompass a larger area, given the results of the initial follow-up in Package R2500, this additional work was performed under Package R2501.

SUMMARY OF CHARACTERIZATION ACTIVITIES

(1) Three survey units were established; survey unit 01 consisted of twenty 5-meter by 5-meter grids surrounding grid 130 from Package R0900, survey unit 02 consisted of twenty 5-meter by 5-meter grids surrounding grid 122 from package R1000, and survey unit 03 consisted of twenty-five 5-meter by 5-meter grids surrounding elevated area #2 in survey unit 06 in package R2500. Grid locations are provided in Figures 8 and 9.

(2) Surface soil samples were collected from the 20 grids established in survey unit 01. All 20 samples showed Cs-137 activity greater than 2 pCi/g. The proximity of the Forebay prevented extension of the survey to the south. The surveys were extended west along the fence line for 12 more 5-meter by 5-meter grids (grid numbers 21-33), and surface samples collected from each grid. Elevated Cs-137 activities were found in most of these grids, especially adjacent to the fence line. Four additional 5-meter by 5-meter grids were established (grids 33-36), and surface soil samples collected. Cs-137 activity fell to below 2 pCi/g in these grids.

(3) Surface soil samples were collected from the 20 grids established in survey unit 02 (grids 1-20 on Figure 8). Several of these grids showed Cs-137 activity greater than 2 pCi/g. Six more grids were established east along the fence line (grids 21-26), and surface soil samples collected. Cs-137 activity fell to or below 2 pCi/g in these grids.

(4) In order to determine the depth of contamination in survey units 01 and 02, subsurface (6-12 inch) samples were collected from 3 grids; grids 21, 11, and 15 in survey unit 01. Six to 12 inch depth samples from grids 21 and 11 both showed activity greater than 2 pCi/g. Bedrock was encountered at 12 inches under grid 21, a 12 to 18-inch sample collected from grid 11 showed Cs-137 activity was less than 2 pCi/g. An asphalt

DBACORR Documentation
aProgDBACORR C_HSTRY.RSL
OUTPUT BATCH SN = 226

Survey Package   R2501 ENVIRONS

Evaluation of Soil Samples Exhibiting Elevated Activities

UNIT : 01   SURFACE : OA1   REASON : C01   ANALYSIS TYPE CODE : LAB06

SAMPLE TYPE OR SURFACE SAMPLED: Surface Soil Sample @ 0"-6" Depth
      SAMPLE LOCATOR: 00001

LAB ID	SPECTRUM	MASS (grams)	COUNT TIME (seconds)	NUCLIDE	ACTIVITY (pCi/g)	MDA (pCi/g)	ERROR (± pCi/g)
MY1272	ENV00320	1,600.00	2400	Co-57	< .06	0.06	0.00
				Co-60	.05	0.04	0.02
				Cs-134	< .05	0.05	0.00
				Cs-137	17.80	0.08	1.24
				K-40	21.50	0.38	1.64
				Mn-54	< .04	0.04	0.00

SAMPLE TYPE OR SURFACE SAMPLED: Surface Soil Sample @ 0"-6" Depth
      SAMPLE LOCATOR: 00002

LAB ID	SPECTRUM	MASS (grams)	COUNT TIME (seconds)	NUCLIDE	ACTIVITY (pCi/g)	MDA (pCi/g)	ERROR (± pCi/g)
MY1273	ENV00305	1,720.00	1200	Co-57	< .08	0.08	0.00
				Co-60	< .08	0.08	0.00
				Cs-134	< .08	0.08	0.00
				Cs-137	11.30	0.07	0.86
				K-40	19.60	0.48	1.86
				Mn-54	< .06	0.06	0.00

SAMPLE TYPE OR SURFACE SAMPLED: Surface Soil Sample @ 0"-6" Depth
      SAMPLE LOCATOR: 00003

LAB ID	SPECTRUM	MASS (grams)	COUNT TIME (seconds)	NUCLIDE	ACTIVITY (pCi/g)	MDA (pCi/g)	ERROR (± pCi/g)
MY1274	ENV00314	1,760.00	1200	Co-57	< .07	0.07	0.00
				Co-60	< .06	0.06	0.00
				Cs-134	< .06	0.06	0.00
				Cs-137	8.80	0.09	0.64
				K-40	21.50	0.43	1.80
				Mn-54	< .05	0.05	0.00

DBACORR Documentation
aProgDBACORR.R_GSPEC.RSL

PAJ

347

Characterization Summary for Survey Package Number: R0500
Bailey Point Drive-Over Gamma Scan

Discussion of this portion of the report can be found online in RADNET Section 12: Maine Yankee: Decommissioning Chronicle Continued: March 5, 1999 and the Hide the Evidence Contest. There is also an analysis in RADNET Section 12: Maine Yankee: *Patterns of Noncompliance* Part II. C. 3. c. Survey Package Summaries. The contamination discovered on Bailey Point is particularly significant for a variety of reasons. First of all, it's a great distance from Bailey Point to the reactor vessel where the discovery of a hot particle would be more likely. Secondly, this is a very hot, hot particle as the reporting unit is pCi/gm of ^{60}Co and the reporting level is 33,600 pCi/gm or, to use standard reporting units, 33,600,000 pCi/kg. This survey summary does not tell us how large the particle was (a tenth of a gram or 10 grams?), nor can we really trust the site characterization survey given the large amount of disinformation surrounding the characterization process that additional areas of Bailey Point are also contaminated. The characterization summary notes that all other samples were showing contamination levels at or below (expected) background levels. Also see Secret Survey 2500.

DURATEK

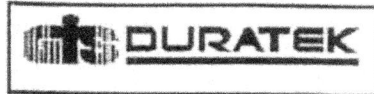

Maine Yankee Atomic Power Plant Site Characterization

04/10/98

CHARACTERIZATION SUMMARY

SURVEY PACKAGE NUMBER :R0500

PACKAGE DESCRIPTION

ENVIRONS

Bailey Point

SURVEY AREA DESCRIPTION
Bailey Point

GENERAL HISTORICAL INFORMATION (Operational history, etc.)

Soil, gravel and asphalt were removed from the Protected Area and placed on Bailey Point. Subsequently, it was found to be contaminated, with dose rates as high as 50 mR/hr. The material was then moved to the contractor trailer area in the Protected Area (refer to package #R0200 for details.).

SUMMARY OF CHARACTERIZATION ACTIVITIES

For Bailey Point (all):

(1) An approximate 100% gamma scan of accessible areas was performed using a large plastic scintillator detector. The locations of the scanning results were identified using a global positioning system (GPS). One location gave an elevated count rate, and it was marked in the field with a flag.

(2) The marked location was investigated with a NaI(Tl) detector to measure the dose rate of the elevated location and to identify the apparent extent of the potential contamination.

(3) Soil samples and exposure rate measurements from this marked location were obtained under Package R2500.

For Bailey Point (Grassy Knoll):

(4) The perimeter of the grassy knoll was marked in approximately 10 meter increments to assist in performing steps (5) and (6).

(5) Thirty grid locations (approximately 10 x 10 meter) were identified by random selection. The location of the 30 random sample points were in addition to the one marked elevated area.

(6) Twenty-nine 6" depth soil samples were obtained from the random sample locations established in step (5), using a sample point offset value of 3 meters north and 7 meters west from the southeast grid corners. The analysis of one random sample was not reported.

(7) At each soil sample location, a 1-meter gamma exposure rate (micro-R/hr) measurement was performed.

(8) Duplicates of surface and subsurface soil samples were obtained under Package 2400 from the two Geoprobe test bore locations in this area.

For Bailey Point (except Grassy Knoll):

(9) The perimeter of the remainder of Bailey Point was marked in approximately 25 meter increments to establish the grids to be sampled. A total of 16 grids were identified.

(10) Sixteen 6" depth soil samples were obtained from the grids identified in step (9).

(11) At each soil sample location, a 1-meter gamma exposure rate (micro-R/hr) measurement was performed.

CHARACTERIZATION SURVEY RESULTS

(1) The Bailey Point drive over gamma scan

349

Figure 3. The follow-up manual gamma scan survey did not find any areas with higher readings. Of the seven 0-6" soil samples taken at this location under Package R2500, the gamma spectroscopic results showed three samples with 8.06, 47.50 and 33,600 pCi/g Co-60. The MicroSpec *in situ* gamma spectroscopic analysis of the flagged location also identified the presence of Co-60. The other surface samples did not show Co-60. The Cs-137 concentrations for these other samples were below typical background values. The one subsurface (6-12") soil sample taken did not show either Co-60 or Cs-137. At the location of the surface soil samples, the exposure rate measurements gave a mean value of 12.4 µR/hr and a range of 9.8 to 22.3 µR/hr.

(2) Forty-five random surface soil samples were obtained from Bailey Point. These samples were in addition to the samples taken to evaluate the elevated location identified by drive-over gamma scanning. The gamma spectroscopic analyses of these 45 samples did not show the presence of Co-60, and the Cs-137 concentrations ranged from 0.00 to 1.03 pCi/g.

(3) Forty-six 1-meter exposure rate measurements from the random grids gave a mean value of 13.3 uR/hr with a range of 10.6 to 19.8 uR/hr.

REFERENCES (Documents, Interviews)

Atherton, P. J. (March 1, 1978). *Maine Yankee fire protection evaluation*. Prepared for the United States Nuclear Regulatory Commission, Washington, D.C.

- "In general plant areas, redundant divisions of safe shutdown/safeguards cables are routed in the same open ladder type aluminum cable trays with an aluminum partition separator. This layout is contrary to all Nuclear Regulatory Commission safety requirements, especially those within Regulatory Guide 1.75." (pg. 1).
- "Equipment required for safe shutdown is located in the turbine building, a non-safety related area." (pg. 1).
- "Redundant divisions of equipment and cabling are located in the same fire area, making them vulnerable to a design basis fire." (pg. 1-2).
- "The use of highly combustible and explosive chemicals throughout the plant appears to be commonplace." (pg. 2).
- "A seismically qualified dedicated safe shutdown system completely independent of all plant areas outside containment is required." (pg. 2).
- "The fire protection system within the control room is judged to be inadequate to prevent functional loss of redundant safe shutdown systems ...The walk-through instrument tunnels and the cable tray risers contain redundant divisions of the same equipment or cables. A fire in these places if not extinguished early may prevent safe shutdown of the reactor." (Control Room, pg. 2).
- "A design basis fire in this room would eliminate the safe plant shutdown capability. Without the low pressure safety injection pumps which also serve as the residual heat removal pumps the plant is unable to achieve cold shutdown." (Containment Spray Pump Area, pg. 1-2).

350

- "A design basis fire in this room could become large enough to damage redundant divisions of electric cable and collapse the aluminum cable trays. This fire will damage cabling essential to safely shutdown the reactor." (Protected Cable Tray Room, pg. 1).
- "Most of the combustibles are located on the ground floor. Some of these combustibles are lube oil, drums of cotton clothing and rubber wear, wood, oxygen-acetylene units, cabling, wax, wax stripper, sealant, cleaner, waste oil and hydrogen gas ... The turbine building contains a high heat load with a potential for collapsing the building. The complete loss of the component cooling water pumps and service water pump cabling would leave no way of achieving safe shutdown." (Turbine Building, pg. 1-2).
- This recently rediscovered report from 1978 addresses the need to separate redundant cables essential to safe plant shutdown and other safety issues. The failure of the NRC to address these issues in 1978 emphasizes the long duration of unsafe reactor operation which continued unnoticed during the recent Independent Safety Assessment Team analysis.

Christine, K. (March 8, 1997). *1978 Dangerous Year for Maine Residents*. Personal communications to an unidentified journalist.

- The following email message from Kris Christine to a Maine journalist and the Center for Biological Monitoring is reprinted (without permission) because it provides a concise summary of a key component of the collapse of the MYAPC pyramid scheme: the unreliability of NRC and state of Maine assertions about the past and present safety of this aging facility.

> I was just going through some documents and suddenly realized what an extraordinarily dangerous year 1978 was for the citizens of Maine. In March 1978, Peter Atherton identified and reported significant cable separation issues throughout Maine Yankee. Maine Yankee did not reroute these cables, coat them with fire suppressive sealant, or install the Protectowire detection system they proposed themselves.
>
> Mr. Atherton was fired from NRC. Then Maine Yankee was rewarded by NRC with a power upgrade on May 10, 1978 allowing the plant to operate at 2560 MWt. If you recall the ISAT findings, the emergency core cooling system equipment was not demonstrated to be operable at power levels above 2440 MWt. So, in its mandated duty to regulate licensees and protect the public health and welfare, NRC not only allowed Maine Yankee to operate with a fire hazard [that] could wipe out primary and redundant safety-related cables, but they granted them a power upgrade

allowing them to exceed the margins for the containment spray system, the high pressure safety injection system, residual heat removal, service water and component cooling water systems -- all of which are necessary to mitigate the consequences of an accident.

For nearly 19 years, NRC has allowed Maine Yankee to operate with an inadequate emergency core cooling system. They also approved and licensed this facility to operate with improperly routed safety-related cabling. Maine Yankee has posed a significant and undue risk to public safety since the first day of operation!

Maine Yankee Radioactive Waste Reclassification

Please note, the following dates from 1999.

The recent closure of the Maine Yankee Atomic Power Company in Wiscasset, Maine, and the Connecticut Yankee facility at Haddam Neck, Connecticut, and the ongoing decommissioning of the Trojan Nuclear Power Plant near Portland, Oregon highlight the urgent necessity for ending the deceptive radioactive waste classification systems now used by the NRC, DOE, and EPA. These systems allow disposal of highly radioactive reactor vessel internal components and operational reactor-derived resins as "low-level" wastes. These class B, C, and GTCC wastes should not be disposed of in an uncontained formats in landfills in any location including Sierra Blanca, Texas (the Texas LLW Compact), Ward Valley, California, and Barnwell, South Carolina (the mother of all uncontained landfills). A further controversial and deceptive component is misleading and obfuscating federal regulations which allow plutonium laced transuranic wastes (TRU) and other long-lived wastes to be sited as low-level waste.

It is now time to reformulate radioactive wastes classifications because of the imminent decommissioning of the reactors in New England and the plan to site GTCC wastes in the Trojan Washington reactor vessel in an uncontained landfill at Hanford. Other events that make it time to reformulate radioactive waste classifications include the recent recommendations from the Institute for Energy and Environmental Research for nuclear waste management and classification reform and the growing opposition in California and elsewhere to the deceptions involved in burial of these long-lived and high-level wastes. The first priority is to separate class A low-level wastes especially including radiopharmaceuticals and other non-nuclear reactor-derived short-lived wastes from the more dangerous class B, C, GTCC, and long-lived wastes and allow the shorter-lived wastes to be disposed of as low-level wastes in either a landfill situation or a monitored retrievable storage facility.

The deceptive inclusion of commercial nuclear reactor and weapons production-derived wastes in the current nuclear waste management disposal scheme threatens the legitimate use of radioisotopes by research institutes, hospitals, universities, and other laboratories. These medical and research waste generators are, in essence, being held hostage to an obsolete and misleading waste classification system, which makes no sense, and which results in a lack of available facilities to accept low risk, short-lived radioactive wastes that do not pose any significant health physics threat to the public. As a result, radioactive waste disposal options are now limited to a few active landfills in Barnwell, SC, Betty, NV, and Hanford, WA, where class A wastes and the more radiotoxic class B, C, and GTCC wastes may still be sited. As decommissioning of a number of commercial reactors begins and the public becomes more aware of the "hot C scam," it is unlikely that these sites will remain as long-term options for radioactive waste disposal. The continued use of this deceptive radioactive waste classification system threatens the future legitimate monitored retrievable storage and/or disposal of class A low-level wastes at Barnwell, Betty, and possibly Ward Valley, which would be a viable site only if retrievable storage technology were implemented.

For an interesting, informative, and up-to-date discussion of nuclear waste management and disposal issues, including a complete listing of all federal categories of radioactive wastes, readers are urged to review Appendix B of *Containing the Cold War Mess* by Marc Fioravanti, Arjun Makhijani, and Steve Hopkins published by IEER (1999). This is by far the most comprehensive description of the waste classification debacle available from any source.

Specific suggestions for a five tiered classification system have also been submitted to the California legislature by PARDNERS (People Against Radioactive Dumping) who are responding to the deceptive attempt to dump uncontained class B, C, and possibly GTCC wastes at the proposed Ward Valley low-level waste site in southern California. PARD suggests classifying wastes as:

- Type A: medical wastes and short decay life (180 days) wastes.
- Type B: unused recyclable wastes with storage limited to one year after which it is reclassified as C, D, or E waste.
- Type C: 30 year wastes, that is, wastes with a decay life of 180 days to 30 years.
- Type D: 100 year wastes, that is, waste with a 30 to 100 year decay life.
- Type E: high-level wastes including spent fuel and other long-lived wastes.

This organization also suggests a limit on a number of curies per gram without mixing or diluting. The PARD proposal does not clarify what is meant by decay life, but it is usually considered that after 10 half-lives an isotope has decayed to a point where it is no longer a health physics threat. Therefore, an isotope such as Cs-137 (1/2 T = 30 years) would have a decay life of 300 years, and, therefore, would be considered a high-level waste type E.

It is not yet clear how these various proposals can be combined into a comprehensive reform of the radioactive waste classification system. However, growing public awareness of the ongoing and upcoming decommissioning debacles at various federal remediation sites and commercial reactor facilities mandate this waste classification reform. As of the present time, legitimate class A medical and research waste disposal is being held hostage to an obsolete potpourri of federal waste classifications.

November 21:

Decommissioning cost analysis for the Maine Yankee Atomic Power Station (Maine Yankee Atomic Power Company October 1997a) is the third in a series of three decommissioning cost analyses made by TLG for MYAPC. Decommissioning cost estimates remain essentially unchanged from the previous estimates made in 1993, with the exception of new provisions made for onsite storage of spent fuel in the form of an independent spent fuel storage installation (ISFSI). ISFSI siting and construction costs are estimated at $52,249,000.00, but the TLG study contains no details about the design of the ISFSI or a separate cost estimate for the dry casks which would be used as a component of this facility. Major questions still remain about whether current designs for dry cask storage of spent fuel are compatible with the monitored retrievable storage facility now proposed for the Nevada desert as a temporary site for commercial spent fuel. The TLG report not only does not provide any detail about the number and costs of dry casks, there is no mention as to whether the ISFSI will be in the form of ready-to-transport multipurpose canisters (MPC). This is consistent with current indecision by federal regulators (NRC and the DOE) as to what will be the best cask design suitable for both onsite and offsite spent fuel storage. Previous efforts by the Department of Energy to sponsor the development of a Westinghouse-designed MPC were terminated by congressional mandate in 1995. MPC development has now been "privatized" and the lack of detail in the TLG report reflects the uncertain status of multipurpose canister design. Existing dry cask may not be the most suitable for transportation of spent fuel to a monitored retrievable storage facility or for final disposal in a geological repository.

Section 2 of this report provides a disingenuous description of the process of segmenting GTCC reactor vessel internals from the reactor vessel and their storage with spent fuel in the ISFSI after they have been packed in fuel bundle canisters (FBC). This is the "worst case scenario" for GTCC reactor vessel internal components, according to MYAPC officials at the recent November 6, public hearing in Wiscasset, ME. The more likely alternative to reactor vessel internals segmentation, siting the entire reactor vessel as one class C "package" in uncontained land burial in Barnwell, SC, will save millions

354

of dollars in labor and transportation costs. The TLG report makes no explicit reference to this cheaper scenario for GTCC waste disposal. In the 1987 TLG decommissioning estimate, the greater than class C reactor vessel internals (229 cu ft, +4 million Ci) were to be divided into 100 shipments and sited as low-level waste, also at the Barnwell facility. The proposal for the Texas Compact evolved after Barnwell was closed for a period of years; the re-opening of the Barnwell facility has undermined the viability of the Texas low-level waste site, which is too distant to have much practical use for the *large component removal projects* (steam generators, reactor vessel). In the event that the reactor vessel is sited as one intact unit, the 706 ton weight will be augmented by the necessity of welding a 2 inch thick steel protective shield around the entire reactor vessel to reduce radiation exposure from the millions of curies of GTCC internal components inside. The TLG report provides no discussion of this alternative method of reactor vessel disposal, which is referenced in the MYAPC PSDAR (see discussion above). The most glaring difference between this cost analysis and the 1987 TLG report is the lack of an equipment specific radiological inventory. The 1987 TLG report contained a very specific description of the reactor vessel and its internal components as well as the radioactivity and weight of these components. The only reason why this more specific description of the reactor vessel is available to the general public is that it was accidentally reproduced as an appendix in an obscure State of Maine report (Vanags 1992). The 1997 TLG report represents a continuation of the deceptions and the deceitfulness that has characterized MYAPC operations from their inception. It is unlikely that the extremely modest decommissioning cost estimates in this report, based in part on antiquated worker duration schedule estimates, can be executed as described in Table 5.1, which is reproduced below. One of many flies in the ointment is the fact that MYAPC has yet to receive the Duratek site characterization of existing radiological contamination, the preliminary plan for which is discussed in the citation.

- "The objective of the study is to prepare a comprehensive estimate of the cost, a detailed schedule of the associated activities, and the resulting volume of low-level radioactive waste generated in decommissioning the Maine Yankee plant." (Section 1, pg. 1).

- "Following the transfer of the spent fuel inventory from the fuel pool, the ISFSI will continue to operate independently. Transfer of spent fuel to a DOE or interim facility will be exclusively from the ISFSI once the spent fuel storage pool has been emptied and the structure released for decommissioning. Assuming initiation of the Federal Waste Management System in 2010, this study assumes that the DOE will be able to complete the transfer of spent fuel from the Maine Yankee site by the year 2023." (Section 2, pg. 12).

- "Low-level radioactive waste generated in the decontamination and dismantling of the Maine Yankee plant is assumed to be destined for the Barnwell facility. Current rates of disposal were used. These include a unit disposal cost ranging from $4.50 to $7.50 per pound of waste and a $0.30 per millicurie surcharge. The waste stream is assumed to be conditioned to the maximum extent possible, e.g., through decontamination, volume reduction, incineration, etc., so as to avoid the high cost of direct disposal. Contaminated soil is assumed to be sent to the Envirocare disposal facility in Clive, Utah at a disposal charge of $87 per cubic foot." (Section 3, pg. 1).

- "The dismantling of reactor internal components at the Maine Yankee plant will generate radioactive waste generally unsuitable for shallow land disposal. This waste is generally referred to as 'Greater-than-Class-C' (GTCC). Although the material is not classified as high-level waste, DOE has indicated it will accept title to this waste for disposal at the future high-level waste repository (Ref. 8). However, the DOE has not yet established an acceptance criteria or a disposition schedule for this material, and numerous questions remain as to the ultimate disposal cost and waste form requirements. As such, for purposes of this study, the GTCC waste will be packaged and disposed of as high-level waste at a cost equivalent to that envisioned for the spent fuel." (Section 3, pg. 10).

- "Once at the storage area, each generator will have a two-inch thick carbon steel membrane welded to its outside surface for shielding during transport. The generators will be moved to a barge loading area where the generators on the multi-wheeled transporter will be barged down the east coast to the Barnwell facility. The barge can move up the Columbia River in South Carolina to a point approximately 30 miles from the disposal facility. From there, the generators will be moved [to] the burial facility with the multi-wheeled transporter." (Section 3, pg. 11).

- "All low-level radioactive waste generated in the decontamination and dismantling of [the] Maine Yankee plant, with the exception of contaminated soil, is assumed destined for disposal at the Barnwell facility. Current base disposal rates were used. This consists of a sliding scale based on the packaging density of the waste as follows:

  - Greater than 75 lbs/ft^3 and less than 120 lbs/ft^3 density $4.50 per pound
  - Between 60.1 lbs/ft^3 and 75 lbs/ft^3 density $5.50 per pound
  - Between 45 lbs/ft^3 and 60 lbs/ft^3 density $7.50 per pound

A surcharge of $0.30 per millicurie of waste is also assessed, with a maximum charge of $120,000 per shipment." (Section 3, pg. 12).

- "The reactor vessel and internal components are expected to be transported in accordance with §71, as Type B. *It is conceivable that the reactor, due to its limited specific activity, could qualify as LSA II or III.* However, the high radiation levels on the outer surface would require that additional shielding be incorporated with the packaging so as to attenuate the dose to levels acceptable for transport." (Section 3, pg. 15).

- "The cost to remove and dispose of the spent fuel from the site is not reflected within the estimate to decommission the Maine Yankee plant. Ultimate disposition of the spent fuel is the province of the DOE's Waste Management System, as defined by the Nuclear Waste Policy Act. Any delay in the transfer of spent fuel would increase the on-site management costs." (Section 3, pg. 16).

- The tables reproduced below are from Section 3, pg. 20; Section 5, pg. 3 and pg. xii. The columns may not add exactly due to rounding.

Table 3.1
SCHEDULE OF DECOMMISSIONING EXPENDITURES
(thousands of 1997 dollars)

Year	Period 1 Preparations	Period 2 Decommissioning	Period 3 Site Restoration	Post Period 3 Dry Fuel Storage	Totals
1997	0				0
1998	31,843				31,843
1999	45,542	29,190			74,732
2000		81,494			81,494
2001		80,299			80,299
2002		69,630			69,630
2003		66,026			66,026
2004		2,896	19,647		22,543
2005			6,249	2,576	8,825
2006 to 2022				3,702 or 3,712	3,702 or 3,712
2023				9,861	9,861
Total	77,385	329,536	25,896	75,404	508,221

357

Table 5.1
DECOMMISSIONING RADIOACTIVE WASTE BURIAL VOLUMES

	Waste Class[1]	Volume[2] (Cubic feet)
	A	179,263
	B	14,561
	C	1,564
	>C	227
Total		195,615

[1] Waste is classified according to the requirements as delineated in Title 10 of the code of Federal Regulations, Part 61.55

[2] Columns may not add due to rounding.

X. Links

ANS (American Nuclear Society): "A not-for-profit, international, scientific and educational organization. It was established by a group of individuals who recognized the need to unify the professional activities within the diverse fields of nuclear science and technology." (www.new.ans.org/)

CAN (Citizens Awareness Network): Here you can find information on New England area nuclear power plants: Yankee Rowe decommissioning, CT Yankee closure, VT Yankee shroud repair; NRC related hearings, FSAR petition, etc. (www.nukebusters.org/)

CCNR (Canadian Coalition for Nuclear Responsibility): This group is dedicated to education and research on all issues related to nuclear energy, whether civilian or military -- including non-nuclear alternatives -- especially those pertaining to Canada. The website contains an extensive publications list and information on the Candu reactors. (www.ccnr.org/index.html)

CCNS (Concerned Citizens for Nuclear Safety): An organization focused on increasing public awareness concerning the issues posed by radioactivity and the nuclear industry, this is a premier information source providing information about and links to WIPP, Los Alamos National Laboratory, and Interstate Nuclear Services. Following a three year federal lawsuit, CCNS became the sole authorized citizen monitoring group of a national laboratory (LANL). CRCPD (Conference of Radiation Control Program Directors, Inc.): A "nonprofit professional organization whose primary membership is made up of individuals in state and local government who regulate the use of radiation sources." (www.nuclearactive.org)

Chernobyl Information:

> **ICRIN (International Chernobyl Research and Information Network)**: "This Portal is the pilot version of international Internet-resource, which presents information on the radiological aspects of safe and secure dwelling of population on territories affected by the Chernobyl NPP catastrophe." (www.chernobyl.info/Default.aspx?tabid=120)

> - Report on the Chernobyl accident: (www.un.org.ua/files/ICRIN_Report-2004.pdf)

> **Chernobyl Radioactive Disaster (Belarus)**: "This file is an attempt to compile information on Chernobyl disaster and its influence on Belarus." (http://popcorn-km.blogspot.com/2011/03/belarusian-chernobyl-tragedy.html)

CND (Campaign for Nuclear Disarmament): Venerable British organization working for a nuclear-free Great Britain. It is dedicated to campaigning for the abolition of nuclear weapons and contains no information about source points of anthropogenic radioactivity. (www.cnduk.org/)

CNR (Committee for Nuclear Responsibility): "CNR is a non-profit, educational group organized to provide independent analyses of the health effects and sources of ionizing radiation." One of the largest and most complicated environmental information websites. Sponsored by the RatHaus, this site includes a section on the health costs of low-level ionizing radiation including the pioneering work on radiation hazards conducted by the committee chairman, Dr. John Gofman M.D., Ph.D., author of Radiation and Human Health, the most important book ever published on the health impact of anthropogenic radioactivity. This site also has extensive information on nuclear technology and links to nuclear, renewable energy, and other information sources and publications of every description. (www.ratical.org/radiation/CNR/)

CRCPD (Conference of Radiation Control Program Directors, Inc.): A "nonprofit professional organization whose primary membership is made up of individuals in state and local government who regulate the use of radiation sources." (www.crcpd.org/Radon.asp)

Deadly Nuclear Radiation Hazards USA: This site includes an important database listing all US anthropogenic radioactivity source points, small and large. (www.prop1.org/prop1/radiated/drh.htm)

EFN (Environmentalists For Nuclear Energy): "A totally independent *(sic)* environmental non-profit organization which aims at developing information to the public on energy and the environment, promoting the benefits of nuclear energy for a cleaner world, & uniting people in favor of clean nuclear energy." (www.ecolo.org/)

EnviroWatch: This is an organization that researches and exposes environmental abuse. (www.envirowatch.org/)

FAS (Federation of American Scientists): The Federation of American Scientists is engaged in analysis and advocacy on science, technology and public policy for global security. Their projects include Arms Sales Monitoring Project, Biological and Toxin Weapons Project, Military Analysis Network, Space Policy Project, Nuclear Non-Proliferation and Disarmament, North Korea and Non-Proliferation as well as many non-nuclear topics. (www.fas.org)

FedWorld Information Network: Available in either FTP or TELNET formats, this US government information directory is a gateway to the most important US government databases. Sponsored by the National Technology Information Service, an

360

agency of the US Department of Commerce, it provides access to over 130 government dial-up bulletin boards. (www.fedworld.gov/index.html)

FERN (Food Emergency Response Network): "FERN integrates the nation's food-testing laboratories at the local, state, and federal levels into a network that is able to respond to emergencies involving biological, chemical, or radiological contamination of food." (http://www.fernlab.org/)

FOE (Friends of the Earth): "Friends of the Earth is fighting to defend the environment and create a more healthy and just world." (www.foe.org)

> **FOEI (Friends of the Earth International)**: The Eastern European nuclear power plants (Mochovce, etc.) are one among FOE's numerous areas of interest. The website contains links to the Siemens campaign and many European anti-nuclear organizations. (www.foei.org)

Greenpeace: The world's largest and most important environmental organization. Alerts, articles, audio, "one of the best websites of any kind anywhere." (NIRS). (www.greenpeace.org/usa)

> **Greenpeace International**: (www.greenpeace.org/international)

IEER (Institute for Energy and Environmental Research): Another important nuclear information source, the primary focus of IEER is on environmental and security issues relating to nuclear weapons production and testing. (www.ieer.org/) (www.ieer.org/fctsheet/radiationhealthfactsheet_2011.pdf)

Indigenous Environmental Network: This site includes information about Ward Valley, depleted uranium, and other nuclear hot topics. (www.ienearth.org/)

INSC (International Nuclear Safety Center of Ukraine): "A comprehensive resource database for safety analysis and risk evaluation of nuclear power plants and facilities... sponsored by the US Department of Energy (DOE). Although the scope of the database is world-wide, the current focus is on Soviet-designed nuclear power plants in Russia and Eastern Europe, and on reactor types in China and India." (www.insc.gov.ua/)

IRSN (Institut de Radioprotection Nucleaire): "Following the earthquake and tsunami that hit Japan on March 11, 2011, IRSN has published information on its French website regarding the status of the nuclear reactors at the Fukushima Daiichi nuclear plant, and the consequences of the crisis." (www.irsn.fr/EN/Pages/home.aspx)

ISIS (Institute for Science and International Security): "Employing Science in the Pursuit of International Peace." (www.isis-online.org/)

NECONA (National Environmental Coalition of Native Americans): This website contains information on radioactive waste and transportation, especially as it affects Native American communities with lots of links to Native American groups. (http://necona.indigenousnative.org/)

NILU (Norwegian Institute for Air Research): "Special forecast products for Fukushima produced by NILU-ATMOS." (http://transport.nilu.no/products/fukushima)

NIRS (Nuclear Information Resource Service): One of the most important nuclear information sources, the NIRS site includes access to their important newsletter, alerts, NIRS net campaigns, NIRS net fact sheets, press releases, information on nuclear issues as well as alternative energy, and extensive links to other sites. (www.nirs.org/)

NRC (Nuclear Regulatory Commission): The US government oversight for all things nuclear in the US; includes a handy map of all nuclear facilities in the US. (www.nrc.gov/)

NRDC (Natural Resources Defense Council): One of the largest Internet environmental information sources, this page is divided into four sites: environmental news and information, *Amicus Journal*, technical information, and links to "green spots" on the net. It includes extensive information on nuclear weapons and energy, with a report on the environmental legislation of the 104th Congress and a sub-section on nuclear program publications. Also includes the very important work by Thomas Cochran, *Nuclear Weapons Databook* series, among many other topics. (www.nrdc.org/)

PhysicalGeography.Net: This link is to a map that shows the usual direction of ocean currents. (www.physicalgeography.net/fundamentals/8q_1.html)

PSR (Physicians for Social Responsibility): "Physicians for Social Responsibility is committed to the elimination of nuclear and other weapons of mass destruction, the achievement of a sustainable environment, and the reduction of violence and its causes." (www.psr.org)

> PSR is the US affiliate of **IPPNW (International Physicians for the Prevention of Nuclear War)**. (www.ippnw.org/)

Public Citizen's Critical Mass Energy Project: Information on nuclear power, reactor safety, radioactive waste, and other energy issues, this organization is one of the most politically active energy information organizations and the publisher of the *Annual Report on Nuclear Lemons*. (www.citizen.org/Page.aspx?pid=183)

RadWaste.org: "WasteLink, your guide to radioactive waste related material on the Internet... provid[ing] a reference source for radioactive waste management professionals." (www.radwaste.org/index.html)

Rathaus: "Rathaus is the German word for city hall. Here, the sense of city as community, is blended with library, a gathering place for people to come and explore issues and illuminate the space we live within." (www.ratical.com/rat_haus.html)

- **Ratitor's corner**: "Independent Watchdogs: An Antidote to Nuclear Pollution" (www.ratical.com/ratitorsCorner/)

Redwood Alliance: Originally formed in 1978 to shut down the Humboldt Bay Nuclear Power Plant, this alliance now covers issues ranging from nuclear waste to renewable energy, with a focus on political action. (www.redwoodalliance.org/)

UCS (Union of Concerned Scientists): This venerable organization was the first to raise important safety issues about US nuclear plants. Current topics of this site include: agriculture, arms control, energy, global resources, and transportation; current publications include important case studies by Robert Pollard on U. S. nuclear power plants and the *Bulletin of the Atomic Scientists*. (www.ucsusa.org/)

US DOE (US Department of Energy): This department has many sub-agencies. (http://www.energy.gov)

- DOE report on why the US called for a 50 mile evacuation around Fukushima Daiichi (currently, the subject of FOIA request by FOE, NIRS, and PSR due to suspicion that it is incomplete). (http://www.energy.gov/news/10194.htm)
  - FOIA request on DOE report. (www.foe.org/sites/default/files/FOE-NIRS-PSR-RadiationFOIA-3-22-2011.pdf)
- DOE blog on Japanese accident. (blog.energy.gov/content/situation-japan/)
- National Nuclear Security Administration. (www.nnsa.energy.gov/)
- The Office of Health, Safety, and Security (HSS). (http://www.hss.energy.gov/)
  - A Guide to Archival Collections Relating to Radioactive Fallout from Nuclear Weapon Testing. (http://www.hss.doe.gov/healthsafety/ohre/new/findingaids/radioactive/index.html)

WISE (World Information Service on Energy): An international switchboard for local and national safe energy groups around the world who want to exchange information and support one another. The primary focus is on topics related to the nuclear industry, such as new developments in policy and technology, nuclear proliferation, accidents, and activities of local movements. (www10.antenna.nl/wise/index.html)

YouTube.com: This video shows a model of the radioactive dispersion from the airborne plume (wind was moving east, obviously) beginning on March 12, 2011. (www.youtube.com/watch?v=qHbQZQygrag&feature=player_embedded)

XI. General Bibliography

Aarkrog, A., Buch, E., Chen, Q. J., Christensen, G. C., Dahlgaard, H., Hansen, H., Holm, E. and Nielsen, S. P. (July 1989). *Environmental radioactivity in the North Atlantic region. The Faroe Islands and Greenland included. 1987*. Riso-R-564. Riso National Laboratory, Roskilde, Denmark.

Aarkrog, A., Botter-Jensen, L., Jiang, Chen Quing, Dahlgaard, H., Hansen, H., Holm, E., Lauridsen, B., Nielsen, S.P., Strandberg, M. and Sogaard-Hansen, J. (1992). *Environmental radioactivity in Denmark in 1990 and 1991*. Riso National Laboratory, Roskilde, Denmark.

Aarkrog, A. (1994). *Source terms and inventories of anthropogenic radionuclides*. Riso National Laboratory, Roskilde, Denmark.

Aarkrog, A., Botter-Jensen, L., Chen, Q. J., Clausen, J., Dahlgaard, H., Hansen, H., Holm, E., Lauridsen, B., Nielsen, S. P., Strandberg, M. and Sogaard-Hansen, J. (February 1995). *Environmental radioactivity in Denmark in 1992 and 1993*. Riso-R-756(EN). Riso National Laboratory, Roskilde, Denmark.

Andrews, Anthony. (2004). *Spent nuclear fuel storage locations and inventory*. CRS Report for Congress. http://ncseonline.org/nle/crsreports/04dec/RS22001.pdf.

BB-Gibson, Pamela Reed. (2000). *Multiple chemical sensitivity: A survival guide*. New Harbinger Publications, Oakland, CA.

Bailik, Carl. (March 23, 2011). Radiation math: How do we count the rays? *Wall Street Journal*.

Baldash, Lawrence. (1979). *Radioactivity in America: Growth and decay of a science*. Johns Hopkins University Press, Baltimore, MD.

Belson, Ken and Myers, Steven Lee. (April 18, 2011). Tokyo utility lays out plan for reactors. *The New York Times*. pg. A1.

Bertell, Rosalie. (2000). *Host response to depleted uranium*. International Institute of Concern for Public Health, Toronto, Canada. http://iicph.org/host_response_to_du.

Blumenthal, Susan. (March 23, 2011). Radiation and health: The aftershocks of Japan's nuclear disaster. *Huffington Post*. http://www.nytimes.com/2011/03/18/world/asia/18spent.html?scp=1&sq=3/18/2011&st=cse.

Brack, H. G. (1984). *RADSCAN: Information sampler on long-lived radionuclides*. Pennywheel Press, Hulls Cove, ME.

Brack, H. G. (1986). *A review of radiological surveillance reports of waste effluents in marine pathways at the Maine Yankee Atomic Power Company at Wiscasset, Maine--- 1970-1984: An annotated bibliography.* Pennywheel Press, Hulls Cove, ME.

Brack, H. G. (1993). *Legacy for our children: The unfunded costs of decommissioning the Maine Yankee Atomic Power Station.* Pennywheel Press, Hulls Cove, ME.

Brack, H. G. (1998). *Patterns of noncompliance: The Nuclear Regulatory Commission and the Maine Yankee Atomic Power Company: Generic and site-specific deficiencies in radiological surveillance programs.* Center for Biological Monitoring, Hulls Cove, ME.

Brack, H. G., Ed. (2009). *Chernobyl fallout data: Annotated bibliography.* Extracted from Section 10 of RADNET. www.davistownmuseum.org/cbm/Rad7.html.

Bradley, David. (1983). *No place to hide: 1946/1984.* University Press of New England, Hanover, NH.

Bryant, P. M. and Jones, J. A. (December 1972). *The future implications of some long-lived fission product nuclides discharged to the environment in fuel reprocessing wastes.* NRPB-R8. National Radiological Protection Board, Harwell, Didcot, Berkshire, England.

Caldicott, Helen. (1978). *Nuclear madness: What you can do!* Bantam Books, Inc., NY.

Camplin, W. C. and Aarkrog, A. (1989). *Radioactivity in north European waters: Report of Working Group II of CEC Project MARINA.* Fisheries Research Data Report No 20. Directorate of Fisheries Research, Ministry of Agriculture, Fisheries and Food, Lowestoft, England.

Center for Biological Monitoring. (1998). *Nuclear dada and the traffic in nuclear waste: The Nuclear Regulatory Commission and the Maine Yankee Atomic Power Company: Patterns of noncompliance: Generic and site-specific deficiencies in radiological surveillance programs.* Pennywheel Press, Hulls Cove, ME.

Centers for Disease Control. (February 1999). *Savannah River Site (SRS) dose reconstruction.* Radiation Studies Branch, National Center for Environmental Health, CDC, Atlanta, GA.

Crowther, James Arnold. (1934). *Ions, electrons, and ionizing radiations.* Edward Arnold & Co., London.

D'Agata, John. (2010). *About a mountain.* W. W. Norton & Company, NY.

Dotto, Lydia. (1986). *Planet earth in jeopardy: Environmental consequences of nuclear war.* John Wiley and Sons, NY.

Eisenbud, Merril. (1987). *Environmental radioactivity*. Fourth edition. Academic Press, Orlando, FL.

European Commission. (1996). *The radiological consequences of the Chernobyl accident.* EUR 16544 EN. European Commission, Brussels, Belgium.

European Commission. (1998). *Atlas of caesium deposition on Europe after the Chernobyl accident*. European Commission. EUR 19810 EN RU. European Commission, Brussels, Belgium.

Fairlie, Ian and Sumner, David. (2006). *The other report on Chernobyl (TORCH): An independent scientific evaluation of the health-related effects of the Chernobyl nuclear disaster with critical analyses of recent IAEA/WHO reports*. The Greens/EFA Party in the European Parliament and the Altner-Combecher Foundation. http://www.chernobylreport.org/torch.pdf.

Farber, S. A. and Hodgdon, A. D. (July 25, 1991). Cesium-137 in wood ash: Results of nationwide survey. PP 91-015. *Presented at the Annual Meeting of the Health Physics Society*. Yankee Atomic Electric Company, Boston, MA.

Federal Radiation Council. (1960-61). *Background material for the development of radiation protection standards: Reports 1 and 2*. US Government Printing Office, Washington, D.C.

Ford, Daniel F. (1982). *The cult of the atom: The secret papers of the Atomic Energy Commission*. Simon and Schuster, NY.

Ford, Daniel F. and Kendall, Henry W. (1974). *An assessment of the emergency core cooling systems rulemaking hearings.* Union of Concerned Scientists, Cambridge, MA.

Gedikoglu, A. and Sipahi, B. L. (January 1989). Chernobyl radioactivity in Turkish tea. *Health Physics*. 56(1). pg. 97-101.

Glanz, James and Broad, William J. (April 5, 2011). US sees new threats at Japan's nuclear plant. *The New York Times.* http://www.nytimes.com/2011/04/06/world/asia/06nuclear.html?scp=1&sq=%22U.S.%20Sees%20New%20Threats%20at%20Japan's%20Nuclear%20Plant%22&st=cse.

Glasstone, Samuel and Jordan, Walter H. (1980). *Nuclear power and its environmental effects.* American Nuclear Society, La Grange Park, IL.

Gofman, J.W. (1981). *Radiation and human health; a comprehensive investigation of the evidence relating low-level radiation to cancer and other diseases*. Sierra Club Books, San Francisco, CA.

Goldman. (1987). *Health and environmental consequences of the Chernobyl nuclear power plant accident*. Report ER-0332. Prepared by the Interlaboratory Task Group, US Department of Energy, Washington, DC.

Grossman, Karl. (1997). *The wrong stuff: The space program's nuclear threat to our planet*. Common Courage Press, Monroe, ME.

Hanson, W. G., Ed. (1980). *Transuranic elements in the environment*. DOE TIC 22800. US Dept. of Energy, Washington, D.C.

Hardy, E. P., Jr. (July 1, 1980). Environmental Measurements Laboratory: *Environmental Quarterly*. EML-374. Department of Energy, NY.

Hardy, E. P., Jr. (October 1, 1980). Environmental Measurements Laboratory: *Environmental Quarterly*. EML-381. Department of Energy, NY.

Hardy, E. P., Krey, P. W. and Volchok, H. L. (February 1973). Global inventory and distribution of fallout plutonium. *Nature*. 241. pg. 444-445.

Helus, Frank, Ed. (1983). *Radionuclides production: Volume I*. CRC Press, Inc., Boca Raton, FL.

Hoopes, Roy. (1962). *A report on fallout in your food with tables and illustrations*. The New American Library, NY.

Hunt, G. J. (1987). *Aquatic environment monitoring report number 18: Radioactivity in surface and coastal waters of the British Isles, 1986*. Ministry of Agriculture, Fisheries and Food, Directorate of Fisheries Research, Lowestoft, England.

Ikaheimonen, T. K., Ilus, E. I. and Saxen, R. (1988). *Finnish studies on radioactivity in the Baltic Sea in 1987: Supplement 8 to Annual Report 1987*. No. STUK-A74. Report No. STUK-A82. Finnish Centre for Radiation and Nuclear Safety, Helsinki, Finland.

Ilus, E., Klemola, S., Sjoblom, K. L. and Ikaheimonen, T. K. (1988). *Radioactivity of Fucus vesiculosus along the Finnish coast in 1987: Supplement 9 to Annual Report 1987 (STUK-A74)*. Report No. STUK-A83. Finnish Centre for Radiation and Nuclear Safety, Helsinki, Finland.

International Commission on Radiological Protection. (1977). Recommendations of the International Commission on Radiological Protection. ICRP Publication 26. *Annals of the ICRP*. 1(3).

International Commission on Radiological Protection. (1989). *Age dependent doses to members of the public from intake of radionuclides: Part I*. ICRP publication No. 56. Pergamon Press, Oxford.

International Physicians for the Prevention of Nuclear War and the Institute for Energy and Environmental Research. (1991). *Radioactive heaven and earth: The health and environmental effects of nuclear weapons testing in, on, and above the earth.* The Apex Press, NY.

Lash, Terry R., Bryson, John E. and Cotton, Richard. (November 1975). *Citizens' guide: The national debate on the handling of radioactive wastes from nuclear power plants.* National Resources Defense Council, Inc., Palo Alto, CA.

Linsley, G. S., Simmonds, J. R. and Kelly, G. N. (December 1978). *An evaluation of the food chain pathway for transuranium elements dispersed in soils.* NRPB-R81. National Radiological Protection Board, Harwell, Didcot, Oxfordshire, England.

Lochbaum, David A. (1996). *Nuclear waste disposal crisis.* PennWell Books, Tulsa, OK.

Long, Michael E. (July 2002). Half-life: The lethal legacy of America's nuclear waste. *National Geographic.* pg. 2-33.

MacKenzie, A. B. (April 2000). Environmental radioactivity: Experience from the 20th century – trends and issues for the 21st century. *The Science of the Total Environment.* 249(1-3). pg. 313-29.

Maddock, Shane J. (2010). *Nuclear apartheid: The quest for American atomic supremacy from World War II to the present.* University of North Carolina Press, Chapel Hill, NC.

Makhijani, A. and Fioravanti, M. (January 1999). Cleaning up the Cold War mess. S*cience for Democratic Action.* IEER. 7(2). pg. 1-24.

Maine Yankee Atomic Power Company. (October 1997a). *Decommissioning cost analysis for the Maine Yankee Atomic Power Station.* Document No. M01-1258-002, prepared for the Maine Yankee Atomic Power Company by TLG Services, Inc., Bridgewater, CT.

Maine Yankee Atomic Power Company. (October 1997b). *Site characterization management plan.* Prepared by GTS Duratek, Inc., for the Maine Yankee Atomic Power Plant.

Maine Yankee Atomic Power Company. (April 16, 1998). *Appendix A: Spent fuel and other radioactive material stored in the Maine Yankee spent fuel pool.* MYPS-101, Rev. 0. Maine Yankee Atomic Power Company, Wiscasset, ME.

Maine Yankee Atomic Power Company. (April 1998). *GTS Duratek characterization survey report for the Maine Yankee Atomic Power Plant, revision 1*. 9 vols. Prepared by GTS Duratek, Inc. for the Maine Yankee Atomic Power Plant, Wiscasset, ME.

Markey, Edward J. and Waller, Douglas. (1982). *Nuclear peril: The politics of proliferation*. Ballinger Publishing Company, Cambridge, MA.

Ministry of Agriculture, Fisheries and Food and Scottish Environment Protection Agency. (September 1998). *Radioactivity in food and the environment, 1997*. RIFE-1. MAFF, London.

Moulder, John E. *Frequently asked questions about static electromagnetic fields and cancer*. Stason.org. http://stason.org/TULARC/health/static-fields-cancer/. Accessed March 2010.

National Council of Churches of Christ in the USA. (September 1975). *The plutonium economy*. National Council of Churches of Christ in the USA, NY, NY.

Nelkin, Dorothy. (1971). *Nuclear power and its critics: The Cayuga Lake controversy*. Cornell University Press, Ithaca, NY.

Ng, Kwan-Hoong. (October 20-22, 2003). Non-ionizing radiations–sources, biological effects, emissions and exposures. *Proceedings of the International Conference on Non-Ionizing Radiation at UNITEN (ICNIR2003) Electromagnetic Fields and Our Health*. http://www.who.int/peh-emf/meetings/archive/en/keynote3ng.pdf.

Nuclear Energy Agency. (April 1980). *Review of the continued suitability of the dumping site for radioactive waste in the north-east Atlantic*. NEA, Organisation for Economic Co-operation and Development, Paris, France.

Nuclear Energy Agency. (September 1981). *The environmental and biological behaviour of plutonium and some other transuranium elements*. NEA, Organisation for Economic Co-operation and Development, Paris, France.

Oak Ridge National Laboratory. (December 1997). *Integrated Data Base Report, 1996: U.S. Spent Nuclear Fuel and Radioactive Waste Inventories, projections, and characteristics*. Report No. DOE/RW-0006, Rev. 13. Oak Ridge National Laboratory, Oak Ridge, TN. 305

Physicians for Social Responsibility. (2011). *Lessons from Fukushima and Chernobyl for U.S. public health*. Physicians for Social Responsibility, Washington, DC. http://www.psr.org/assets/pdfs/fukushima-and-chernobyl.pdf.

Platt, R. B., Palms, J. M., Ragsdale, H. L., Shure, D. J., Mayer, P. G. and Mohrbacher, J. A. (1974). Empirical benefits derived from an ecosystem approach to environmental monitoring of a nuclear fuel reprocessing plant. *Proceedings: Environmental behavior of radionuclides released in the nuclear industry*. International Atomic Energy Agency, Stationery Office Books, Vienna, Austria.

Puhakainen, M., Rahola, T. and Suomela, M. (1987). *Radioactivity of sludge after the Chernobyl accident in 1986: Supplement 13 to Annual Report STUK-A55*. Report No. STUK-A68. Finnish Centre for Radiation and Nuclear Safety, Helsinki, Finland.

Reuters. (March 17, 2011). Japan's race against time: Getting to the core of the stricken Fukushima nuclear plant. *ThomsonReuters.com*. http://graphics.thomsonreuters.com/AS/pdf/JapanReactors1603_mv.pdf.

Shannon, Sara. (1987). *Diet for the atomic age: How to protect yourself from low-level radiation*. Avery Publishing Group Inc., Wayne, NJ.

Shapiero, Fred C. (1981). *Radwaste: A reporter's investigation of a growing nuclear menace*. Random House, NY.

Shapiro, Jacob. (1972). *Radiation protection: A guide for scientists and physicians*. Harvard University Press, Cambridge, MA.

Shleien, B., Pharm, D., Schmidt, G.D. and Chiacchiernini, R.P. (1982). *Background for protective action recommendations: Accidental radioactive contamination of food and animal feeds*. HHS Publication FDA 82-8196. US Department of Health and Human Services, Public Health Service, Washington D.C.

Spotts, Pete. (March 15, 2011). Meltdown 101: What are spent-fuel pools and why are they a threat. *Christian Science Monitor*. http://www.csmonitor.com/USA/2011/0315/Meltdown-101-What-are-spent-fuel-pools-and-why-are-they-a-threat.

Sternglass, Ernest. (1972). *Secret fallout: Low-level radiation from Hiroshima to Three-Mile Island*. Ballantine Books, NY.

The New York Times. (March 18, 2011). Greater danger lies in spent fuel than reactors; Radiation spread seen; frantic repairs go on; Easy fixes at reactors in the long run are elusive; Japan offers little response to US assessment; Radiation fears and distrust push thousands from homes. *The New York Times*. http://www.nytimes.com/2011/03/18/world/asia/18spent.html?scp=1&sq=3/18/2011&st=cse.

Toombs, George L., Martin, Sylvia L., Culter, Peter B. and Dibblee, Martha G. (n.d.). *Environmental radiological surveillance report on Oregon surface waters 1961 - 1983: Volume I*. Radiation Control Section, Health Division, Oregon Department of Human Resources, Portland, OR.

United Nations. (1972). *Ionizing radiation: Levels and effects: A report of the United Nations Scientific Committee on the effects of atomic radiation to the General Assembly, with annexes: Volume I: Levels*. UN, NY.

United Nations. (1996). *Sources and effects of ionizing radiation: United Nations Scientific Committee on the Effects of Atomic Radiation UNSCEAR 1996 report to the General Assembly, with scientific annex*. UN, NY.

United Nations Scientific Committee on the Effects of Atomic Radiation. (1982). *Ionizing radiation: Sources and biological effects*. UNSCEAR, Report to the General Assembly, United Nations, New York, NY. http://www.unscear.org/unscear/en/publications/1982.html.

United Nations Environment Programme, the International Labour Organization, and the World Health Organization. (1983). *Selected radionuclides: Tritium, carbon-14, krypton-85, strontium-90, iodine, caesium-137, radon, plutonium*. Environmental Health Criteria 25. World Health Organization, Geneva.

United States Department of Energy. (October 1992). *Integrated data base for 1992: U.S. spent fuel and radioactive waste inventories, projections, and characteristics*. DOE/RW-006, Rev. 8. Oak Ridge National Laboratory, Oak Ridge, Tennessee.

United States Department of Energy. (1996). *BEMR: Baseline environmental management report*. Secretary of Energy, Washington, DC. http://www.em.doe.gov/bemr/pages/bemr96.aspx.

United States Department of Energy, Environmental Protection Agency, Nuclear Regulatory Commission and Department of Defense. (December 6, 1996). *Multi-Agency radiation survey and site investigation manual (MARSSIM): Draft for public comment*. NUREG-1575. EPA 402-R-96-018. NTIS-PB97-117659. Washington, D.C. http://www.epa.gov/radiation/marssim/index.html. The final copy was released in December 1997.

United States Department of Energy, Environmental Protection Agency, Nuclear Regulatory Commission and Department of Defense. (August 2000). *Multi-Agency radiation survey and site investigation manual (MARSSIM): Revision 1*. NUREG-1575, REV 1. EPA 402-R-97-016, Rev 1. DOE/EH-0624, Rev 1. Washington, D.C. http://www.epa.gov/radiation/marssim/obtain.html.

United States Department of Health & Human Services. (September 1997). *Toxicological profile for ionizing radiation: Draft for public comment*. Agency for Toxic Substances and Disease Registry, Public Health Service, US Dept. of Health &

Human Services, Washington, D.C.
http://www.atsdr.cdc.gov/ToxProfiles/tp.asp?id=484&tid=86.

United States Environmental Protection Agency. (1987). Radiation protection guidance to federal agencies for occupational exposure. *Federal Register*. 52.

United States Environmental Protection Agency. (May 1992). *Manual of protective action guides and protective actions for nuclear incidents*. Office of Radiation Programs, US EPA, Washington, DC. http://www.epa.gov/rpdweb00/docs/er/400-r-92-001.pdf.

United States Federal Energy Regulatory Commission. (September 15, 1997). *Connecticut Yankee Atomic Power Company: Rebuttal testimony of James K. Joosten*. Docket No. ER97-913-000. Office of the Attorney General, State of Connecticut.

United States Food and Drug Administration. (March 5, 1997). *Draft: Accidental radioactive contamination of human food and animal feeds: Recommendations for state and local agencies*. Center for Devices and Radiological Health, US FDA, Washington, D.C.

United States Nuclear Regulatory Commission. (December 1998). *Draft regulatory guide DG-1074: Steam generator tube integrity*. Office of Nuclear Regulatory Research, US NRC, Washington, D.C.

Van Unnik, J. G., Broerse, J. J., Geleijns, J., Jansen, J. T., Zoetelief, J. and Zweers, D. (April 1997). Survey of CT techniques and absorbed dose in various Dutch hospitals. *Br J Radiol*. 70(832). pg. 367-71.

Wasserman, Harvey and Soloman, Norman. (1982). *Killing our own: The disaster of America's experience with atomic radiation*. Dell Publishing, NY.

World Health Organization. (1982). *Nuclear power: Health implications of transuranium elements*. European Series No. 11, WHO Regional Publications, Copenhagen.

Yablokov, Alexey V., Nesterenko, Vassily B. and Nesterenko, Alexey V. (December 2009). Chernobyl: Consequences of the catastrophe for people and the environment. *Annals of the New York Academy of Sciences*. 1181.

Subject Index by Heading

Table of Contents